Mathematical Sciences

From its pre-historic roots in simple counting to the algorithms powering modern desktop computers, from the genius of Archimedes to the genius of Einstein, advances in mathematical understanding and numerical techniques have been directly responsible for creating the modern world as we know it. This series will provide a library of the most influential publications and writers on mathematics in its broadest sense. As such, it will show not only the deep roots from which modern science and technology have grown, but also the astonishing breadth of application of mathematical techniques in the humanities and social sciences, and in everyday life.

An Investigation of the Laws of Thought

Self-taught mathematician and father of Boolean algebra, George Boole (1815-1864) published An Investigation of the Laws of Thought in 1854. In this highly original investigation of the fundamental laws of human reasoning, a sequel to ideas he had explored in earlier writings, Boole uses the symbolic language of mathematics to establish a method to examine the nature of the human mind using logic and the theory of probabilities. Boole considers language not just as a mode of expression, but as a system one can use to understand the human mind. In the first 12 chapters, he sets down the rules necessary to represent logic in this unique way. Then he analyses a variety of arguments and propositions of various writers from Aristotle to Spinoza. One of history's most insightful mathematicians, Boole is compelling reading for today's student of intellectual history and the science of the mind.

Cambridge University Press has long been a pioneer in the reissuing of out-of-print titles from its own backlist, producing digital reprints of books that are still sought after by scholars and students but could not be reprinted economically using traditional technology. The Cambridge Library Collection extends this activity to a wider range of books which are still of importance to researchers and professionals, either for the source material they contain, or as landmarks in the history of their academic discipline.

Drawing from the world-renowned collections in the Cambridge University Library, and guided by the advice of experts in each subject area, Cambridge University Press is using state-of-the-art scanning machines in its own Printing House to capture the content of each book selected for inclusion. The files are processed to give a consistently clear, crisp image, and the books finished to the high quality standard for which the Press is recognised around the world. The latest print-on-demand technology ensures that the books will remain available indefinitely, and that orders for single or multiple copies can quickly be supplied.

The Cambridge Library Collection will bring back to life books of enduring scholarly value across a wide range of disciplines in the humanities and social sciences and in science and technology.

An Investigation of the Laws of Thought

On Which Are Founded the Mathematical Theories of Logic and Probabilities

George Boole

CAMBRIDGE UNIVERSITY PRESS

Cambridge New York Melbourne Madrid Cape Town Singapore São Paolo Delhi

Published in the United States of America by Cambridge University Press, New York

www.cambridge.org
Information on this title: www.cambridge.org/9781108001533

This edition first published 1854
This digitally printed version 2009

ISBN 978-1-108-00153-3

AN INVESTIGATION

OF

THE LAWS OF THOUGHT,

ON WHICH ARE FOUNDED

THE MATHEMATICAL THEORIES OF LOGIC AND PROBABILITIES.

BY

GEORGE BOOLE, LL.D.

PROFESSOR OF MATHEMATICS IN QUEEN'S COLLEGE, CORK.

LONDON:

WALTON AND MABERLY,

UPPER GOWER-STREET, AND IVY-LANE, PATERNOSTER-ROW.

CAMBRIDGE: MACMILLAN AND CO.

1854.

TO

JOHN RYALL, LL.D.,

VICE-PRESIDENT AND PROFESSOR OF GREEK

IN QUEEN'S COLLEGE, CORK,

THIS WORK IS INSCRIBED

IN TESTIMONY OF FRIENDSHIP AND ESTEEM.

cially of the doctrine of Elimination, and of the solution of Equations containing more than one unknown quantity. Preliminary information upon the subject-matter will be found in the special treatises on Probabilities in "Lardner's Cabinet Cyclopædia," and the "Library of Useful Knowledge," the former of these by Professor De Morgan, the latter by Sir John Lubbock; and in an interesting series of Letters translated from the French of M. Quetelet. Other references will be given in the work. On a first perusal the reader may omit at his discretion, Chapters x., xiv., and xix., together with any of the applications which he may deem uninviting or irrelevant.

In different parts of the work, and especially in the notes to the concluding chapter, will be found references to various writers, ancient and modern, chiefly designed to illustrate a certain view of the history of philosophy. With respect to these, the Author thinks it proper to add, that he has in no instance given a citation which he has not believed upon careful examination to be supported either by parallel authorities, or by the general tenor of the work from which it was taken. While he would gladly have avoided the introduction of anything which might by possibility be construed into the parade of learning, he felt it to be due both to his subject and to the truth, that the statements in the text should be accompanied by the means of verification. And if now, in bringing to its close a labour, of the extent of which few persons will be able to judge from its apparent fruits, he may be permitted to speak for a single moment of the feelings with which he has pursued, and with which he now lays aside, his task, he would say, that he never doubted that it was worthy of his best efforts; that he felt that whatever of truth it might bring to light was not a private or arbitrary thing, not dependent, as to its essence, upon any human opinion. He was fully aware that learned and able men maintained opinions upon the subject of

PREFACE.

THE following work is not a republication of a former treatise by the Author, entitled, "The Mathematical Analysis of Logic." Its earlier portion is indeed devoted to the same object, and it begins by establishing the same system of fundamental laws, but its methods are more general, and its range of applications far wider. It exhibits the results, matured by some years of study and reflection, of a principle of investigation relating to the intellectual operations, the previous exposition of which was written within a few weeks after its idea had been conceived.

That portion of this work which relates to Logic presupposes in its reader a knowledge of the most important terms of the science, as usually treated, and of its general object. On these points there is no better guide than Archbishop Whately's "Elements of Logic," or Mr. Thomson's "Outlines of the Laws of Thought." To the former of these treatises, the present revival of attention to this class of studies seems in a great measure due. Some acquaintance with the principles of Algebra is also requisite, but it is not necessary that this application should have been carried beyond the solution of simple equations. For the study of those chapters which relate to the theory of probabilities, a somewhat larger knowledge of Algebra is required, and espe-

Logic directly opposed to the views upon which the entire argument and procedure of his work rested. While he believed those opinions to be erroneous, he was conscious that his own views might insensibly be warped by an influence of another kind. He felt in an especial manner the danger of that intellectual bias which long attention to a particular aspect of truth tends to produce. But he trusts that out of this conflict of opinions the same truth will but emerge the more free from any personal admixture; that its different parts will be seen in their just proportion; and that none of them will eventually be too highly valued or too lightly regarded because of the prejudices which may attach to the mere form of its exposition.

To his valued friend, the Rev. George Stephens Dickson, of Lincoln, the Author desires to record his obligations for much kind assistance in the revision of this work, and for some important suggestions.

5, Grenville-place, Cork,
Nov. 30th, 1853.

CONTENTS.

AN INVESTIGATION

OF

THE LAWS OF THOUGHT.

CHAPTER I.

NATURE AND DESIGN OF THIS WORK.

1. THE design of the following treatise is to investigate the fundamental laws of those operations of the mind by which reasoning is performed; to give expression to them in the symbolical language of a Calculus, and upon this foundation to establish the science of Logic and construct its method; to make that method itself the basis of a general method for the application of the mathematical doctrine of Probabilities; and, finally, to collect from the various elements of truth brought to view in the course of these inquiries some probable intimations concerning the nature and constitution of the human mind.

2. That this design is not altogether a novel one it is almost needless to remark, and it is well known that to its two main practical divisions of Logic and Probabilities a very considerable share of the attention of philosophers has been directed. In its ancient and scholastic form, indeed, the subject of Logic stands almost exclusively associated with the great name of Aristotle. As it was presented to ancient Greece in the partly technical, partly metaphysical disquisitions of the Organon, such, with scarcely any essential change, it has continued to the present day. The stream of original inquiry has rather been directed towards questions of general philosophy, which, though they

have arisen among the disputes of the logicians, have outgrown their origin, and given to successive ages of speculation their peculiar bent and character. The eras of Porphyry and Proclus, of Anselm and Abelard, of Ramus, and of Descartes, together with the final protests of Bacon and Locke, rise up before the mind as examples of the remoter influences of the study upon the course of human thought, partly in suggesting topics fertile of discussion, partly in provoking remonstrance against its own undue pretensions. The history of the theory of Probabilities, on the other hand, has presented far more of that character of steady growth which belongs to science. In its origin the early genius of Pascal,—in its maturer stages of development the most recondite of all the mathematical speculations of Laplace,—were directed to its improvement; to omit here the mention of other names scarcely less distinguished than these. As the study of Logic has been remarkable for the kindred questions of Metaphysics to which it has given occasion, so that of Probabilities also has been remarkable for the impulse which it has bestowed upon the higher departments of mathematical science. Each of these subjects has, moreover, been justly regarded as having relation to a speculative as well as to a practical end. To enable us to deduce correct inferences from given premises is not the only object of Logic; nor is it the sole claim of the theory of Probabilities that it teaches us how to establish the business of life assurance on a secure basis; and how to condense whatever is valuable in the records of innumerable observations in astronomy, in physics, or in that field of social inquiry which is fast assuming a character of great importance. Both these studies have also an interest of another kind, derived from the light which they shed upon the intellectual powers. They instruct us concerning the mode in which language and number serve as instrumental aids to the processes of reasoning; they reveal to us in some degree the connexion between different powers of our common intellect; they set before us what, in the two domains of demonstrative and of probable knowledge, are the essential standards of truth and correctness,—standards not derived from without, but deeply founded in the constitution of the human faculties. These ends of speculation yield neither in interest nor in dignity, nor yet, it may be added, in importance, to the practical objects, with the pursuit of which they have been historically associated. To unfold the secret laws and relations of those high faculties of thought by which all beyond the merely perceptive knowledge of the world and of ourselves is attained or matured, is an object which does not stand in need of commendation to a rational mind.

3. But although certain parts of the design of this work have been entertained by others, its general conception, its method, and, to a considerable extent, its results, are believed to be original. For this reason I shall offer, in the present chapter, some preparatory statements and explanations, in order that the real aim of this treatise may be understood, and the treatment of its subject facilitated.

It is designed, in the first place, to investigate the fundamental laws of those operations of the mind by which reasoning is performed. It is unnecessary to enter here into any argument to prove that the operations of the mind are in a certain real sense subject to laws, and that a science of the mind is therefore *possible*. If these are questions which admit of doubt, that doubt is not to be met by an endeavour to settle the point of dispute *à priori*, but by directing the attention of the objector to the evidence of actual laws, by referring him to an actual science. And thus the solution of that doubt would belong not to the introduction to this treatise, but to the treatise itself. Let the assumption be granted, that a science of the intellectual powers is possible, and let us for a moment consider how the knowledge of it is to be obtained.

4. Like all other sciences, that of the intellectual operations must primarily rest upon observation,—the subject of such observation being the very operations and processes of which we desire to determine the laws. But while the necessity of a foundation in experience is thus a condition common to all sciences, there are some special differences between the modes in which this principle becomes available for the determination of general truths when the subject of inquiry is the mind, and when the subject is external nature. To these it is necessary to direct attention.

The general laws of Nature are not, for the most part, immediate objects of perception. They are either inductive inferences from a large body of facts, the common truth in which they express, or, in their origin at least, physical hypotheses of a causal nature serving to explain phænomena with undeviating precision, and to enable us to predict new combinations of them. They are in all cases, and in the strictest sense of the term, *probable* conclusions, approaching, indeed, ever and ever nearer to certainty, as they receive more and more of the confirmation of experience. But of the character of probability, in the strict and proper sense of that term, they are never wholly divested. On the other hand, the knowledge of the laws of the mind does not require as its basis any extensive collection of observations. The general truth is seen in the particular instance, and it is not confirmed by the repetition of instances. We may illustrate this position by an obvious example. It may be a question whether that formula of reasoning, which is called the *dictum* of Aristotle, *de omni et nullo*, expresses a primary law of human reasoning or not; but it is no question that it expresses a general truth in Logic. Now that truth is made manifest in all its generality by reflection upon a single instance of its application. And this is both an evidence that the particular principle or formula in question is founded upon some general law or laws of the mind, and an illustration of the doctrine that the perception of such general truths is not derived from an induction from many instances, but is involved in the clear apprehension of a single instance. In connexion with this truth is seen the not less important one that our knowledge of the laws upon which the science of the intellectual powers rests, whatever may be its extent or its deficiency, is not probable knowledge. For we not only see in the particular example the general truth, but we see it also as a certain truth,—a truth, our confidence in which will not continue to increase with increasing experience of its practical verifications.

5. But if the general truths of Logic are of such a nature that when presented to the mind they at once command assent, wherein consists the difficulty of constructing the Science of Logic? Not, it may be answered, in collecting the materials of knowledge, but in discriminating their nature, and determining

their mutual place and relation. All sciences consist of general truths, but of those truths some only are primary and fundamental, others are secondary and derived. The laws of elliptic motion, discovered by Kepler, are general truths in astronomy, but they are not its fundamental truths. And it is so also in the purely mathematical sciences. An almost boundless diversity of theorems, which are known, and an infinite possibility of others, as yet unknown, rest together upon the foundation of a few simple axioms; and yet these are all *general* truths. It may be added, that they are truths which to an intelligence sufficiently refined would shine forth in their own unborrowed light, without the need of those connecting links of thought, those steps of wearisome and often painful deduction, by which the knowledge of them is actually acquired. Let us define as fundamental those laws and principles from which all other general truths of science may be deduced, and into which they may all be again resolved. Shall we then err in regarding that as the true science of Logic which, laying down certain elementary laws, confirmed by the very testimony of the mind, permits us thence to deduce, by uniform processes, the entire chain of its secondary consequences, and furnishes, for its practical applications, methods of perfect generality? Let it be considered whether in any science, viewed either as a system of truth or as the foundation of a practical art, there can properly be any other test of the completeness and the fundamental character of its laws, than the completeness of its system of derived truths, and the generality of the methods which it serves to establish. Other questions may indeed present themselves. Convenience, prescription, individual preference, may urge their claims and deserve attention. But as respects the question of what constitutes science in its abstract integrity, I apprehend that no other considerations than the above are properly of any value.

6. It is designed, in the next place, to give expression in this treatise to the fundamental laws of reasoning in the symbolical language of a Calculus. Upon this head it will suffice to say, that those laws are such as to suggest this mode of expression, and to give to it a peculiar and exclusive fitness for the ends in view.

There is not only a close analogy between the operations of the mind in general reasoning and its operations in the particular science of Algebra, but there is to a considerable extent an exact agreement in the laws by which the two classes of operations are conducted. Of course the laws must in both cases be determined independently; any formal agreement between them can only be established *a posteriori* by actual comparison. To borrow the notation of the science of Number, and then assume that in its new application the laws by which its use is governed will remain unchanged, would be mere hypothesis. There exist, indeed, certain general principles founded in the very nature of language, by which the use of symbols, which are but the elements of scientific language, is determined. To a certain extent these elements are arbitrary. Their interpretation is purely conventional: we are permitted to employ them in whatever sense we please. But this permission is limited by two indispensable conditions,—first, that from the sense once conventionally established we never, in the same process of reasoning, depart; secondly, that the laws by which the process is conducted be founded exclusively upon the above fixed sense or meaning of the symbols employed. In accordance with these principles, any agreement which may be established between the laws of the symbols of Logic and those of Algebra can but issue in an agreement of processes. The two provinces of interpretation remain apart and independent, each subject to its own laws and conditions.

Now the actual investigations of the following pages exhibit Logic, in its practical aspect, as a system of processes carried on by the aid of symbols having a definite interpretation, and subject to laws founded upon that interpretation alone. But at the same time they exhibit those laws as identical in form with the laws of the general symbols of algebra, with this single addition, viz., that the symbols of Logic are further subject to a special law (Chap. II.), to which the symbols of quantity, as such, are not subject. Upon the nature and the evidence of this law it is not purposed here to dwell. These questions will be fully discussed in a future page. But as constituting the essential ground of difference between those forms of inference with which Logic is conversant, and those which present themselves in the particular science of Number, the law in question is deserving of more than a passing notice. It may be said that it lies at the very foundation of general reasoning,—that it governs those intellectual acts of conception or of imagination which are preliminary to the processes of logical deduction, and that it gives to the processes themselves much of their actual form and expression. It may hence be affirmed that this law constitutes the germ or seminal principle, of which every approximation to a general method in Logic is the more or less perfect development.

7. The principle has already been laid down (5) that the sufficiency and truly fundamental character of any assumed system of laws in the science of Logic must partly be seen in the perfection of the methods to which they conduct us. It remains, then, to consider what the requirements of a general method in Logic are, and how far they are fulfilled in the system of the present work.

Logic is conversant with two kinds of relations,—relations among things, and relations among facts. But as facts are expressed by propositions, the latter species of relation may, at least for the purposes of Logic, be resolved into a relation among propositions. The assertion that the fact or event A is an invariable consequent of the fact or event B may, to this extent at least, be regarded as equivalent to the assertion, that the truth of the proposition affirming the occurrence of the event B always implies the truth of the proposition affirming the occurrence of the event A. Instead, then, of saying that Logic is conversant with relations among things and relations among facts, we are permitted to say that it is concerned with relations among things and relations among propositions. Of the former kind of relations we have an example in the proposition—"All men are mortal;" of the latter kind in the proposition—"If the sun is totally eclipsed, the stars will become visible." The one expresses a relation between "men" and "mortal beings," the other between the elementary propositions—"The sun is totally eclipsed;" "The stars will become visible." Among such relations I suppose to be included those which affirm or deny existence with respect to things, and those which affirm or deny truth with respect to propositions. Now let those things or those propositions among which relation is expressed be termed the elements of the propositions by which such relation is expressed. Proceeding from this definition, we may then say that the *premises* of any logical argument express *given* relations among certain elements, and that the *conclusion* must express an *implied* relation among those elements, or among a part of them, i. e. a relation implied by or inferentially involved in the premises.

8. Now this being premised, the requirements of a general method in Logic seem to be the following:—

1st. As the conclusion must express a relation among the whole or among a part of the elements involved in the premises, it is requisite that we should possess the means of eliminating those elements which we desire not to appear in the conclusion, and of determining the whole amount of relation implied by the premises among the elements which we wish to retain. Those elements which do not present themselves in the conclusion are, in the language of the common Logic, called middle terms; and the species of elimination exemplified in treatises on Logic consists in deducing from two propositions, containing a common element or middle term, a conclusion connecting the two remaining terms. But the problem of elimination, as contemplated in this work, possesses a much wider scope. It proposes not merely the elimination of one middle term from two propositions, but the elimination generally of middle terms from propositions, without regard to the number of either of them, or to the nature of their connexion. To this object neither the processes of Logic nor those of Algebra, in their actual state, present any strict parallel. In the latter science the problem of elimination is known to be limited in the following manner:—From two equations we can eliminate one symbol of quantity; from three equations two symbols; and, generally, from n equations $n - 1$ symbols. But though this condition, necessary in Algebra, seems to prevail in the existing Logic also, it has no essential place in Logic as a science. There, no relation whatever can be proved to prevail between the number of terms to be eliminated and the number of propositions from which the elimination is to be effected. From the equation representing a single proposition, any number of symbols representing terms or elements in Logic may be eliminated; and from any number of equations representing propositions, one or any other number of symbols of this kind may be eliminated in a similar manner. For such elimination there exists one general process applicable to all cases. This is one of the many remarkable consequences of that distinguishing law of the symbols of Logic, to which attention has been already directed.

2ndly. It should be within the province of a general method in Logic to express the final relation among the elements of the conclusion by any admissible *kind* of proposition, or in any selected *order* of terms. Among varieties of kind we may reckon those which logicians have designated by the terms categorical, hypothetical, disjunctive, &c. To a choice or selection in the order of the terms, we may refer whatsoever is dependent upon the appearance of particular elements in the subject or in the predicate, in the antecedent or in the consequent, of that proposition which forms the "conclusion." But waiving the language of the schools, let us consider what really distinct species of problems may present themselves to our notice. We have seen that the elements of the final or inferred relation may either be *things* or *propositions*. Suppose the former case; then it might be required to deduce from the premises a definition or description of some one thing, or class of things, constituting an element of the conclusion in terms of the other things involved in it. Or we might form the conception of some thing or class of things, involving more than one of the elements of the conclusion, and require its expression in terms of the other elements. Again, suppose the elements retained in the conclusion to be propositions, we might desire to ascertain such points as the following, viz., Whether, in virtue of the premises, any of those propositions, taken singly, are true or false?—Whether particular combinations of them are true or false?—Whether, assuming a particular proposition to be true, any consequences will follow, and if so, what consequences, with respect to the other propositions?—Whether any particular condition being assumed with reference to certain of the propositions, any consequences, and what consequences, will follow with respect to the others? and so on. I say that these are general questions, which it should fall within the scope or province of a general method in Logic to solve. Perhaps we might include them all under this one statement of the final problem of practical Logic. Given a set of premises expressing relations among certain elements, whether things or propositions: required explicitly the whole relation consequent among *any* of those elements under any proposed conditions, and in any proposed form. That this problem, under all its aspects, is resolvable, will hereafter appear. But it is not for the sake of noticing this fact, that the above inquiry into the nature and the functions of a general method in Logic has been introduced. It is necessary that the reader should apprehend what are the specific ends of the investigation upon which we are entering, as well as the principles which are to guide us to the attainment of them.

9. Possibly it may here be said that the Logic of Aristotle, in its rules of syllogism and conversion, sets forth the elementary processes of which all reasoning consists, and that beyond these there is neither scope nor occasion for a general method. I have no desire to point out the defects of the common Logic, nor do I wish to refer to it any further than is necessary, in order to place in its true light the nature of the present treatise. With this end alone in view, I would remark:—1st. That syllogism, conversion, &c., are not the ultimate processes of Logic. It will be shown in this treatise that they are founded upon, and are resolvable into, ulterior and more simple processes which constitute the real elements of method in Logic. Nor is it true in fact that all inference is reducible to the particular forms of syllogism and conversion.—*Vide* Chap. xv. 2ndly. If all inference were reducible to these two processes (and it has been maintained that it is reducible to syllogism alone), there would still exist the same necessity for a general method. For it would still be requisite to determine in what order the processes should succeed each other, as well as their particular nature, in order that the desired relation should be obtained. By the desired relation I mean that full relation which, in virtue of the premises, connects any elements selected out of the premises at will, and which, moreover, expresses that relation in any desired form and order.

If we may judge from the mathematical sciences, which are the most perfect examples of method known, this *directive* function of Method constitutes its chief office and distinction. The fundamental processes of arithmetic, for instance, are in themselves but the elements of a possible science. To assign their nature is the first business of its method, but to arrange their succession is its subsequent and higher function. In the more complex examples of logical deduction, and especially in those which form a basis for the solution of difficult questions in the theory of Probabilities, the aid of a directive method, such as a Calculus alone can supply, is indispensable.

10. Whence it is that the ultimate laws of Logic are mathematical in their form; why they are, except in a single point, identical with the general laws of Number; and why in that particular point they differ;—are questions upon which it might not be very remote from presumption to endeavour to pronounce a positive judgment. Probably they lie beyond the reach of our limited faculties. It may, perhaps, be permitted to the mind to attain a knowledge of the laws to which it is itself subject, without its being also given to it to understand their ground and origin, or even, except in a very limited degree, to comprehend their fitness for their end, as compared with other and conceivable systems of law. Such knowledge is, indeed, unnecessary for the ends of science, which properly concerns itself with what is, and seeks not for grounds of preference or reasons of appointment. These considerations furnish a sufficient answer to all protests against the exhibition of Logic in the form of a Calculus. It is not because we choose to assign to it such a mode of manifestation, but because the ultimate laws of thought render that mode possible, and prescribe its character, and forbid, as it would seem, the perfect manifestation of the science in any other form, that such a mode demands adoption. It is to be remembered that it is the business of science not to create laws, but to discover them. We do not originate the constitution of our own minds, greatly as it may be in our power to modify their character. And as the laws of the human intellect do not depend upon our will, so the forms of the science, of which they constitute the basis, are in all essential regards independent of individual choice.

11. Beside the general statement of the principles of the above method, this treatise will exhibit its application to the analysis of a considerable variety of propositions, and of trains of propositions constituting the premises of demonstrative arguments. These examples have been selected from various writers, they differ greatly in complexity, and they embrace a wide range of subjects. Though in this particular respect it may appear to some that too great a latitude of choice has been exercised, I do not deem it necessary to offer any apology upon this account. That Logic, as a science, is susceptible of very wide applications is admitted; but it is equally certain that its ultimate forms and processes are mathematical. Any objection *à priori* which may therefore be supposed to lie against the adoption of such forms and processes in the discussion of a problem of morals or of general philosophy must be founded upon misapprehension or false analogy. It is not of the essence of mathematics to be conversant with the ideas of number and quantity. Whether as a general habit of mind it would be desirable to apply symbolical processes to moral argument, is another question. Possibly, as I have elsewhere observed,* the perfection of the method of Logic may be chiefly valuable as an evidence of the speculative truth of its principles. To supersede the employment of common reasoning, or to subject it to the rigour of technical forms, would be the last desire of one who knows the value of that intellectual toil and warfare which imparts to the mind an athletic vigour, and teaches it to contend with difficulties, and to rely upon itself in emergencies. Nevertheless, cases may arise in which the value of a scientific procedure, even in those things which fall confessedly under the ordinary dominion of the reason, may be felt and acknowledged. Some examples of this kind will be found in the present work.

12. The general doctrine and method of Logic above explained form also the basis of a theory and corresponding method of Probabilities. Accordingly, the development of such a theory and method, upon the above principles, will constitute a distinct object of the present treatise. Of the nature of this application it may be desirable to give here some account, more especially as

* Mathematical Analysis of Logic. London: G. Bell. 1847.

regards the character of the solutions to which it leads. In connexion with this object some further detail will be requisite concerning the forms in which the results of the logical analysis are presented.

The ground of this necessity of a prior method in Logic, as the basis of a theory of Probabilities, may be stated in a few words. Before we can determine the mode in which the expected frequency of occurrence of a particular event is dependent upon the known frequency of occurrence of any other events, we must be acquainted with the mutual dependence of the events themselves. Speaking technically, we must be able to express the event whose probability is sought, as a function of the events whose probabilities are given. Now this explicit determination belongs in all instances to the department of Logic. Probability, however, in its mathematical acceptation, admits of numerical measurement. Hence the subject of Probabilities belongs equally to the science of Number and to that of Logic. In recognising the co-ordinate existence of both these elements, the present treatise differs from all previous ones; and as this difference not only affects the question of the possibility of the solution of problems in a large number of instances, but also introduces new and important elements into the solutions obtained, I deem it necessary to state here, at some length, the peculiar consequences of the theory developed in the following pages.

13. The measure of the probability of an event is usually defined as a fraction, of which the numerator represents the number of cases favourable to the event, and the denominator the whole number of cases favourable and unfavourable; all cases being supposed equally likely to happen. That definition is adopted in the present work. At the same time it is shown that there is another aspect of the subject (shortly to be referred to) which might equally be regarded as fundamental, and which would actually lead to the same system of methods and conclusions. It may be added, that so far as the received conclusions of the theory of Probabilities extend, and so far as they are consequences of its fundamental definitions, they do not differ from the results (supposed to be equally correct in inference) of the method of this work.

Again, although questions in the theory of Probabilities present themselves under various aspects, and may be variously modified by algebraical and other conditions, there seems to be one general type to which all such questions, or so much of each of them as truly belongs to the theory of Probabilities, may be referred. Considered with reference to the *data* and the *quæsitum*, that type may be described as follows:—1st. The data are the probabilities of one or more given events, each probability being either that of the absolute fulfilment of the event to which it relates, or the probability of its fulfilment under given supposed conditions. 2ndly. The *quæsitum*, or object sought, is the probability of the fulfilment, absolutely or conditionally, of some other event differing in expression from those in the data, but more or less involving the same elements. As concerns the data, they are either *causally given*,—as when the probability of a particular throw of a die is deduced from a knowledge of the constitution of the piece,—or they are derived from observation of repeated instances of the success or failure of events. In the latter case the probability of an event may be defined as the limit toward which the ratio of the favourable to the whole number of observed cases approaches (the uniformity of nature being presupposed) as the observations are indefinitely continued. Lastly, as concerns the nature or relation of the events in question, an important distinction remains. Those events are either *simple* or *compound*. By a compound event is meant one of which the expression in language, or the conception in thought, depends upon the expression or the conception of other events, which, in relation to it, may be regarded as *simple* events. To say "it rains," or to say "it thunders," is to express the occurrence of a simple event; but to say "it rains and thunders," or to say "it either rains or thunders," is to express that of a compound event. For the expression of that event depends upon the elementary expressions, "it rains," "it thunders." The criterion of simple events is not, therefore, any supposed simplicity in their nature. It is founded solely on the mode of their expression in language or conception in thought.

14. Now one general problem, which the existing theory of Probabilities enables us to solve, is the following, viz.:—Given

the probabilities of any simple events: required the probability of a given compound event, i. e. of an event compounded in a given manner out of the given simple events. The problem can also be solved when the compound event, whose probability is required, is subjected to given conditions, i. e. to conditions dependent also in a given manner on the given simple events. Beside this general problem, there exist also particular problems of which the principle of solution is known. Various questions relating to *causes* and *effects* can be solved by known methods under the particular hypothesis that the causes are mutually exclusive, but apparently not otherwise. Beyond this it is not clear that any advance has been made toward the solution of what may be regarded as the general problem of the science, viz.: Given the probabilities of any events, simple or compound, conditioned or unconditioned: required the probability of any other event equally arbitrary in expression and conception. In the statement of this question it is not even postulated that the events whose probabilities are given, and the one whose probability is sought, should involve some common elements, because it is the office of a method to determine whether the data of a problem are sufficient for the end in view, and to indicate, when they are not so, wherein the deficiency consists.

This problem, in the most unrestricted form of its statement, is resolvable by the method of the present treatise; or, to speak more precisely, its theoretical solution is completely given, and its practical solution is brought to depend only upon processes purely mathematical, such as the resolution and analysis of equations. The order and character of the general solution may be thus described.

15. In the first place it is always possible, by the preliminary method of the Calculus of Logic, to express the event whose probability is sought as a logical function of the events whose probabilities are given. The result is of the following character: Suppose that X represents the event whose probability is sought, A, B, C, &c. the events whose probabilities are given, those events being either simple or compound. Then the *whole* relation of the event X to the events A, B, C, &c. is deduced in the form of what mathematicians term a *development*, consisting, in

the most general case, of four distinct classes of terms. By the first class are expressed those combinations of the events A, B, C, which both necessarily accompany and necessarily indicate the occurrence of the event X; by the second class, those combinations which necessarily accompany, but do not necessarily imply, the occurrence of the event X; by the third class, those combinations whose occurrence in connexion with the event X is impossible, but not otherwise impossible; by the fourth class, those combinations whose occurrence is impossible under any circumstances. I shall not dwell upon this statement of the result of the logical analysis of the problem, further than to remark that the elements which it presents are precisely those by which the expectation of the event X, as dependent upon our knowledge of the events A, B, C, is, or alone can be, affected. General reasoning would verify this conclusion; but general reasoning would not usually avail to disentangle the complicated web of events and circumstances from which the solution above described must be evolved. The attainment of this object constitutes the first step towards the complete solution of the question proposed. It is to be noted that thus far the process of solution is logical, i. e. conducted by symbols of logical significance, and resulting in an equation interpretable into a *proposition*. Let this result be termed the *final logical equation.*

The second step of the process deserves attentive remark. From the final logical equation to which the previous step has conducted us, are deduced, by inspection, a series of algebraic equations implicitly involving the complete solution of the problem proposed. Of the mode in which this transition is effected let it suffice to say, that there exists a definite relation between the laws by which the probabilities of events are expressed as algebraic functions of the probabilities of other events upon which they depend, and the laws by which the logical connexion of the events is itself expressed. This relation, like the other coincidences of formal law which have been referred to, is not founded upon hypothesis, but is made known to us by observation (I. 4), and reflection. If, however, its reality were assumed *à priori* as the basis of the very definition of Probability, strict deduction would thence lead us to the received numerical definition as a necessary consequence. The Theory of Probabilities stands, as it has already been remarked (I. 12), in equally close relation to Logic and to Arithmetic; and it is indifferent, so far as results are concerned, whether we regard it as springing out of the latter of these sciences, or as founded in the mutual relations which connect the two together.

16. There are some circumstances, interesting perhaps to the mathematician, attending the general solutions deduced by the above method, which it may be desirable to notice.

1st. As the method is independent of the number and the nature of the data, it continues to be applicable when the latter are insufficient to render determinate the value sought. When such is the case, the final expression of the solution will contain terms with arbitrary constant coefficients. To such terms there will correspond terms in the final logical equation (I. 15), the interpretation of which will inform us what new data are requisite in order to determine the values of those constants, and thus render the numerical solution complete. If such data are not to be obtained, we can still, by giving to the constants their limiting values 0 and 1, determine the limits within which the probability sought must lie independently of all further experience. When the event whose probability is sought is *quite* independent of those whose probabilities are given, the limits thus obtained for its value will be 0 and 1, as it is evident that they ought to be, and the interpretation of the constants will only lead to a re-statement of the original problem.

2ndly. The expression of the final solution will in all cases involve a particular element of quantity, determinable by the solution of an algebraic equation. Now when that equation is of an elevated degree, a difficulty may seem to arise as to the selection of the proper root. There are, indeed, cases in which both the elements given and the element sought are so obviously restricted by necessary conditions that no choice remains. But in complex instances the discovery of such conditions, by unassisted force of reasoning, would be hopeless. A distinct method is requisite for this end,—a method which might not inappropriately be termed the Calculus of Statistical Conditions. Into the nature of this method I shall not here further enter than to say, that, like the previous method, it is based upon the employment of the "final logical equation," and that it definitely assigns, 1st, the conditions which must be fulfilled among the numerical elements of the data, in order that the problem may be real, i. e. derived from a *possible experience ;* 2ndly, the numerical limits, within which the probability sought must have been confined, if, instead of being determined by theory, it had been deduced directly by observation from the same system of phænomena from which the data were derived. It is clear that these limits will be actual limits of the probability sought. Now, on supposing the data subject to the conditions above assigned to them, it appears in every instance which I have examined that there exists one root, and only one root, of the final algebraic equation which is subject to the required limitations. Every source of ambiguity is thus removed. It would even seem that new truths relating to the theory of algebraic equations are thus incidentally brought to light. It is remarkable that the special element of quantity, to which the previous discussion relates, depends only upon the *data*, and not at all upon the *quæsitum* of the problem proposed. Hence the solution of each particular problem unties the knot of difficulty for a system of problems, viz., for that system of problems which is marked by the possession of common data, independently of the nature of their *quæsita*. This circumstance is important whenever from a particular system of data it is required to deduce a series of connected conclusions. And it further gives to the solutions of particular problems that character of relationship, derived from their dependence upon a central and fundamental unity, which not unfrequently marks the application of general methods.

17. But though the above considerations, with others of a like nature, justify the assertion that the method of this treatise, for the solution of questions in the theory of Probabilities, is a general method, it does not thence follow that we are relieved in all cases from the necessity of recourse to hypothetical grounds. It has been observed that a solution may consist entirely of terms affected by arbitrary constant coefficients,—may, in fact, be wholly indefinite. The application of the method of this work to some of the most important questions within its range would—

were the data of experience alone employed—present results of this character. To obtain a *definite* solution it is necessary, in such cases, to have recourse to hypotheses possessing more or less of independent probability, but incapable of exact verification. Generally speaking, such hypotheses will differ from the immediate results of experience in partaking of a logical rather than of a numerical character; in prescribing the conditions under which phænomena occur, rather than assigning the relative frequency of their occurrence. This circumstance is, however, unimportant. Whatever their nature may be, the hypotheses assumed must thenceforth be regarded as belonging to the actual data, although tending, as is obvious, to give to the solution itself somewhat of a hypothetical character. With this understanding as to the possible sources of the data actually employed, the method is perfectly general, but for the correctness of the hypothetical elements introduced it is of course no more responsible than for the correctness of the numerical data derived from experience.

In illustration of these remarks we may observe that the theory of the reduction of astronomical observations* rests, in part, upon hypothetical grounds. It assumes certain positions as to the nature of error, the equal probabilities of its occurrence in the form of excess or defect, &c., without which it would be impossible to obtain any *definite* conclusions from a system of conflicting observations. But granting such positions as the above, the residue of the investigation falls strictly within the province of the theory of Probabilities. Similar observations apply to the important problem which proposes to deduce from the records of the majorities of a deliberative assembly the mean probability of correct judgment in one of its members. If the method of this treatise be applied to the mere numerical data, the solution obtained is of that wholly indefinite kind above described. And to show in a more eminent degree the insufficiency of those data by themselves, the interpretation of the arbitrary constants (I. 16) which appear in the solution, merely produces

* The author designs to treat this subject either in a separate work or in a future Appendix. In the present treatise he avoids the use of the integral calculus.

a re-statement of the original problem. Admitting, however, the hypothesis of the independent formation of opinion in the individual mind, either absolutely, as in the speculations of Laplace and Poisson, or under limitations imposed by the actual data, as will be seen in this treatise, Chap. XXI., the problem assumes a far more definite character. It will be manifest that the ulterior value of the theory of Probabilities must depend very much upon the correct formation of such mediate hypotheses, where the purely experimental data are insufficient for *definite* solution, and where that further experience indicated by the interpretation of the final logical equation is unattainable. Upon the other hand, an undue readiness to form hypotheses in subjects which from their very nature are placed beyond human ken, must re-act upon the credit of the theory of Probabilities, and tend to throw doubt in the general mind over its most legitimate conclusions.

18. It would, perhaps, be premature to speculate here upon the question whether the methods of abstract science are likely at any future day to render service in the investigation of social problems at all commensurate with those which they have rendered in various departments of physical inquiry. An attempt to resolve this question upon pure *à priori* grounds of reasoning would be very likely to mislead us. For example, the consideration of human free-agency would seem at first sight to preclude the idea that the movements of the social system should ever manifest that character of orderly evolution which we are prepared to expect under the reign of a physical necessity. Yet already do the researches of the statist reveal to us facts at variance with such an anticipation. Thus the records of crime and pauperism present a degree of regularity unknown in regions in which the disturbing influence of human wants and passions is unfelt. On the other hand, the distemperature of seasons, the eruption of volcanoes, the spread of blight in the vegetable, or of epidemic maladies in the animal kingdom, things apparently or chiefly the product of natural causes, refuse to be submitted to regular and apprehensible laws. "Fickle as the wind," is a proverbial expression. Reflection upon these points teaches us in some degree to correct our earlier judgments. We learn that we are not to

expect, under the dominion of necessity, an order perceptible to human observation, unless the play of its producing causes is sufficiently simple; nor, on the other hand, to deem that free agency in the individual is inconsistent with regularity in the motions of the system of which he forms a component unit. Human freedom stands out as an apparent fact of our consciousness, while it is also, I conceive, a highly probable deduction of analogy (Chap. XXII.) from the nature of that portion of the mind whose scientific constitution we are able to investigate. But whether accepted as a fact reposing on consciousness, or as a conclusion sanctioned by the reason, it must be so interpreted as not to conflict with an established result of observation, viz.: that phænomena, in the production of which large masses of men are concerned, do actually exhibit a very remarkable degree of regularity, enabling us to collect in each succeeding age the elements upon which the estimate of its state and progress, so far as manifested in outward results, must depend. There is thus no sound objection *à priori* against the possibility of that species of data which is requisite for the experimental foundation of a science of social statistics. Again, whatever other object this treatise may accomplish, it is presumed that it will leave no doubt as to the existence of a system of abstract principles and of methods founded upon those principles, by which any collective body of social data may be made to yield, in an explicit form, whatever information they implicitly involve. There may, where the data are exceedingly complex, be very great difficulty in obtaining this information,—difficulty due not to any imperfection of the theory, but to the laborious character of the analytical processes to which it points. It is quite conceivable that in many instances that difficulty may be such as only united effort could overcome. But that we possess theoretically in all cases, and practically, so far as the requisite labour of calculation may be supplied, the means of evolving from statistical records the seeds of general truths which lie buried amid the mass of figures, is a position which may, I conceive, with perfect safety be affirmed.

19. But beyond these general positions I do not venture to speak in terms of assurance. Whether the results which might be expected from the application of scientific methods to statis-

tical records, over and above those the discovery of which requires no such aid, would so far compensate for the labour involved as to render it worth while to institute such investigations upon a proper scale of magnitude, is a point which could, perhaps, only be determined by experience. It is to be desired, and it might without great presumption be expected, that in this, as in other instances, the abstract doctrines of science should minister to more than intellectual gratification. Nor, viewing the apparent order in which the sciences have been evolved, and have successively contributed their aid to the service of mankind, does it seem very improbable that a day may arrive in which similar aid may accrue from departments of the field of knowledge yet more intimately allied with the elements of human welfare. Let the speculations of this treatise, however, rest at present simply upon their claim to be regarded as true.

20. I design, in the last place, to endeavour to educe from the scientific results of the previous inquiries some general intimations respecting the nature and constitution of the human mind. Into the grounds of the possibility of this species of inference it is not necessary to enter here. One or two general observations may serve to indicate the track which I shall endeavour to follow. It cannot but be admitted that our views of the science of Logic must materially influence, perhaps mainly determine, our opinions upon the nature of the intellectual faculties. For example, the question whether reasoning consists merely in the application of certain first or necessary truths, with which the mind has been originally imprinted, or whether the mind is itself a seat of law, whose operation is as manifest and as conclusive in the particular as in the general formula, or whether, as some not undistinguished writers seem to maintain, all reasoning is of particulars; this question, I say, is one which not merely affects the science of Logic, but also concerns the formation of just views of the constitution of the intellectual faculties. Again, if it is concluded that the mind is by original constitution a seat of law, the question of the nature of its subjection to this law,—whether, for instance, it is an obedience founded upon necessity, like that which sustains the revolutions of the heavens, and preserves the order of Nature,—or whether

it is a subjection of some quite distinct kind, is also a matter of deep speculative interest. Further, if the mind is truly determined to be a subject of law, and if its laws also are truly assigned, the question of their probable or necessary influence upon the course of human thought in different ages is one invested with great importance, and well deserving a patient investigation, as matter both of philosophy and of history. These and other questions I propose, however imperfectly, to discuss in the concluding portion of the present work. They belong, perhaps, to the domain of probable or conjectural, rather than to that of positive, knowledge. But it may happen that where there is not sufficient warrant for the certainties of science, there may be grounds of analogy adequate for the suggestion of highly probable opinions. It has seemed to me better that this discussion should be entirely reserved for the sequel of the main business of this treatise,—which is the investigation of scientific truths and laws. Experience sufficiently instructs us that the proper order of advancement in all inquiries after truth is to proceed from the known to the unknown. There are parts, even of the philosophy and constitution of the human mind, which have been placed fully within the reach of our investigation. To make a due acquaintance with those portions of our nature the basis of all endeavours to penetrate amid the shadows and uncertainties of that conjectural realm which lies beyond and above them, is the course most accordant with the limitations of our present condition.

CHAPTER II.

OF SIGNS IN GENERAL, AND OF THE SIGNS APPROPRIATE TO THE SCIENCE OF LOGIC IN PARTICULAR; ALSO OF THE LAWS TO WHICH THAT CLASS OF SIGNS ARE SUBJECT.

1. THAT Language is an instrument of human reason, and not merely a medium for the expression of thought, is a truth generally admitted. It is proposed in this chapter to inquire what it is that renders Language thus subservient to the most important of our intellectual faculties. In the various steps of this inquiry we shall be led to consider the constitution of Language, considered as a system adapted to an end or purpose; to investigate its elements; to seek to determine their mutual relation and dependence; and to inquire in what manner they contribute to the attainment of the end to which, as co-ordinate parts of a system, they have respect.

In proceeding to these inquiries, it will not be necessary to enter into the discussion of that famous question of the schools, whether Language is to be regarded as an *essential* instrument of reasoning, or whether, on the other hand, it is possible for us to reason without its aid. I suppose this question to be beside the design of the present treatise, for the following reason, viz., that it is the business of Science to investigate laws; and that, whether we regard signs as the representatives of things and of their relations, or as the representatives of the conceptions and operations of the human intellect, in studying the laws of signs, we are in effect studying the manifested laws of reasoning. If there exists a difference between the two inquiries, it is one which does not affect the scientific expressions of formal law, which are the object of investigation in the present stage of this work, but relates only to the mode in which those results are presented to the mental regard. For though in investigating the laws of signs, *à posteriori*, the immediate subject of examination is Language, with the rules which govern its use; while in making the internal processes of thought the direct object of inquiry, we appeal in a more immediate way to our personal consciousness,—it will be found that in both cases the results obtained are formally equivalent. Nor could we easily conceive, that the unnumbered tongues and dialects of the earth should have preserved through a long succession of ages so much that is common and universal, were we not assured of the existence of some deep foundation of their agreement in the laws of the mind itself.

2. The elements of which all language consists are signs or symbols. Words are signs. Sometimes they are said to represent things; sometimes the operations by which the mind combines together the simple notions of things into complex conceptions; sometimes they express the relations of action, passion, or mere quality, which we perceive to exist among the objects of our experience; sometimes the emotions of the perceiving mind. But words, although in this and in other ways they fulfil the office of signs, or representative symbols, are not the only signs which we are capable of employing. Arbitrary marks, which speak only to the eye, and arbitrary sounds or actions, which address themselves to some other sense, are equally of the nature of signs, provided that their representative office is defined and understood. In the mathematical sciences, letters, and the symbols $+$, $-$, $=$, &c., are used as signs, although the term "sign" is applied to the latter class of symbols, which represent operations or relations, rather than to the former, which represent the elements of number and quantity. As the real import of a sign does not in any way depend upon its particular form or expression, so neither do the laws which determine its use. In the present treatise, however, it is with written signs that we have to do, and it is with reference to these exclusively that the term "sign" will be employed. The essential properties of signs are enumerated in the following definition.

Definition.—A sign is an arbitrary mark, having a fixed interpretation, and susceptible of combination with other signs in subjection to fixed laws dependent upon their mutual interpretation.

3. Let us consider the particulars involved in the above definition separately.

(1.) In the first place, a sign is an *arbitrary* mark. It is clearly indifferent what particular word or token we associate with a given idea, provided that the association once made is permanent. The Romans expressed by the word "civitas" what we designate by the word "state." But both they and we might equally well have employed any other word to represent the same conception. Nothing, indeed, in the nature of Language would prevent us from using a mere letter in the same sense. Were this done, the laws according to which that letter would require to be used would be essentially the same with the laws which govern the use of "civitas" in the Latin, and of "state" in the English language, so far at least as the use of those words is regulated by any general principles common to all languages alike.

(2.) In the second place, it is necessary that each sign should possess, within the limits of the same discourse or process of reasoning, a fixed interpretation. The necessity of this condition is obvious, and seems to be founded in the very nature of the subject. There exists, however, a dispute as to the precise nature of the representative office of words or symbols used as names in the processes of reasoning. By some it is maintained, that they represent the conceptions of the mind alone; by others, that they represent things. The question is not of great importance here, as its decision cannot affect the laws according to which signs are employed. I apprehend, however, that the general answer to this and such like questions is, that in the processes of reasoning, signs stand in the place and fulfil the office of the conceptions and operations of the mind; but that as those conceptions and operations represent things, and the connexions and relations of things, so signs represent things with their connexions and relations; and lastly, that as signs stand in the place of the conceptions and operations of the mind, they are subject to the laws of those conceptions and operations. This view will be more fully elucidated in the next chapter; but it here serves to explain the third of those particulars involved in the definition of a sign, viz., its subjection to fixed laws of combination depending upon the nature of its interpretation.

4. The analysis and classification of those signs by which the

operations of reasoning are conducted will be considered in the following Proposition:

PROPOSITION I.

All the operations of Language, as an instrument of reasoning, may be conducted by a system of signs composed of the following elements, viz.:

1st. *Literal symbols, as x, y, &c., representing things as subjects of our conceptions.*

2nd. *Signs of operation, as $+$, $-$, $\times$, standing for those operations of the mind by which the conceptions of things are combined or resolved so as to form new conceptions involving the same elements.*

3rd. *The sign of identity, $=$.*

And these symbols of Logic are in their use subject to definite laws, partly agreeing with and partly differing from the laws of the corresponding symbols in the science of Algebra.

Let it be assumed as a criterion of the true elements of rational discourse, that they should be susceptible of combination in the simplest forms and by the simplest laws, and thus combining should generate all other known and conceivable forms of language; and adopting this principle, let the following classification be considered.

CLASS I.

5. *Appellative or descriptive signs, expressing either the name of a thing, or some quality or circumstance belonging to it.*

To this class we may obviously refer the substantive proper or common, and the adjective. These may indeed be regarded as differing only in this respect, that the former expresses the substantive existence of the individual thing or things to which it refers; the latter implies that existence. If we attach to the adjective the universally understood subject "being" or "thing," it becomes virtually a substantive, and may for all the essential purposes of reasoning be replaced by the substantive. Whether or not, in every particular of the mental regard, it is the same thing to say, "Water is a fluid thing," as to say, "Water is fluid;" it is at least equivalent in the expression of the processes of reasoning.

It is clear also, that to the above class we must refer any sign which may conventionally be used to express some circumstance or relation, the detailed exposition of which would involve the use of many signs. The epithets of poetic diction are very frequently of this kind. They are usually compounded adjectives, singly fulfilling the office of a many-worded description. Homer's "deep-eddying ocean" embodies a virtual description in the single word βαθυδίνης. And conventionally any other description addressed either to the imagination or to the intellect might equally be represented by a single sign, the use of which would in all essential points be subject to the same laws as the use of the adjective "good" or "great." Combined with the subject "thing," such a sign would virtually become a substantive; and by a single substantive the combined meaning both of thing and quality might be expressed.

6. Now, as it has been defined that a sign is an arbitrary mark, it is permissible to replace all signs of the species above described by letters. Let us then agree to represent the class of individuals to which a particular name or description is applicable, by a single letter, as x. If the name is "men," for instance, let x represent "all men," or the class "men." By a class is usually meant a collection of individuals, to each of which a particular name or description may be applied; but in this work the meaning of the term will be extended so as to include the case in which but a single individual exists, answering to the required name or description, as well as the cases denoted by the terms "nothing" and "universe," which as "classes" should be understood to comprise respectively "no beings," "all beings." Again, if an adjective, as "good," is employed as a term of description, let us represent by a letter, as y, all things to which the description "good" is applicable, i. e. "all good things," or the class "good things." Let it further be agreed, that by the combination xy shall be represented that class of things to which the names or descriptions represented by x and y are simultaneously applicable. Thus, if x alone stands for "white things," and y for "sheep," let xy stand for "white sheep;" and in like manner, if z stand for "horned things," and x and y retain their previous interpretations, let zxy represent

"horned white sheep," i.e. that collection of things to which the name "sheep," and the descriptions "white" and "horned" are together applicable.

Let us now consider the laws to which the symbols x, y, &c., used in the above sense, are subject.

7. First, it is evident, that according to the above combinations, the order in which two symbols are written is indifferent. The expressions xy and yx equally represent that class of things to the several members of which the names or descriptions x and y are together applicable. Hence we have,

$$xy = yx. \qquad (1)$$

In the case of x representing white things, and y sheep, either of the members of this equation will represent the class of "white sheep." There may be a difference as to the order in which the conception is formed, but there is none as to the individual things which are comprehended under it. In like manner, if x represent "estuaries," and y "rivers," the expressions xy and yx will indifferently represent "rivers that are estuaries," or "estuaries that are rivers," the combination in this case being in ordinary language that of two substantives, instead of that of a substantive and an adjective as in the previous instance. Let there be a third symbol, as z, representing that class of things to which the term "navigable" is applicable, and any one of the following expressions,

$$zxy, \quad zyx, \quad xyz, \text{ \&c.},$$

will represent the class of "navigable rivers that are estuaries."

If one of the descriptive terms should have some implied reference to another, it is only necessary to include that reference expressly in its stated meaning, in order to render the above remarks still applicable. Thus, if x represent "wise" and y "counsellor," we shall have to define whether x implies wisdom in the absolute sense, or only the wisdom of counsel. With such definition the law $xy = yx$ continues to be valid.

We are permitted, therefore, to employ the symbols x, y, z, &c., in the place of the substantives, adjectives, and descriptive phrases subject to the rule of interpretation, that any expression in which several of these symbols are written together shall represent all the objects or indi-

viduals to which their several meanings are together applicable, and to the law that the order in which the symbols succeed each other is indifferent.

As the rule of interpretation has been sufficiently exemplified, I shall deem it unnecessary always to express the subject "things" in defining the interpretation of a symbol used for an adjective. When I say, let x represent "good," it will be understood that x only represents "good" when a subject for that quality is supplied by another symbol, and that, used alone, its interpretation will be "good things."

8. Concerning the law above determined, the following observations, which will also be more or less appropriate to certain other laws to be deduced hereafter, may be added.

First, I would remark, that this law is a law of thought, and not, properly speaking, a law of things. Difference in the order of the qualities or attributes of an object, apart from all questions of causation, is a difference in conception merely. The law (1) expresses as a general truth, that the same thing may be conceived in different ways, and states the nature of that difference; and it does no more than this.

Secondly, As a law of thought, it is actually developed in a law of Language, the product and the instrument of thought. Though the tendency of prose writing is toward uniformity, yet even there the order of sequence of adjectives absolute in their meaning, and applied to the same subject, is indifferent, but poetic diction borrows much of its rich diversity from the extension of the same lawful freedom to the substantive also. The language of Milton is peculiarly distinguished by this species of variety. Not only does the substantive often precede the adjectives by which it is qualified, but it is frequently placed in their midst. In the first few lines of the invocation to Light, we meet with such examples as the following:

"*Offspring of heaven first-born.*"

"The rising world of *waters dark and deep.*"

"Bright effluence of *bright essence increate.*"

Now these inverted forms are not simply the fruits of a poetic license. They are the natural expressions of a freedom sanctioned by the intimate laws of thought, but for reasons of convenience not exercised in the ordinary use of language.

Thirdly, The law expressed by (1) may be characterized by saying that the literal symbols x, y, z, are *commutative, like the symbols of Algebra*. In saying this, it is not affirmed that the process of multiplication in Algebra, of which the fundamental law is expressed by the equation

$$xy = yx,$$

possesses in itself any analogy with that process of logical combination which xy has been made to represent above; but only that if the arithmetical and the logical process are expressed in the same manner, their symbolical expressions will be subject to the same formal law. The evidence of that subjection is in the two cases quite distinct.

9. As the combination of two literal symbols in the form xy expresses the whole of that class of objects to which the names or qualities represented by x and y are together applicable, it follows that if the two symbols have exactly the same signification, their combination expresses no more than either of the symbols taken alone would do. In such case we should therefore have

$$xy = x.$$

As y is, however, supposed to have the same meaning as x, we may replace it in the above equation by x, and we thus get

$$xx = x.$$

Now in common Algebra the combination xx is more briefly represented by x^2. Let us adopt the same principle of notation here; for the mode of expressing a particular succession of mental operations is a thing in itself quite as arbitrary as the mode of expressing a single idea or operation (II. 3). In accordance with this notation, then, the above equation assumes the form

$$x^2 = x, \qquad (2)$$

and is, in fact, the expression of a second general law of those symbols by which names, qualities, or descriptions, are symbolically represented.

The reader must bear in mind that although the symbols x and y in the examples previously formed received significations distinct from each other, nothing prevents us from attributing to them precisely the same signification. It is evident that the more nearly their actual significations approach to each other, the more nearly does the class of things denoted by the combination xy approach to identity with the class denoted by x, as well as with that denoted by y. The case supposed in the demonstration of the equation (2) is that of *absolute* identity of meaning. The law which it expresses is practically exemplified in language. To say "good, good," in relation to any subject, though a cumbrous and useless pleonasm, is the same as to say "good." Thus "good, good" men, is equivalent to "good" men. Such repetitions of words are indeed sometimes employed to heighten a quality or strengthen an affirmation. But this effect is merely secondary and conventional; it is not founded in the intrinsic relations of language and thought. Most of the operations which we observe in nature, or perform ourselves, are of such a kind that their effect is augmented by repetition, and this circumstance prepares us to expect the same thing in language, and even to use repetition when we design to speak with emphasis. But neither in strict reasoning nor in exact discourse is there any just ground for such a practice.

10. We pass now to the consideration of another class of the signs of speech, and of the laws connected with their use.

CLASS II.

11. *Signs of those mental operations whereby we collect parts into a whole, or separate a whole into its parts.*

We are not only capable of entertaining the conceptions of objects, as characterized by names, qualities, or circumstances, applicable to each individual of the group under consideration, but also of forming the aggregate conception of a group of objects consisting of partial groups, each of which is separately named or described. For this purpose we use the conjunctions "and," "or," &c. "Trees and minerals," "barren mountains, or fertile vales," are examples of this kind. In strictness, the words "and," "or," interposed between the terms descriptive of two or more classes of objects, imply that those classes are quite distinct, so that no member of one is found in another. In this and in all other respects the words "and" "or" are analogous with the sign + in algebra, and their laws are identical. Thus the expression "men and women" is, conventional meanings set aside, equivalent with the expression "women and men." Let x represent "men," y, "women;" and let + stand for "*and*" and "*or*," then we have

$$x + y = y + x, \tag{3}$$

an equation which would equally hold true if x and y represented *numbers*, and + were the sign of arithmetical addition.

Let the symbol z stand for the adjective "European," then since it is, in effect, the same thing to say "European men and women," as to say "European men and European women," we have

$$z(x + y) = zx + zy. \tag{4}$$

And this equation also would be equally true were x, y, and z symbols of number, and were the juxtaposition of two literal symbols to represent their algebraic product, just as in the logical signification previously given, it represents the class of objects to which both the epithets conjoined belong.

The above are the laws which govern the use of the sign +, here used to denote the positive operation of aggregating parts into a whole. But the very idea of an operation effecting some positive change seems to suggest to us the idea of an opposite or negative operation, having the effect of undoing what the former one has done. Thus we cannot conceive it possible to collect parts into a whole, and not conceive it also possible to separate a part from a whole. This operation we express in common language by the sign *except*, as, "All men *except* Asiatics," "All states *except* those which are monarchical." Here it is implied that the things excepted form a part of the things from which they are excepted. As we have expressed the operation of aggregation by the sign +, so we may express the negative operation above described by − minus. Thus if x be taken to represent men, and y, Asiatics, i. e. Asiatic men, then the conception of "All men except Asiatics" will be expressed by $x - y$. And if we represent by x, "states," and by y the descriptive property "having a monarchical form," then the conception of "All states except those which are monarchical" will be expressed by $x - xy$.

As it is indifferent for all the *essential* purposes of reasoning whether we express excepted cases first or last in the order of speech, it is also indifferent in what order we write any series of terms, some of which are affected by the sign $-$. Thus we have, as in the common algebra,

$$x - y = - y + x. \quad (5)$$

Still representing by x the class "men," and by y "Asiatics," let z represent the adjective "white." Now to apply the adjective "white" to the collection of men expressed by the phrase "Men except Asiatics," is the same as to say, "White men, except white Asiatics." Hence we have

$$z(x - y) = zx - zy. \quad (6)$$

This is also in accordance with the laws of ordinary algebra.

The equations (4) and (6) may be considered as exemplification of a single general law, which may be stated by saying, *that the literal symbols, x, y, z, &c. are distributive in their operation.* The general fact which that law expresses is this, viz.:—If any quality or circumstance is ascribed to all the members of a group, formed either by aggregation or exclusion of partial groups, the resulting conception is the same as if the quality or circumstance were first ascribed to each member of the partial groups, and the aggregation or exclusion effected afterwards. That which is ascribed to the members of the whole is ascribed to the members of all its parts, howsoever those parts are connected together.

CLASS III.

12. *Signs by which relation is expressed, and by which we form propositions.*

Though all verbs may with propriety be referred to this class, it is sufficient for the purposes of Logic to consider it as including only the substantive verb *is* or *are*, since every other verb may be resolved into this element, and one of the signs included under Class I. For as those signs are used to express quality or circumstance of every kind, they may be employed to express the active or passive relation of the subject of the verb, considered with reference either to past, to present, or to future time. Thus the Proposition, "Cæsar conquered the Gauls," may be resolved into "Cæsar is he who conquered the Gauls." The ground of this analysis I conceive to be the following:—Unless we understand what is meant by having conquered the Gauls, i. e. by the expression "One who conquered the Gauls," we cannot understand the sentence in question. It is, therefore, truly an element of that sentence; another element is "Cæsar," and there is yet another required, the copula *is*, to show the connexion of these two. I do not, however, affirm that there is no other mode than the above of contemplating the relation expressed by the proposition, "Cæsar conquered the Gauls;" but only that the analysis here given is a correct one for the particular point of view which has been taken, and that it suffices for the purposes of logical deduction. It may be remarked that the passive and future participles of the Greek language imply the existence of the principle which has been asserted, viz.: that the sign *is* or *are* may be regarded as an element of every personal verb.

13. The above sign, *is* or *are*, may be expressed by the symbol $=$. The laws, or as would usually be said, the axioms which the symbol introduces, are next to be considered.

Let us take the Proposition, "The stars are the suns and the planets," and let us represent stars by x, suns by y, and planets by z; we have then

$$x = y + z. \tag{7}$$

Now if it be true that the stars are the suns and the planets, it will follow that the stars, except the planets, are suns. This would give the equation

$$x - z = y, \tag{8}$$

which must therefore be a deduction from (7). Thus a term z has been removed from one side of an equation to the other by changing its sign. This is in accordance with the algebraic rule of transposition.

But instead of dwelling upon particular cases, we may at once affirm the general axioms :—

1st. If equal things are added to equal things, the wholes are equal.

2nd. If equal things are taken from equal things, the remainders are equal.

And it hence appears that we may add or subtract equations, and employ the rule of transposition above given just as in common algebra.

Again : If two classes of things, x and y, be identical, that is, if all the members of the one are members of the other, then those members of the one class which possess a given property z will be identical with those members of the other which possess the same property z. Hence if we have the equation

$$x = y ;$$

then whatever class or property z may represent, we have also

$$zx = zy.$$

This is formally the same as the algebraic law :—If both members of an equation are multiplied by the same quantity, the products are equal.

In like manner it may be shown that if the corresponding members of two equations are multiplied together, the resulting equation is true.

14. Here, however, the analogy of the present system with that of algebra, as commonly stated, appears to stop. Suppose it true that those members of a class x which possess a certain property z are identical with those members of a class y which possess the same property z, it does not follow that the members of the class x universally are identical with the members of the class y. Hence it cannot be inferred from the equation

$$zx = zy,$$

that the equation

$$x = y$$

is also true. In other words, the axiom of algebraists, that both sides of an equation may be divided by the same quantity, has no formal equivalent here. I say no *formal equivalent*, because, in accordance with the general spirit of these inquiries, it is not even sought to determine whether the mental operation which is represented by removing a logical symbol, z, from a combination zx, is in itself analogous with the operation of division in Arithmetic. That mental operation is indeed identical with what is commonly termed Abstraction, and it will hereafter appear that its laws are dependent upon the laws already deduced in this chapter. What has now been shown is, that there does not exist among those laws anything analogous in *form* with a commonly received axiom of Algebra.

But a little consideration will show that even in common algebra that axiom does not possess the generality of those other axioms which have been considered. The deduction of the equation $x = y$ from the equation $zx = zy$ is only valid when it is known that z is not equal to 0. If then the value $z = 0$ is supposed to be admissible in the algebraic system, the axiom above stated ceases to be applicable, and the analogy before exemplified remains at least unbroken.

15. However, it is not with the symbols of quantity generally that it is of any importance, except as a matter of speculation, to trace such affinities. We have seen (II. 9) that the symbols of Logic are subject to the special law,

$$x^2 = x.$$

Now of the symbols of Number there are but two, viz. 0 and 1, which are subject to the same formal law. We know that $0^2 = 0$, and that $1^2 = 1$; and the equation $x^2 = x$, considered as algebraic, has no other roots than 0 and 1. Hence, instead of determining the measure of formal agreement of the symbols of Logic with those of Number generally, it is more immediately suggested to us to compare them with symbols of quantity *admitting only of the values* 0 *and* 1. Let us conceive, then, of an Algebra in which the symbols x, y, z, &c. admit indifferently of the values 0 and 1, and of these values alone. The laws, the axioms, and the processes, of such an Algebra will be identical in their whole extent with the laws, the axioms, and the processes of an Algebra of Logic. Difference of interpretation will alone divide them. Upon this principle the method of the following work is established.

16. It now remains to show that those constituent parts of ordinary language which have not been considered in the previous sections of this chapter are either resolvable into the same elements as those which have been considered, or are subsidiary to those elements by contributing to their more precise definition.

The substantive, the adjective, and the verb, together with the particles *and*, *except*, we have already considered. The pronoun may be regarded as a particular form of the substantive or the adjective. The adverb modifies the meaning of the verb, but does not affect its nature. Prepositions contribute to the expression of circumstance or relation, and thus tend to give precision and detail to the meaning of the literal symbols. The conjunctions *if*, *either*, *or*, are used chiefly in the expression of relation among propositions, and it will hereafter be shown that the same relations can be completely expressed by elementary symbols analogous in interpretation, and identical in form and law with the symbols whose use and meaning have been explained in this Chapter. As to any remaining elements of speech, it will, upon examination, be found that they are used either to give a more definite significance to the terms of discourse, and thus enter into the interpretation of the literal symbols already considered, or to express some emotion or state of feeling accompanying the utterance of a proposition, and thus do not belong to the province of the understanding, with which alone our present concern lies. Experience of its use will testify to the sufficiency of the classification which has been adopted.

CHAPTER III.

DERIVATION OF THE LAWS OF THE SYMBOLS OF LOGIC FROM THE LAWS OF THE OPERATIONS OF THE HUMAN MIND.

1. THE object of science, properly so called, is the knowledge of laws and relations. To be able to distinguish what is essential to this end, from what is only accidentally associated with it, is one of the most important conditions of scientific progress. I say, to *distinguish* between these elements, because a consistent devotion to science does not require that the attention should be altogether withdrawn from other speculations, often of a metaphysical nature, with which it is not unfrequently connected. Such questions, for instance, as the existence of a sustaining ground of phænomena, the reality of cause, the propriety of forms of speech implying that the successive states of things are connected by *operations*, and others of a like nature, may possess a deep interest and significance in relation to science, without being essentially scientific. It is indeed scarcely possible to express the conclusions of natural science without borrowing the language of these conceptions. Nor is there necessarily any practical inconvenience arising from this source. They who believe, and they who refuse to believe, that there is more in the relation of cause and effect than an invariable order of succession, agree in their interpretation of the conclusions of physical astronomy. But they only agree because they recognise a common element of scientific truth, which is independent of their particular views of the nature of causation.

2. If this distinction is important in physical science, much more does it deserve attention in connexion with the science of the intellectual powers. For the questions which this science presents become, in expression at least, almost necessarily mixed up with modes of thought and language, which betray a metaphysical origin. The idealist would give to the laws of reasoning

one form of expression; the sceptic, if true to his principles, another. They who regard the phænomena with which we are concerned in this inquiry as the mere successive *states* of the thinking subject devoid of any causal connexion, and they who refer them to the *operations* of an active intelligence, would, if consistent, equally differ in their modes of statement. Like difference would also result from a difference of classification of the mental faculties. Now the principle which I would here assert, as affording us the only ground of confidence and stability amid so much of seeming and of real diversity, is the following, viz., that if the laws in question are really deduced from observation, they have a real existence as laws of the human mind, independently of any metaphysical theory which may seem to be involved in the mode of their statement. They contain an element of truth which no ulterior criticism upon the nature, or even upon the reality, of the mind's operations, can essentially affect. Let it even be granted that the mind is but a succession of states of consciousness, a series of fleeting impressions uncaused from without or from within, emerging out of nothing, and returning into nothing again,—the last refinement of the sceptic intellect,—still, as laws of succession, or at least of a past succession, the results to which observation had led would remain true. They would require to be interpreted into a language from whose vocabulary all such terms as cause and effect, operation and subject, substance and attribute, had been banished; but they would still be valid as scientific truths.

Moreover, as any statement of the laws of thought, founded upon actual observation, must thus contain scientific elements which are independent of metaphysical theories of the nature of the mind, the practical application of such elements to the construction of a system or method of reasoning must also be independent of metaphysical distinctions. For it is upon the scientific elements involved in the statement of the laws, that any practical application will rest, just as the practical conclusions of physical astronomy are independent of any theory of the cause of gravitation, but rest only on the knowledge of its phænomenal effects. And, therefore, as respects both the determination of the laws of thought, and the practical use of them when discovered, we are, for all really scientific ends, unconcerned with the truth or falsehood of any metaphysical speculations whatever.

3. The course which it appears to me to be expedient, under these circumstances, to adopt, is to avail myself as far as possible of the language of common discourse, without regard to any theory of the nature and powers of the mind which it may be thought to embody. For instance, it is agreeable to common usage to say that we converse with each other by the communication of ideas, or conceptions, such communication being the office of words; and that with reference to any particular ideas or conceptions presented to it, the mind possesses certain powers or faculties by which the mental regard may be fixed upon some ideas, to the exclusion of others, or by which the given conceptions or ideas may, in various ways, be combined together. To those faculties or powers different names, as Attention, Simple Apprehension, Conception or Imagination, Abstraction, &c., have been given,—names which have not only furnished the titles of distinct divisions of the philosophy of the human mind, but passed into the common language of men. Whenever, then, occasion shall occur to use these terms, I shall do so without implying thereby that I accept the theory that the mind possesses such and such powers and faculties as distinct elements of its activity. Nor is it indeed necessary to inquire whether such powers of the understanding have a distinct existence or not. We may merge these different titles under the one generic name of *Operations* of the human mind, define these operations so far as is necessary for the purposes of this work, and then seek to express their ultimate laws. Such will be the general order of the course which I shall pursue, though reference will occasionally be made to the names which common agreement has assigned to the particular states or operations of the mind which may fall under our notice.

It will be most convenient to distribute the more definite results of the following investigation into distinct Propositions.

PROPOSITION I.

4. *To deduce the laws of the symbols of Logic from a consideration of those operations of the mind which are implied in the strict use of language as an instrument of reasoning.*

In every discourse, whether of the mind conversing with its own thoughts, or of the individual in his intercourse with others, there is an assumed or expressed limit within which the subjects of its operation are confined. The most unfettered discourse is that in which the words we use are understood in the widest possible application, and for them the limits of discourse are co-extensive with those of the universe itself. But more usually we confine ourselves to a less spacious field. Sometimes, in discoursing of men we imply (without expressing the limitation) that it is of men only under certain circumstances and conditions that we speak, as of civilized men, or of men in the vigour of life, or of men under some other condition or relation. Now, whatever may be the extent of the field within which all the objects of our discourse are found, that field may properly be termed the universe of discourse.

5. Furthermore, this universe of discourse is in the strictest sense the ultimate *subject* of the discourse. The office of any name or descriptive term employed under the limitations supposed is not to raise in the mind the conception of all the beings or objects to which that name or description is applicable, but only of those which exist within the supposed universe of discourse. If that universe of discourse is the actual universe of things, which it always is when our words are taken in their real and literal sense, then by men we mean *all men that exist*; but if the universe of discourse is limited by any antecedent implied understanding, then it is of men under the limitation thus introduced that we speak. It is in both cases the business of the word *men* to direct a certain operation of the mind, by which, from the proper universe of discourse, we select or fix upon the individuals signified.

6. Exactly of the same kind is the mental operation implied by the use of an adjective. Let, for instance, the universe of discourse be the actual Universe. Then, as the word *men* directs

us to select mentally from that Universe all the beings to which the term "men" is applicable; so the adjective "good," in the combination "good men," directs us still further to select mentally from the class of *men* all those who possess the further quality "good;" and if another adjective were prefixed to the combination "good men," it would direct a further operation of the same nature, having reference to that further quality which it might be chosen to express.

It is important to notice carefully the real nature of the operation here described, for it is conceivable, that it might have been different from what it is. Were the adjective simply *attributive* in its character, it would seem, that when a particular set of beings is designated by *men*, the prefixing of the adjective *good* would direct us to attach mentally to all those beings the quality of goodness. But this is not the real office of the adjective. The operation which we really perform is one of *selection according to a prescribed principle or idea.* To what faculties of the mind such an operation would be referred, according to the received classification of its powers, it is not important to inquire, but I suppose that it would be considered as dependent upon the two faculties of Conception or Imagination, and Attention. To the one of these faculties might be referred the formation of the general conception; to the other the fixing of the mental regard upon those individuals within the prescribed universe of discourse which answer to the conception. If, however, as seems not improbable, the power of Attention is nothing more than the power of continuing the exercise of any other faculty of the mind, we might properly regard the whole of the mental process above described as referrible to the mental faculty of Imagination or Conception, the first step of the process being the conception of the Universe itself, and each succeeding step limiting in a definite manner the conception thus formed. Adopting this view, I shall describe each such step, or any definite combination of such steps, as a *definite act of conception.* And the use of this term I shall extend so as to include in its meaning not only the conception of classes of objects represented by particular names or simple attributes of quality, but also the combination of such conceptions in any manner consistent with the powers and limitations of the human mind; indeed, any intellectual operation short of that which is involved in the structure of a sentence or proposition. The general laws to which such operations of the mind are subject are now to be considered.

7. Now it will be shown that the laws which in the preceding chapter have been determined *à posteriori* from the constitution of language, for the use of the literal symbols of Logic, are in reality the laws of that definite mental operation which has just been described. We commence our discourse with a certain understanding as to the limits of its subject, i. e. as to the limits of its Universe. Every name, every term of description that we employ, directs him whom we address to the performance of a certain mental operation upon that subject. And thus is thought communicated. But as each name or descriptive term is in this view but the representative of an intellectual operation, that operation being also prior in the order of nature, it is clear that the laws of the name or symbol must be of a derivative character,—must, in fact, originate in those of the operation which they represent. That the laws of the symbol and of the mental process are identical in expression will now be shown.

8. Let us then suppose that the universe of our discourse is the actual universe, so that words are to be used in the full extent of their meaning, and let us consider the two mental operations implied by the words "white" and "men." The word "men" implies the operation of selecting in thought from its subject, the universe, all men; and the resulting conception, *men*, becomes the subject of the next operation. The operation implied by the word "white" is that of selecting from its subject, "men," all of that class which are white. The final resulting conception is that of "white men." Now it is perfectly apparent that if the operations above described had been performed in a converse order, the result would have been the same. Whether we begin by forming the conception of "*men*," and then by a second intellectual act limit that conception to "white men," or whether we begin by forming the conception of "white objects," and then limit it to such of that class as are "men," is perfectly indifferent so far as the result is concerned. It is obvious that the order of the mental processes would be equally

indifferent if for the words "white" and "men" we substituted any other descriptive or appellative terms whatever, provided only that their meaning was fixed and absolute. And thus the indifference of the order of two successive acts of the faculty of Conception, the one of which furnishes the subject upon which the other is supposed to operate, is a general condition of the exercise of that faculty. It is a law of the mind, and it is the real origin of that law of the literal symbols of Logic which constitutes its formal expression (1) Chap. II.

9. It is equally clear that the mental operation above described is of such a nature that its effect is not altered by repetition. Suppose that by a definite act of conception the attention has been fixed upon men, and that by another exercise of the same faculty we limit it to those of the race who are white. Then any further repetition of the latter mental act, by which the attention is limited to white objects, does not in any way modify the conception arrived at, viz., that of white men. This is also an example of a general law of the mind, and it has its formal expression in the law ((2) Chap. II.) of the literal symbols.

10. Again, it is manifest that from the conceptions of two distinct classes of things we can form the conception of that collection of things which the two classes taken together compose; and it is obviously indifferent in what order of position or of priority those classes are presented to the mental view. This is another general law of the mind, and its expression is found in (3) Chap. II.

11. It is not necessary to pursue this course of inquiry and comparison. Sufficient illustration has been given to render manifest the two following positions, viz.:

First, That the operations of the mind, by which, in the exercise of its power of imagination or conception, it combines and modifies the simple ideas of things or qualities, not less than those operations of the reason which are exercised upon truths and propositions, are subject to general laws.

Secondly, That those laws are mathematical in their form, and that they are actually developed in the essential laws of human language. Wherefore the laws of the symbols of Logic are deducible from a consideration of the operations of the mind in reasoning.

12. The remainder of this chapter will be occupied with questions relating to that law of thought whose expression is $x^2 = x$ (II. 9), a law which, as has been implied (II. 15), forms the characteristic distinction of the operations of the mind in its ordinary discourse and reasoning, as compared with its operations when occupied with the general algebra of quantity. An important part of the following inquiry will consist in proving that the symbols 0 and 1 occupy a place, and are susceptible of an interpretation, among the symbols of Logic; and it may first be necessary to show how particular symbols, such as the above, may with propriety and advantage be employed in the representation of distinct systems of thought.

The ground of this propriety cannot consist in any community of interpretation. For in systems of thought so truly distinct as those of Logic and Arithmetic (I use the latter term in its widest sense as the science of Number), there is, properly speaking, no community of subject. The one of them is conversant with the very conceptions of things, the other takes account solely of their numerical relations. But inasmuch as the forms and methods of any system of reasoning depend immediately upon the laws to which the symbols are subject, and only mediately, through the above link of connexion, upon their interpretation, there may be both propriety and advantage in employing the same symbols in different systems of thought, provided that such interpretations can be assigned to them as shall render their formal laws identical, and their use consistent. The ground of that employment will not then be community of interpretation, but the community of the formal laws, to which in their respective systems they are subject. Nor must that community of formal laws be established upon any other ground than that of a careful observation and comparison of those results which are seen to flow independently from the interpretations of the systems under consideration.

These observations will explain the process of inquiry adopted in the following Proposition. The literal symbols of Logic are universally subject to the law whose expression is $x^2 = x$. Of the symbols of Number there are two only, 0 and 1, which satisfy this law. But each of these symbols is also subject to a law peculiar to itself in the system of numerical magnitude, and this suggests the inquiry, what interpretations must be given to the literal symbols of Logic, in order that the same peculiar and formal laws may be realized in the logical system also.

Proposition II.

13. *To determine the logical value and significance of the symbols* 0 *and* 1.

The symbol 0, as used in Algebra, satisfies the following formal law,

$$0 \times y = 0, \text{ or } 0y = 0, \qquad (1)$$

whatever *number* y may represent. That this formal law may be obeyed in the system of Logic, we must assign to the symbol 0 such an interpretation that the *class* represented by $0y$ may be identical with the class represented by 0, whatever the class y may be. A little consideration will show that this condition is satisfied if the symbol 0 represent Nothing. In accordance with a previous definition, we may term Nothing a class. In fact, Nothing and Universe are the two limits of class extension, for they are the limits of the possible interpretations of general names, none of which can relate to fewer individuals than are comprised in Nothing, or to more than are comprised in the Universe. Now whatever the class y may be, the individuals which are common to it and to the class "Nothing" are identical with those comprised in the class "Nothing," for they are none. And thus by assigning to 0 the interpretation Nothing, the law (1) is satisfied; and it is not otherwise satisfied consistently with the perfectly general character of the class y.

Secondly, The symbol 1 satisfies in the system of Number the following law, viz.,

$$1 \times y = y, \text{ or } 1y = y,$$

whatever number y may represent. And this formal equation being assumed as equally valid in the system of this work, in which 1 and y represent classes, it appears that the symbol 1 must represent such a class that all the individuals which are found in *any* proposed class y are also all the individuals $1y$ that are common to that class y and the class represented by 1. A little consideration will here show that the class represented by 1 must be "the Universe," since this is the only class in which are found *all* the individuals that exist in *any* class. Hence the respective interpretations of the symbols 0 and 1 in the system of Logic are *Nothing* and *Universe*.

14. As with the idea of any class of objects as "men," there is suggested to the mind the idea of the contrary class of beings which are not men; and as the whole Universe is made up of these two classes together, since of every individual which it comprehends we may affirm either that it is a man, or that it is not a man, it becomes important to inquire how such contrary names are to be expressed. Such is the object of the following Proposition.

Proposition III.

If x represent any class of objects, then will $1 - x$ *represent the contrary or supplementary class of objects*, i. e. *the class including all objects which are not comprehended in the class x.*

For greater distinctness of conception let x represent the class *men*, and let us express, according to the last Proposition, the Universe by 1; now if from the conception of the Universe, as consisting of "men" and "not-men," we exclude the conception of "men," the resulting conception is that of the contrary class, "not-men." Hence the class "not-men" will be represented by $1 - x$. And, in general, whatever class of objects is represented by the symbol x, the contrary class will be expressed by $1 - x$.

15. Although the following Proposition belongs in strictness to a future chapter of this work, devoted to the subject of *maxims* or *necessary truths*, yet, on account of the great importance of that law of thought to which it relates, it has been thought proper to introduce it here.

PROPOSITION IV.

That axiom of metaphysicians which is termed the principle of contradiction, and which affirms that it is impossible for any being to possess a quality, and at the same time not to possess it, is a consequence of the fundamental law of thought, whose expression is $x^2 = x$.

Let us write this equation in the form

$$x - x^2 = 0,$$

whence we have

$$x(1 - x) = 0; \qquad (1)$$

both these transformations being justified by the axiomatic laws of combination and transposition (II. 13). Let us, for simplicity of conception, give to the symbol x the particular interpretation of *men*, then $1 - x$ will represent the class of "not-men" (Prop. III.) Now the formal product of the expressions of two classes represents that class of individuals which is common to them both (II. 6). Hence $x(1 - x)$ will represent the class whose members are at once "men," and "not men," and the equation (1) thus express the principle, *that a class whose members are at the same time men and not men does not exist.* In other words, that *it is impossible for the same individual to be at the same time a man and not a man.* Now let the meaning of the symbol x be extended from the representing of "men," to that of any class of beings characterized by the possession of any quality whatever; and the equation (1) will then express that it is impossible for a being to possess a quality and not to possess that quality at the same time. But this is identically that "principle of contradiction" which Aristotle has described as the fundamental axiom of all philosophy. "It is impossible that the same quality should both belong and not belong to the same thing. . . This is the most certain of all principles. . . Wherefore they who demonstrate refer to this as an ultimate opinion. For it is by nature the source of all the other axioms."*

* *Τὸ γὰρ αὐτὸ ἅμα ὑπάρχειν τε καὶ μὴ ὑπάρχειν ἀδύνατον τῷ αὐτῷ καὶ κατὰ τὸ αὐτό. . . Αὕτη δὴ πασῶν ἐστὶ βεβαιοτάτη τῶν ἀρχῶν. . . Διὸ πάντες οἱ ἀποδεικνύντες εἰς ταύτην ἀνάγουσιν ἐσχάτην δόξαν· φύσει γὰρ ἀρχὴ καὶ τῶν ἄλλων ἀξιωμάτων αὕτη πάντων.—Metaphysica*, III. 3.

The above interpretation has been introduced not on account of its immediate value in the present system, but as an illustration of a significant fact in the philosophy of the intellectual powers, viz., that what has been commonly regarded as the fundamental axiom of metaphysics is but the consequence of a law of thought, mathematical in its form. I desire to direct attention also to the circumstance that the equation (1) in which that fundamental law of thought is expressed is an equation of the second degree.* Without speculating at all in this chapter upon the question, whether that circumstance is necessary in its own nature, we may venture to assert that if it had not existed, the whole procedure of the understanding would have been different from what it is. Thus it is a consequence of the fact that the fundamental equation of thought is of the second degree, that we perform the operation of analysis and classification, by division into pairs of

* Should it here be said that the existence of the equation $x^2 = x$ necessitates also the existence of the equation $x^3 = x$, which is of the third degree, and then inquired whether that equation does not indicate a process of *trichotomy*; the answer is, that the equation $x^3 = x$ is not interpretable in the system of logic. For writing it in either of the forms

$$x(1-x)(1+x) = 0, \quad (2)$$
$$x(1-x)(-1-x) = 0, \quad (3)$$

we see that its interpretation, if possible at all, must involve that of the factor $1 + x$, or of the factor $-1 - x$. The former is not interpretable, because we cannot conceive of the addition of any class x to the universe 1; the latter is not interpretable, because the symbol -1 is not subject to the law $x(1-x) = 0$, to which all class symbols are subject. Hence the equation $x^3 = x$ admits of no interpretation analogous to that of the equation $x^2 = x$. Were the former equation, however, true independently of the latter, i. e. were that act of the mind which is denoted by the symbol x, such that its second repetition should reproduce the result of a single operation, but not its first or mere repetition, it is presumable that we should be able to interpret one of the forms (2), (3), which under the actual conditions of thought we cannot do. There exist operations, known to the mathematician, the law of which may be adequately expressed by the equation $x^3 = x$. But they are of a nature altogether foreign to the province of general reasoning.

In saying that it is conceivable that the law of thought might have been different from what it is, I mean only that we can frame such an hypothesis, and study its consequences. The possibility of doing this involves no such doctrine as that the actual law of human reason is the product either of chance or of arbitrary will.

opposites, or, as it is technically said, by *dichotomy*. Now if the equation in question had been of the third degree, still admitting of interpretation as such, the mental division must have been threefold in character, and we must have proceeded by a species of *trichotomy*, the real nature of which it is impossible for us, with our existing faculties, adequately to conceive, but the laws of which we might still investigate as an object of intellectual speculation.

16. The law of thought expressed by the equation (1) will, for reasons which are made apparent by the above discussion, be occasionally referred to as the "law of duality."

CHAPTER IV.

OF THE DIVISION OF PROPOSITIONS INTO THE TWO CLASSES OF "PRIMARY" AND "SECONDARY;" OF THE CHARACTERISTIC PROPERTIES OF THOSE CLASSES, AND OF THE LAWS OF THE EXPRESSION OF PRIMARY PROPOSITIONS.

1. THE laws of those mental operations which are concerned in the processes of Conception or Imagination having been investigated, and the corresponding laws of the symbols by which they are represented explained, we are led to consider the practical application of the results obtained: first, in the expression of the complex terms of propositions; secondly, in the expression of propositions; and lastly, in the construction of a general method of deductive analysis. In the present chapter we shall be chiefly concerned with the first of these objects, as an introduction to which it is necessary to establish the following Proposition:

PROPOSITION I.

All logical propositions may be considered as belonging to one or the other of two great classes, to which the respective names of "Primary" or "Concrete Propositions," and "Secondary" or "Abstract Propositions," may be given.

Every assertion that we make may be referred to one or the other of the two following kinds. Either it expresses a relation among *things*, or it expresses, or is equivalent to the expression of, a relation among *propositions*. An assertion respecting the properties of things, or the phænomena which they manifest, or the circumstances in which they are placed, is, properly speaking, the assertion of a relation among things. To say that "snow is white," is for the ends of logic equivalent to saying, that "snow is a white thing." An assertion respecting facts or events, their mutual connexion and dependence, is, for the same ends, generally equivalent to the assertion, that such and such propositions concerning those events have a certain relation to each other as respects their mutual truth or falsehood. The former class of propositions, relating to *things*, I call "Primary;" the latter class, relating to *propositions*, I call "Secondary." The distinction is in practice nearly but not quite co-extensive with the common logical distinction of propositions as categorical or hypothetical.

For instance, the propositions, "The sun shines," "The earth is warmed," are primary; the proposition, "If the sun shines the earth is warmed," is secondary. To say, "The sun shines," is to say, "The sun is that which shines," and it expresses a relation between two classes of things, viz., "the sun" and "things which shine." The secondary proposition, however, given above, expresses a relation of dependence between the two primary propositions, "The sun shines," and "The earth is warmed." I do not hereby affirm that the relation between these propositions is, like that which exists between the facts which they express, a relation of causality, but only that the relation among the propositions so implies, and is so implied by, the relation among the facts, that it may for the ends of logic be used as a fit representative of that relation.

2. If instead of the proposition, "The sun shines," we say, "It is true that the sun shines," we then speak not directly of things, but of a proposition concerning things, viz., of the proposition, "The sun shines." And, therefore, the proposition in which we thus speak is a secondary one. Every primary proposition may thus give rise to a secondary proposition, viz., to that secondary proposition which asserts its truth, or declares its falsehood.

It will usually happen, that the particles *if*, *either*, *or*, will indicate that a proposition is secondary; but they do not necessarily imply that such is the case. The proposition, "Animals are either rational or irrational," is primary. It cannot be resolved into "Either animals are rational or animals are irrational," and it does not therefore express a relation of dependence between the two propositions connected together in the latter disjunctive sentence. The particles, *either*, *or*, are in fact no *criterion* of the nature of propositions, although it happens that they are more frequently found in secondary propositions. Even the conjunction *if* may be found in primary propositions. "Men are, if wise, then temperate," is an example of the kind. It cannot be resolved into "If all men are wise, then all men are temperate."

3. As it is not my design to discuss the merits or defects of the ordinary division of propositions, I shall simply remark here, that the principle upon which the present classification is founded is clear and definite in its application, that it involves a real and fundamental distinction in propositions, and that it is of essential importance to the development of a general method of reasoning. Nor does the fact that a primary proposition may be put into a form in which it becomes secondary at all conflict with the views here maintained. For in the case thus supposed, it is not of the things connected together in the primary proposition that any direct account is taken, but only of the proposition itself considered as *true* or as *false*.

4. In the expression both of primary and of secondary propositions, the same symbols, subject, as it will appear, to the same laws, will be employed in this work. The difference between the two cases is a difference not of form but of interpretation. In both cases the actual relation which it is the object of the proposition to express will be denoted by the sign $=$. In the expression of primary propositions, the members thus connected will usually represent the "terms" of a proposition, or, as they are more particularly designated, its subject and predicate.

Proposition II.

5. *To deduce a general method, founded upon the enumeration of possible varieties, for the expression of any class or collection of things, which may constitute a "term" of a Primary Proposition.*

First, If the class or collection of things to be expressed is defined only by names or qualities common to all the individuals of which it consists, its expression will consist of a single term, in which the symbols expressive of those names or qualities will be combined without any connecting sign, as if by the algebraic process of multiplication. Thus, if x represent opaque substances, y polished substances, z stones, we shall have,

xyz = opaque polished stones ;
$xy(1-z)$ = opaque polished substances which are not stones;
$x(1-y)(1-z)$ = opaque substances which are not polished, and are not stones ;

and so on for any other combination. Let it be observed, that each of these expressions satisfies the same law of duality, as the individual symbols which it contains. Thus,

$$xyz \times xyz = xyz;$$
$$xy(1-z) \times xy(1-z) = xy(1-z);$$

and so on. Any such term as the above we shall designate as a "class term," because it expresses a class of things by means of the common properties or names of the individual members of such class.

Secondly, If we speak of a collection of things, different portions of which are defined by different properties, names, or attributes, the expressions for those different portions must be separately formed, and then connected by the sign +. But if the collection of which we desire to speak has been formed by excluding from some wider collection a defined portion of its members, the sign − must be prefixed to the symbolical expression of the excluded portion. Respecting the use of these symbols some further observations may be added.

6. Speaking generally, the symbol + is the equivalent of the conjunctions "and," "or," and the symbol −, the equivalent of the preposition "except." Of the conjunctions "and" and "or," the former is usually employed when the collection to be described forms the subject, the latter when it forms the predicate, of a proposition. "The scholar *and* the man of the world desire happiness," may be taken as an illustration of one of these cases. "Things possessing utility are *either* productive of pleasure *or* preventive of pain," may exemplify the other. Now whenever an expression involving these particles presents itself in a primary proposition, it becomes very important to know whether the groups or classes separated in thought by them are intended to be quite distinct from each other and mutually exclusive, or not. Does the expression, "Scholars and men of the world," include or exclude those who are both? Does the ex-

pression, "Either productive of pleasure or preventive of pain," include or exclude things which possess both these qualities? I apprehend that in strictness of meaning the conjunctions "and," "or," do possess the power of separation or exclusion here referred to; that the formula, "All x's are either y's or z's," rigorously interpreted, means, "All x's are either y's, but not z's," or, "z's but not y's." But it must at the same time be admitted, that the "jus et norma loquendi" seems rather to favour an opposite interpretation. The expression, "Either y's or z's," would generally be understood to include things that are y's and z's at the same time, together with things which come under the one, but not the other. Remembering, however, that the symbol + does possess the separating power which has been the subject of discussion, we must resolve any disjunctive expression which may come before us into elements really separated in thought, and then connect their respective expressions by the symbol +.

And thus, according to the meaning implied, the expression, "Things which are either x's or y's," will have two different symbolical equivalents. If we mean, "Things which are x's, but not y's, or y's, but not x's," the expression will be

$$x(1-y)+y(1-x);$$

the symbol x standing for x's, y for y's. If, however, we mean, "Things which are either x's, or, if not x's, then y's," the expression will be

$$x+y(1-x).$$

This expression supposes the admissibility of things which are both x's and y's at the same time. It might more fully be expressed in the form

$$xy+x(1-y)+y(1-x);$$

but this expression, on addition of the two first terms, only reproduces the former one.

Let it be observed that the expressions above given satisfy the fundamental law of duality (III. 16). Thus we have

$$\{x(1-y)+y(1-x)\}^2 = x(1-y)+y(1-x),$$
$$\{x+y(1-x)\}^2 = x+y(1-x).$$

It will be seen hereafter, that this is but a particular manifestation of a general law of expressions representing "classes or collections of things."

7. The results of these investigations may be embodied in the following rule of expression.

RULE.—*Express simple names or qualities by the symbols* x, y, z, *&c., their contraries by* $1-x$, $1-y$, $1-z$, *&c.; classes of things defined by common names or qualities, by connecting the corresponding symbols as in multiplication; collections of things, consisting of portions different from each other, by connecting the expressions of those portions by the sign* $+$. *In particular, let the expression, "Either* x*'s or* y*'s," be expressed by* $x(1-y)+y(1-x)$, *when the classes denoted by* x *and* y *are exclusive, by* $x+y(1-x)$ *when they are not exclusive. Similarly let the expression, "Either* x*'s, or* y*'s, or* z*'s," be expressed by* $x(1-y)(1-z)+y(1-x)(1-z)+z(1-x)(1-y)$, *when the classes denoted by* x, y, *and* z, *are designed to be mutually exclusive, by* $x+y(1-x)+z(1-x)(1-y)$, *when they are not meant to be exclusive, and so on.*

8. On this rule of expression is founded the converse rule of interpretation. Both these will be exemplified with, perhaps, sufficient fulness in the following instances. Omitting for brevity the universal subject "things," or "beings," let us assume

$$x = \text{hard}, \quad y = \text{elastic}, \quad z = \text{metals};$$

and we shall have the following results:

"Non-elastic metals," will be expressed by $z(1-y)$;

"Elastic substances with non-elastic metals," by $y+z(1-y)$;

"Hard substances, except metals," by $x-y$;

"Metallic substances, except those which are neither hard nor elastic," by $z-z(1-x)(1-y)$, or by $z\{1-(1-x)(1-y)\}$, *vide* (6), Chap. II.

In the last example, what we had really to express was "Metals, except not hard, not elastic, metals." Conjunctions used between *adjectives* are usually superfluous, and, therefore, must not be expressed symbolically.

Thus, "Metals hard and elastic," is equivalent to "Hard elastic metals," and expressed by xyz.

Take next the expression, "Hard substances, except those

which are metallic and non-elastic, and those which are elastic and non-metallic." Here the word *those* means hard substances, so that the expression really means, *Hard substances except hard substances, metallic, non-elastic, and hard substances non-metallic, elastic;* the word *except* extending to both the classes which follow it. The complete expression is

$$x - \{xz(1-y) + xy(1-z)\};$$

or,

$$x - xz(1-y) - xy(1-z).$$

9. The preceding Proposition, with the different illustrations which have been given of it, is a necessary preliminary to the following one, which will complete the design of the present chapter.

Proposition III.

To deduce from an examination of their possible varieties a general method for the expression of Primary or Concrete Propositions.

A primary proposition, in the most general sense, consists of two terms, between which a relation is asserted to exist. These terms are not necessarily single-worded names, but may represent any collection of objects, such as we have been engaged in considering in the previous sections. The mode of expressing those terms is, therefore, comprehended in the general precepts above given, and it only remains to discover how the relations between the terms are to be expressed. This will evidently depend upon the nature of the relation, and more particularly upon the question whether, in that relation, the terms are understood to be universal or particular, i. e. whether we speak of the whole of that collection of objects to which a term refers, or indefinitely of the whole or of a part of it, the usual signification of the prefix, "some."

Suppose that we wish to express a relation of identity between the two classes, "Fixed Stars" and "Suns," i. e. to express that "All fixed stars are suns," and "All suns are fixed stars." Here, if x stand for fixed stars, and y for suns, we shall have

$$x = y$$

for the equation required.

In the proposition, "All fixed stars are suns," the term "all fixed stars" would be called the *subject*, and "suns" the *predicate*. Suppose that we extend the meaning of the terms *subject* and *predicate* in the following manner. By *subject* let us mean the first term of any affirmative proposition, i. e. the term which precedes the copula *is* or *are*; and by *predicate* let us agree to mean the second term, i. e. the one which follows the copula; and let us admit the assumption that either of these may be universal or particular, so that, in either case, the whole class may be implied, or only a part of it. Then we shall have the following Rule for cases such as the one in the last example:—

10. RULE.—*When both Subject and Predicate of a Proposition are universal, form the separate expressions for them, and connect them by the sign* =.

This case will usually present itself in the expression of the definitions of science, or of subjects treated after the manner of pure science. Mr. Senior's definition of wealth affords a good example of this kind, viz.:

"Wealth consists of things transferable, limited in supply, and either productive of pleasure or preventive of pain."

Before proceeding to express this definition symbolically, it must be remarked that the conjunction *and* is superfluous. Wealth is really defined by its possession of three properties or qualities, not by its composition out of three classes or collections of objects. Omitting then the conjunction *and*, let us make

w = wealth.
t = things transferable.
s = limited in supply.
p = productive of pleasure.
r = preventive of pain.

Now it is plain from the nature of the subject, that the expression, "Either productive of pleasure or preventive of pain," in the above definition, is meant to be equivalent to "Either productive of pleasure; or, if not productive of pleasure, preventive of pain." Thus the class of things which the above expression, taken alone, would define, would consist of all things productive of pleasure, together with all things not productive of pleasure, but preventive of pain, and its symbolical expression would be

$$p + (1 - p) r.$$

If then we attach to this expression placed in brackets to denote that both its terms are referred to, the symbols s and t limiting its application to things "transferable" and "limited in supply," we obtain the following symbolical equivalent for the original definition, viz.:

$$w = st \{p + r (1 - p)\}. \qquad (1)$$

If the expression, "Either productive of pleasure or preventive of pain," were intended to point out merely those things which are productive of pleasure without being preventive of pain, $p (1 - r)$, or preventive of pain, without being productive of pleasure, $r (1 - p)$ (exclusion being made of those things which are both productive of pleasure and preventive of pain), the expression in symbols of the definition would be

$$w = st \{p (1 - r) + r (1 - p)\}. \qquad (2)$$

All this agrees with what has before been more generally stated.

The reader may be curious to inquire what effect would be produced if we literally translated the expression, "Things productive of pleasure or preventive of pain," by $p + r$, making the symbolical equation of the definition to be

$$w = st (p + r). \qquad (3)$$

The answer is, that this expression would be equivalent to (2), with the additional implication that the classes of things denoted by stp and str are quite distinct, so that of things transferable and limited in supply there exist none in the universe which are at the same time both productive of pleasure and preventive of pain. How the full import of any equation may be determined will be explained hereafter. What has been said may show that before attempting to translate our data into the rigorous language of symbols, it is above all things necessary to ascertain the *intended* import of the words we are using. But this necessity cannot be regarded as an evil by those who value correctness of thought, and regard the right employment of language as both its instrument and its safeguard.

11. Let us consider next the case in which the predicate of the proposition is particular, e. g. "All men are mortal."

In this case it is clear that our meaning is, "All men are some mortal beings," and we must seek the expression of the predicate, "some mortal beings." Represent then by v, a class indefinite in every respect but this, viz., that some of its members are mortal beings, and let x stand for "mortal beings," then will vx represent "some mortal beings." Hence if y represent men, the equation sought will be

$$y = vx.$$

From such considerations we derive the following Rule, for expressing an affirmative universal proposition whose predicate is particular:

RULE.—*Express as before the subject and the predicate, attach to the latter the indefinite symbol v, and equate the expressions.*

It is obvious that v is a symbol of the same kind as x, y, &c., and that it is subject to the general law,

$$v^2 = v, \text{ or } v(1 - v) = 0.$$

Thus, to express the proposition, "The planets are either primary or secondary," we should, according to the rule, proceed thus:

Let x represent planets (the subject);
y = primary bodies;
z = secondary bodies;

then, assuming the conjunction "or" to separate absolutely the class of "primary" from that of "secondary" bodies, so far as they enter into our consideration in the proposition given, we find for the equation of the proposition

$$x = v\{y(1 - z) + z(1 - y)\}. \qquad (4)$$

It may be worth while to notice, that in this case the *literal* translation of the premises into the form

$$x = v(y + z) \qquad (5)$$

would be exactly equivalent, v being an indefinite class symbol. The form (4) is, however, the better, as the expression

$$y(1-z)+z(1-y)$$

consists of terms representing classes quite distinct from each other, and satisfies the fundamental law of duality.

If we take the proposition, "The heavenly bodies are either suns, or planets, or comets," representing these classes of things by w, x, y, z, respectively, its expression, on the supposition that none of the heavenly bodies belong at once to two of the divisions above mentioned, will be

$$w=v\{x(1-y)(1-z)+y(1-x)(1-z)+z(1-x)(1-y)\}.$$

If, however, it were meant to be implied that the heavenly bodies were either suns, or, if not suns, planets, or, if neither suns nor planets, fixed stars, a meaning which does not exclude the supposition of some of them belonging at once to two or to all three of the divisions of suns, planets, and fixed stars,—the expression required would be

$$w=v\{x+y(1-x)+z(1-x)(1-y)\}. \qquad (6)$$

The above examples belong to the class of descriptions, not definitions. Indeed the predicates of propositions are usually particular. When this is not the case, either the predicate is a singular term, or we employ, instead of the copula "is" or "are," some form of connexion, which implies that the predicate is to be taken universally.

12. Consider next the case of universal negative propositions, e. g. "No men are perfect beings."

Now it is manifest that in this case we do not speak of a class termed "no men," and assert of this class that all its members are "perfect beings." But we virtually make an assertion about "*all men*" to the effect that they are "*not perfect beings.*" Thus the true meaning of the proposition is this:

"All men (subject) are (copula) not perfect (predicate);" whence, if y represent "men," and x "perfect beings," we shall have

$$y=v(1-x),$$

and similarly in any other case. Thus we have the following Rule:

RULE.—*To express any proposition of the form "No x's are y's," convert it into the form "All x's are not y's," and then proceed as in the previous case.*

13. Consider, lastly, the case in which the subject of the proposition is particular, e.g. "Some men are not wise." Here, as has been remarked, the negative *not* may properly be referred, certainly, at least, for the ends of Logic, to the predicate *wise*; for we do not mean to say that it is not true that "Some men are wise," but we intend to predicate of "some men" a want of wisdom. The requisite form of the given proposition is, therefore, "Some men are not-wise." Putting, then, y for "men," x for "wise," i.e. "wise beings," and introducing v as the symbol of a class indefinite in all respects but this, that it contains some individuals of the class to whose expression it is prefixed, we have

$$vy = v(1 - x).$$

14. We may comprise all that we have determined in the following general Rule:

GENERAL RULE FOR THE SYMBOLICAL EXPRESSION OF PRIMARY PROPOSITIONS.

1st. *If the proposition is affirmative, form the expression of the subject and that of the predicate. Should either of them be particular, attach to it the indefinite symbol v, and then equate the resulting expressions.*

2ndly. *If the proposition is negative, express first its true meaning by attaching the negative particle to the predicate, then proceed as above.*

One or two additional examples may suffice for illustration.

Ex.—"No men are placed in exalted stations, and free from envious regards."

Let y represent "men," x, "placed in exalted stations," z, "free from envious regards."

Now the expression of the class described as "placed in

exalted station," and "free from envious regards," is xz. Hence the contrary class, i. e. they to whom this description does not apply, will be represented by $1 - xz$, and to this class all men are referred. Hence we have

$$y = v(1 - xz).$$

If the proposition thus expressed had been placed in the equivalent form, "Men in exalted stations are not free from envious regards," its expression would have been

$$yx = v(1 - z).$$

It will hereafter appear that this expression is really equivalent to the previous one, on the particular hypothesis involved, viz., that v is an indefinite class symbol.

Ex.—"No men are heroes but those who unite self-denial to courage."

Let x = "men," y = "heroes," z = "those who practise self-denial," w, "those who possess courage."

The assertion really is, that "men who do not possess courage and practise self-denial are not heroes."

Hence we have

$$x(1 - zw) = v(1 - y)$$

for the equation required.

15. In closing this Chapter it may be interesting to compare together the great leading types of propositions symbolically expressed. If we agree to represent by X and Y the symbolical expressions of the "terms," or things related, those types will be

$$X = vY,$$
$$X = Y,$$
$$vX = vY.$$

In the first, the predicate only is particular; in the second, both terms are universal; in the third, both are particular. Some minor forms are really included under these. Thus, if $Y = 0$, the second form becomes

$$X = 0;$$

and if $Y = 1$ it becomes

$$X = 1;$$

both which forms admit of interpretation. It is further to be noticed, that the expressions X and Y, if founded upon a sufficiently careful analysis of the meaning of the "terms" of the proposition, will satisfy the fundamental law of duality which requires that we have

$$X^2 = X \text{ or } X(1 - X) = 0,$$
$$Y^2 = Y \text{ or } Y(1 - Y) = 0.$$

CHAPTER V.

OF THE FUNDAMENTAL PRINCIPLES OF SYMBOLICAL REASONING, AND OF THE EXPANSION OR DEVELOPMENT OF EXPRESSIONS INVOLVING LOGICAL SYMBOLS.

1. THE previous chapters of this work have been devoted to the investigation of the fundamental laws of the operations of the mind in reasoning; of their development in the laws of the symbols of Logic; and of the principles of expression, by which that species of propositions called primary may be represented in the language of symbols. These inquiries have been in the strictest sense preliminary. They form an indispensable introduction to one of the chief objects of this treatise—the construction of a system or method of Logic upon the basis of an exact summary of the fundamental laws of thought. There are certain considerations touching the nature of this end, and the means of its attainment, to which I deem it necessary here to direct attention.

2. I would remark in the first place that the generality of a method in Logic must very much depend upon the generality of its elementary processes and laws. We have, for instance, in the previous sections of this work investigated, among other things, the laws of that logical process of *addition* which is symbolized by the sign +. Now those laws have been determined from the study of instances, in all of which it has been a necessary condition, that the classes or things added together in thought should be mutually exclusive. The expression $x + y$ seems indeed uninterpretable, unless it be assumed that the things represented by x and the things represented by y are entirely separate; that they embrace no individuals in common. And conditions analogous to this have been involved in those acts of conception from the study of which the laws of the other symbolical operations have been ascertained. The question then arises, whether

it is necessary to restrict the application of these symbolical laws and processes by the same conditions of interpretability under which the knowledge of them was obtained. If such restriction is necessary, it is manifest that no such thing as a general method in Logic is possible. On the other hand, if such restriction is unnecessary, in what light are we to contemplate processes which appear to be uninterpretable in that sphere of thought which they are designed to aid? These questions do not belong to the science of Logic alone. They are equally pertinent to every developed form of human reasoning which is based upon the employment of a symbolical language.

3. I would observe in the second place, that this apparent failure of correspondency between process and interpretation does not manifest itself in the *ordinary* applications of human reason. For no operations are there performed of which the meaning and the application are not seen; and to most minds it does not suffice that merely formal reasoning should connect their premises and their conclusions; but every step of the connecting train, every mediate result which is established in the course of demonstration, must be intelligible also. And without doubt, this is both an actual condition and an important safeguard, in the reasonings and discourses of common life.

There are perhaps many who would be disposed to extend the same principle to the general use of symbolical language as an instrument of reasoning. It might be argued, that as the laws or axioms which govern the use of symbols are established upon an investigation of those cases only in which interpretation is possible, we have no right to extend their application to other cases in which interpretation is impossible or doubtful, even though (as should be admitted) such application is employed in the intermediate steps of demonstration only. Were this objection conclusive, it must be acknowledged that slight advantage would accrue from the use of a symbolical method in Logic. Perhaps that advantage would be confined to the mechanical gain of employing short and convenient symbols in the place of more cumbrous ones. But the objection itself is fallacious. Whatever our *à priori* anticipations might be, it is an unquestionable fact that the validity of a conclusion arrived at by any symbolical process of reasoning, does not depend upon our ability to interpret the formal results which have presented themselves in the different stages of the investigation. There exist, in fact, certain general principles relating to the use of symbolical methods, which, as pertaining to the particular subject of Logic, I shall first state, and I shall then offer some remarks upon the nature and upon the grounds of their claim to acceptance.

4. The conditions of valid reasoning, by the aid of symbols, are—

1st, That a fixed interpretation be assigned to the symbols employed in the expression of the data; and that the laws of the combination of those symbols be correctly determined from that interpretation.

2nd, That the formal processes of solution or demonstration be conducted throughout in obedience to all the laws determined as above, without regard to the question of the interpretability of the particular results obtained.

3rd, That the final result be interpretable in form, and that it be actually interpreted in accordance with that system of interpretation which has been employed in the expression of the data. Concerning these principles, the following observations may be made.

5. The necessity of a fixed interpretation of the symbols has already been sufficiently dwelt upon (II. 3). The necessity that the fixed result should be in such a form as to admit of that interpretation being applied, is founded on the obvious principle, that the use of symbols is a means towards an end, that end being the knowledge of some intelligible fact or truth. And that this end may be attained, the final result which expresses the symbolical conclusion must be in an interpretable form. It is, however, in connexion with the second of the above general principles or conditions (V. 4), that the greatest difficulty is likely to be felt, and upon this point a few additional words are necessary.

I would then remark, that the principle in question may be considered as resting upon a general law of the mind, the knowledge of which is not given to us *à priori*, i. e. antecedently to experience, but is derived, like the knowledge of the other laws of the mind, from the clear manifestation of the general principle in the particular instance. A single example of reasoning, in which symbols are employed in obedience to laws founded upon their interpretation, but without any sustained reference to that interpretation, the chain of demonstration conducting us through intermediate steps which are not interpretable, to a final result which is interpretable, seems not only to establish the validity of the particular application, but to make known to us the general law manifested therein. No accumulation of instances can properly add weight to such evidence. It may furnish us with clearer conceptions of that common element of truth upon which the application of the principle depends, and so prepare the way for its reception. It may, where the immediate force of the evidence is not felt, serve as a verification, *à posteriori*, of the practical validity of the principle in question. But this does not affect the position affirmed, viz., that the general principle must be seen in the particular instance,—seen to be general in application as well as true in the special example. The employment of the uninterpretable symbol $\sqrt{-1}$, in the intermediate processes of trigonometry, furnishes an illustration of what has been said. I apprehend that there is no mode of explaining that application which does not covertly assume the very principle in question. But that principle, though not, as I conceive, warranted by formal reasoning based upon other grounds, seems to deserve a place among those axiomatic truths which constitute, in some sense, the foundation of the possibility of general knowledge, and which may properly be regarded as expressions of the mind's own laws and constitution.

6. The following is the mode in which the principle above stated will be applied in the present work. It has been seen, that any system of propositions may be expressed by equations involving symbols x, y, z, which, whenever interpretation is possible, are subject to laws identical in form with the laws of a system of quantitative symbols, susceptible only of the values 0 and 1 (II. 15). But as the formal processes of reasoning depend only upon the laws of the symbols, and not upon the nature of their interpretation, we are permitted to treat the above symbols,

x, y, z, as if they were quantitative symbols of the kind above described. *We may in fact lay aside the logical interpretation of the symbols in the given equation; convert them into quantitative symbols, susceptible only of the values* 0 *and* 1; *perform upon them as such all the requisite processes of solution; and finally restore to them their logical interpretation.* And this is the mode of procedure which will actually be adopted, though it will be deemed unnecessary to restate in every instance the nature of the transformation employed. The processes to which the symbols x, y, z, regarded as quantitative and of the species above described, are subject, are not limited by those conditions of thought to which they would, if performed upon purely logical symbols, be subject, and a freedom of operation is given to us in the use of them, without which, the inquiry after a general method in Logic would be a hopeless quest.

Now the above system of processes would conduct us to no intelligible result, unless the final equations resulting therefrom were in a form which should render their interpretation, after restoring to the symbols their logical significance, possible. There exists, however, a general method of reducing equations to such a form, and the remainder of this chapter will be devoted to its consideration. I shall say little concerning the way in which the method renders interpretation possible,—this point being reserved for the next chapter,—but shall chiefly confine myself here to the mere process employed, which may be characterized as a process of "development." As introductory to the nature of this process, it may be proper first to make a few observations.

7. Suppose that we are considering any class of things with reference to this question, viz., the relation in which its members stand as to the possession or the want of a certain property x. As every individual in the proposed class either possesses or does not possess the property in question, we may divide the class into two portions, the former consisting of those individuals which possess, the latter of those which do not possess, the property. This possibility of dividing in thought the whole class into two constituent portions, is antecedent to all knowledge of the constitution of the class derived from any other source; of

which knowledge the effect can only be to inform us, more or less precisely, to what further conditions the portions of the class which possess and which do not possess the given property are subject. Suppose, then, such knowledge is to the following effect, viz., that the members of that portion which possess the property x, possess also a certain property u, and that these conditions united are a sufficient definition of them. We may then represent that portion of the original class by the expression ux (II. 6). If, further, we obtain information that the members of the original class which do not possess the property x, are subject to a condition v, and are thus defined, it is clear, that those members will be represented by the expression $v(1-x)$. Hence the class in its totality will be represented by

$$ux + v(1-x);$$

which may be considered as a general developed form for the expression of any class of objects considered with reference to the possession or the want of a given property x.

The general form thus established upon purely logical grounds may also be deduced from distinct considerations of formal law, applicable to the symbols x, y, z, equally in their logical and in their quantitative interpretation already referred to (V. 6).

8. *Definition.*—Any algebraic expression involving a symbol x is termed a function of x, and may be represented under the abbreviated general form $f(x)$. Any expression involving two symbols, x and y, is similarly termed a function of x and y, and may be represented under the general form $f(x, y)$, and so on for any other case.

Thus the form $f(x)$ would indifferently represent any of the following functions, viz., x, $1-x$, $\frac{1+x}{1-x}$, &c.; and $f(x, y)$ would equally represent any of the forms $x+y$, $x-2y$, $\frac{x+y}{x-2y}$, &c.

On the same principles of notation, if in any function $f(x)$, we change x into 1, the result will be expressed by the form $f(1)$; if in the same function we change x into 0, the result will be expressed by the form $f(0)$. Thus, if $f(x)$ represent the function $\frac{a+x}{a-2x}$, $f(1)$ will represent $\frac{a+1}{a-2}$, and $f(0)$ will represent $\frac{a}{a}$.

9. *Definition.*—Any function $f(x)$, in which x is a logical symbol, or a symbol of quantity susceptible only of the values 0 and 1, is said to be developed, when it is reduced to the form $ax + b(1-x)$, a and b being so determined as to make the result equivalent to the function from which it was derived.

This definition assumes, that it is possible to represent any function $f(x)$ in the form supposed. The assumption is vindicated in the following Proposition.

Proposition I.

10. *To develop any function $f(x)$ in which x is a logical symbol.*

By the principle which has been asserted in this chapter, it is lawful to treat x as a quantitative symbol, susceptible only of the values 0 and 1.

Assume then,

$$f(x) = ax + b(1-x),$$

and making $x = 1$, we have

$$f(1) = a.$$

Again, in the same equation making $x = 0$, we have

$$f(0) = b.$$

Hence the values of a and b are determined, and substituting them in the first equation, we have

$$f(x) = f(1)\,x + f(0)\,(1-x); \qquad (1)$$

as the development sought.* The second member of the equa-

* To some it may be interesting to remark, that the development of $f(x)$ obtained in this chapter, strictly holds, in the logical system, the place of the expansion of $f(x)$ in ascending powers of x in the system of ordinary algebra. Thus it may be obtained by introducing into the expression of Taylor's well-known theorem, viz.:

$$f(x) = f(0) + f'(0)\,x + f''(0)\frac{x^2}{1.2} + f'''(0)\frac{x^3}{1.2.3}, \text{ \&c.} \qquad (1)$$

the condition $x(1-x) = 0$, whence we find $x^2 = x$, $x^3 = x$, &c., and

tion adequately represents the function $f(x)$, whatever the form of that function may be. For x regarded as a quantitative symbol admits only of the values 0 and 1, and for each of these values the development

$$f(1)\,x + f(0)\,(1-x),$$

assumes the same value as the function $f(x)$.

As an illustration, let it be required to develop the function $\frac{1+x}{1+2x}$. Here, when $x=1$, we find $f(1) = \frac{2}{3}$, and when $x=0$, we find $f(0) = \frac{1}{1}$, or 1. Hence the expression required is

$$\frac{1+x}{1+2x} = \frac{2}{3}x + 1 - x;$$

and this equation is satisfied for each of the values of which the symbol x is susceptible.

Proposition II.

To expand or develop a function involving any number of logical symbols.

Let us begin with the case in which there are two symbols, x and y, and let us represent the function to be developed by $f(x, y)$.

First, considering $f(x, y)$ as a function of x alone, and expanding it by the general theorem (1), we have

$$f(x, y) = f(1, y)\,x + f(0, y)\,(1-x); \tag{2}$$

$$f(x) = f(0) + \{f'(0) + \frac{f''(0)}{1.2} + \frac{f'''(0)}{1.2.3} + \&c.\}\,x. \tag{2}$$

But making in (1), $x = 1$, we get

$$f(1) = f(0) + f'(0) + \frac{f''(0)}{1.2} + \frac{f'''(0)}{1.2.3} + \&c.;$$

whence

$$f'(0) + \frac{f''(0)}{1.2} + \&c. = f(1) - f(0),$$

and (2) becomes, on substitution,

$$\begin{aligned} f(x) &= f(0) + \{f(1) - f(0)\}\,x, \\ &= f(1)x + f(0)\,(1-x), \end{aligned}$$

the form in question. This demonstration in supposing $f(x)$ to be developable in a series of ascending powers of x is less general than the one in the text.

wherein $f(1, y)$ represents what the proposed function becomes, when in it for x we write 1, and $f(0, y)$ what the said function becomes, when in it for x we write 0.

Now, taking the coefficient $f(1, y)$, and regarding it as a function of y, and expanding it accordingly, we have

$$f(1, y) = f(1, 1)\, y + f(1, 0)\,(1 - y), \tag{3}$$

wherein $f(1, 1)$ represents what $f(1, y)$ becomes when y is made equal to 1, and $f(1, 0)$ what $f(1, y)$ becomes when y is made equal to 0.

In like manner, the coefficient $f(0, y)$ gives by expansion,

$$f(0, y) = f(0, 1)\, y + f(0, 0)\,(1 - y). \tag{4}$$

Substitute in (2) for $f(1, y)$, $f(0, y)$, their values given in (3) and (4), and we have

$$f(x, y) = f(1, 1)\, xy + f(1, 0)\, x(1 - y) + f(0, 1)\,(1 - x)\, y \\ + f(0, 0)\,(1 - x)(1 - y), \tag{5}$$

for the expansion required. Here $f(1, 1)$ represents what $f(x, y)$ becomes when we make therein $x = 1$, $y = 1$; $f(1, 0)$ represents what $f(x, y)$ becomes when we make therein $x = 1$, $y = 0$, and so on for the rest.

Thus, if $f(x, y)$ represent the function $\frac{1 - x}{1 - y}$, we find

$$f(1, 1) = \frac{0}{0}, \quad f(1, 0) = \frac{0}{1} = 0, \quad f(0, 1) = \frac{1}{0}, \quad f(0, 0) = 1,$$

whence the expansion of the given function is

$$\frac{0}{0}\, xy + 0x(1 - y) + \frac{1}{0}(1 - x)\, y + (1 - x)(1 - y).$$

It will in the next chapter be seen that the forms $\frac{0}{0}$ and $\frac{1}{0}$, the former of which is known to mathematicians as the symbol of indeterminate quantity, admit, in such expressions as the above, of a very important logical interpretation.

Suppose, in the next place, that we have three symbols in the function to be expanded, which we may represent under the general form $f(x, y, z)$. Proceeding as before, we get

$$f(x,y,z) = f(1,1,1)xyz + f(1,1,0)xy(1-z) + f(1,0,1)x(1-y)z$$
$$+ f(1,0,0)x(1-y)(1-z) + f(0,1,1)(1-x)yz$$
$$+ f(0,1,0)(1-x)y(1-z) + f(0,0,1)(1-x)(1-y)z$$
$$+ f(0,0,0)(1-x)(1-y)(1-z),$$

in which $f(1, 1, 1)$ represents what the function $f(x, y, z)$ becomes when we make therein $x = 1$, $y = 1$, $z = 1$, and so on for the rest.

11. It is now easy to see the general law which determines the expansion of any proposed function, and to reduce the method of effecting the expansion to a rule. But before proceeding to the expression of such a rule, it will be convenient to premise the following observations :—

Each form of expansion that we have obtained consists of certain terms, into which the symbols x, y, &c. enter, multiplied by coefficients, into which those symbols do not enter. Thus the expansion of $f(x)$ consists of two terms, x and $1 - x$, multiplied by the coefficients $f(1)$ and $f(0)$ respectively. And the expansion of $f(x, y)$ consists of the four terms xy, $x(1 - y)$, $(1 - x)y$, and $(1 - x)$, $(1 - y)$, multiplied by the coefficients $f(1, 1)$, $f(1, 0)$, $f(0, 1)$, $f(0, 0)$, respectively. The terms x, $1 - x$, in the former case, and the terms xy, $x(1 - y)$, &c., in the latter, we shall call the *constituents* of the expansion. It is evident that they are in form independent of the form of the function to be expanded. Of the constituent xy, x and y are termed the *factors*.

The general rule of development will therefore consist of two parts, the first of which will relate to the formation of the *constituents* of the expansion, the second to the determination of their respective coefficients. It is as follows:

1st. *To expand any function of the symbols x, y, z.*—Form a series of constituents in the following manner: Let the first constituent be the product of the symbols; change in this product any symbol z into $1 - z$, for the second constituent. Then in both these change any other symbol y into $1 - y$, for two more constituents. Then in the four constituents thus obtained change any other symbol x into $1 - x$, for four new constituents, and so on until the number of possible changes is exhausted.

2ndly. *To find the coefficient of any constituent.*—If that constituent involves x as a factor, change in the original function x into 1; but if it involves $1 - x$ as a factor, change in the original function x into 0. Apply the same rule with reference to the symbols y, z, &c.: the final calculated value of the function thus transformed will be the coefficient sought.

The sum of the constituents, multiplied each by its respective coefficient, will be the expansion required.

12. It is worthy of observation, that a function may be developed with reference to symbols which it does not explicitly contain. Thus if, proceeding according to the rule, we seek to develop the function $1 - x$, with reference to the symbols x and y, we have,

When $x = 1$ and $y = 1$ the given function $= 0$.
$x = 1$,, $y = 0$,, ,, $= 0$.
$x = 0$,, $y = 1$,, ,, $= 1$.
$x = 0$,, $y = 0$,, ,, $= 1$.

Whence the development is

$$1 - x = 0\,xy + 0\,x(1-y) + (1-x)y + (1-x)(1-y);$$

and this is a true development. The addition of the terms $(1-x)y$ and $(1-x)(1-y)$ produces the function $1 - x$.

The symbol 1 thus developed according to the rule, with respect to the symbol x, gives

$$x + 1 - x.$$

Developed with respect to x and y, it gives

$$xy + x(1-y) + (1-x)y + (1-x)(1-y).$$

Similarly developed with respect to any set of symbols, it produces a series consisting of all possible constituents of those symbols.

13. A few additional remarks concerning the nature of the general expansions may with propriety be added. Let us take, for illustration, the general theorem (5), which presents the type of development for functions of two logical symbols.

In the first place, that theorem is perfectly true and intelligible when x and y are quantitative symbols of the species considered in this chapter, whatever algebraic form may be assigned to the function $f(x, y)$, and it may therefore be intelligibly employed in any stage of the process of analysis intermediate between the change of interpretation of the symbols from the logical to the quantitative system above referred to, and the final restoration of the logical interpretation.

Secondly. The theorem is perfectly true and intelligible when x and y are logical symbols, provided that the form of the function $f(x, y)$ is such as to represent a *class or collection of things*, in which case the second member is always logically interpretable. For instance, if $f(x, y)$ represent the function $1 - x + xy$, we obtain on applying the theorem

$$1 - x + xy = xy + 0\,x(1-y) + (1-x)y + (1-x)(1-y),$$
$$= xy + (1-x)y + (1-x)(1-y),$$

and this result is intelligible and true.

Thus we may regard the theorem as true and intelligible for quantitative symbols of the species above described, *always*; for logical symbols, *always when interpretable*. Whensoever therefore it is employed in this work it must be understood that the symbols x, y are quantitative and of the particular species referred to, if the expansion obtained is not interpretable.

But though the expansion is not always immediately interpretable, it always conducts us at once to results which are interpretable. Thus the expression $x - y$ gives on development the form

$$x(1-y) - y(1-x),$$

which is not generally interpretable. We cannot take, in thought, from the class of things which are x's and not y's, the class of things which are y's and not x's, because the latter class is not contained in the former. But if the form $x - y$ presented itself as the first member of an equation, of which the second member was 0, we should have on development

$$x(1-y) - y(1-x) = 0.$$

Now it will be shown in the next chapter that the above equation, x and y being regarded as quantitative and of the species described, is resolvable at once into the two equations

$$x(1-y) = 0, \quad y(1-x) = 0,$$

and these equations are directly interpretable in Logic when logical interpretations are assigned to the symbols x and y. And it may be remarked, that though *functions* do not necessarily become interpretable upon development, yet *equations* are always reducible by this process to interpretable forms.

14. The following Proposition establishes some important properties of constituents. In its enunciation the symbol t is employed to represent indifferently any constituent of an expansion. Thus if the expansion is that of a function of two symbols x and y, t represents any of the four forms xy, $x(1-y)$, $(1-x)y$, and $(1-x)(1-y)$. Where it is necessary to represent the constituents of an expansion by single symbols, and yet to distinguish them from each other, the distinction will be marked by suffixes. Thus t_1 might be employed to represent xy, t_2 to represent $x(1-y)$, and so on.

Proposition III.

Any single constituent t of an expansion satisfies the law of duality whose expression is

$$t(1-t)=0.$$

The product of any two distinct constituents of an expansion is equal to 0, *and the sum of all the constituents is equal to* 1.

1st. Consider the particular constituent xy. We have

$$xy \times xy = x^2 y^2.$$

But $x^2 = x$, $y^2 = y$, by the fundamental law of class symbols; hence

$$xy \times xy = xy.$$

Or representing xy by t,

$$t \times t = t,$$

or

$$t(1-t)=0.$$

Similarly the constituent $x(1-y)$ satisfies the same law. For we have

$$x^2 = x, \quad (1-y)^2 = 1-y,$$

$$\therefore \ \{x(1-y)\}^2 = x(1-y), \text{ or } t(1-t)=0.$$

Now every factor of every constituent is either of the form x or of the form $1-x$. Hence the square of each factor is equal to that factor, and therefore the square of the product of the factors, i. e. of the constituent, is equal to the constituent; wherefore t representing any constituent, we have

$$t^2 = t, \text{ or } t(1-t) = 0.$$

2ndly. The product of any two constituents is 0. This is evident from the general law of the symbols expressed by the equation $x(1-x) = 0$; for whatever constituents in the same expansion we take, there will be at least one factor x in the one, to which will correspond a factor $1-x$ in the other.

3rdly. The sum of all the constituents of an expansion is unity. This is evident from addition of the two constituents x and $1-x$, or of the four constituents, xy, $x(1-y)$, $(1-x)y$, $(1-x)(1-y)$. But it is also, and more generally, proved by expanding 1 in terms of any set of symbols (V. 12). The constituents in this case are formed as usual, and all the coefficients are unity.

15. With the above Proposition we may connect the following.

PROPOSITION IV.

If V represent the sum of any series of constituents, the separate coefficients of which are 1, *then is the condition satisfied,*

$$V(1-V) = 0.$$

Let $t_1, t_2 \ldots t_n$ be the constituents in question, then

$$V = t_1 + t_2 \ldots + t_n.$$

Squaring both sides, and observing that $t_1^2 = t_1$, $t_1 t_2 = 0$, &c., we have

$$V^2 = t_1 + t_2 \ldots + t_n;$$

whence

$$V = V^2.$$

Therefore

$$V(1-V) = 0.$$

CHAPTER VI.

OF THE GENERAL INTERPRETATION OF LOGICAL EQUATIONS, AND THE RESULTING ANALYSIS OF PROPOSITIONS. ALSO, OF THE CONDITION OF INTERPRETABILITY OF LOGICAL FUNCTIONS.

1. IT has been observed that the complete expansion of any function by the general rule demonstrated in the last chapter, involves two distinct sets of elements, viz., the constituents of the expansion, and their coefficients. I propose in the present chapter to inquire, first, into the interpretation of constituents, and afterwards into the mode in which that interpretation is modified by the coefficients with which they are connected.

The terms "logical equation," "logical function," &c., will be employed generally to denote any equation or function involving the symbols x, y, &c., which may present itself either in the expression of a system of premises, or in the train of symbolical results which intervenes between the premises and the conclusion. If that function or equation is in a form not immediately interpretable in Logic, the symbols x, y, &c., must be regarded as quantitative symbols of the species described in previous chapters (II. 15), (V. 6), as satisfying the law,

$$x(1 - x) = 0.$$

By the problem, then, of the interpretation of any such logical function or equation, is meant the reduction of it to a form in which, when logical values are assigned to the symbols x, y, &c., it shall become interpretable, together with the resulting interpretation. These conventional definitions are in accordance with the general principles for the conducting of the method of this treatise, laid down in the previous chapter.

PROPOSITION I.

2. *The constituents of the expansion of any function of the logical symbols x, y, &c., are interpretable, and represent the several exclusive divisions of the universe of discourse, formed by the predication and denial in every possible way of the qualities denoted by the symbols x, y, &c.*

For greater distinctness of conception, let it be supposed that the function expanded involves two symbols x and y, with reference to which the expansion has been effected. We have then the following constituents, viz.:

$$xy, \quad x(1-y), \quad (1-x)y, \quad (1-x)(1-y).$$

Of these it is evident, that the first xy represents that class of objects which at the same time possess both the elementary qualities expressed by x and y, and that the second $x(1-y)$ represents the class possessing the property x, but not the property y. In like manner the third constituent represents the class of objects which possess the property represented by y, but not that represented by x; and the fourth constituent $(1-x)(1-y)$, represents that class of objects, the members of which possess neither of the qualities in question.

Thus the constituents in the case just considered represent all the four classes of objects which can be described by affirmation and denial of the properties expressed by x and y. Those classes are distinct from each other. No member of one is a member of another, for each class possesses some property or quality contrary to a property or quality possessed by any other class. Again, these classes together make up the universe, for there is no object which may not be described by the presence or the absence of a proposed quality, and thus each individual thing in the universe may be referred to some one or other of the four classes made by the possible combination of the two given classes x and y, and their contraries.

The remarks which have here been made with reference to the constituents of $f(x, y)$ are perfectly general in character. The constituents of any expansion represent classes—those classes are mutually distinct, through the possession of contrary qualities, and they together make up the universe of discourse.

3. These properties of constituents have their expression in the theorems demonstrated in the conclusion of the last chapter, and might thence have been deduced. From the fact that every constituent satisfies the fundamental law of the individual symbols, it might have been conjectured that each constituent would represent a class. From the fact that the product of any two constituents of an expansion vanishes, it might have been concluded that the classes they represent are mutually exclusive. Lastly, from the fact that the sum of the constituents of an expansion is unity, it might have been inferred, that the classes which they represent, together make up the universe.

4. Upon the laws of constituents and the mode of their interpretation above determined, are founded the analysis and the interpretation of logical equations. That all such equations admit of interpretation by the theorem of development has already been stated. I propose here to investigate the forms of possible solution which thus present themselves in the conclusion of a train of reasoning, and to show how those forms arise. Although, properly speaking, they are but manifestations of a single fundamental type or principle of expression, it will conduce to clearness of apprehension if the minor varieties which they exhibit are presented separately to the mind.

The forms, which are three in number, are as follows:

FORM I.

5. The form we shall first consider arises when any logical equation $V = 0$ is developed, and the result, after resolution into its component equations, is to be interpreted. The function is supposed to involve the logical symbols x, y, &c., in combinations which are not fractional. Fractional combinations indeed only arise in the class of problems which will be considered when we come to speak of the third of the forms of solution above referred to.

Proposition II.

To interpret the logical equation $V = 0$.

For simplicity let us suppose that V involves but two symbols, x and y, and let us represent the development of the given equation by

$$axy + bx(1-y) + c(1-x)y + d(1-x)(1-y) = 0; \quad (1)$$

a, b, c, and d being definite numerical constants.

Now, suppose that any coefficient, as a, does not vanish. Then multiplying each side of the equation by the constituent xy, to which that coefficient is attached, we have

$$axy = 0,$$

whence, as a does not vanish,

$$xy = 0,$$

and this result is quite independent of the nature of the other coefficients of the expansion. Its interpretation, on assigning to x and y their logical significance, is "No individuals belonging at once to the class represented by x, and the class represented by y, exist."

But if the coefficient a *does* vanish, the term axy does not appear in the development (1), and, therefore, the equation $xy = 0$ cannot thence be deduced.

In like manner, if the coefficient b does not vanish, we have

$$x(1-y) = 0,$$

which admits of the interpretation, "There are no individuals which at the same time belong to the class x, and do not belong to the class y."

Either of the above interpretations may, however, as will subsequently be shown, be exhibited in a different form.

The sum of the distinct interpretations thus obtained from the several terms of the expansion whose coefficients do not vanish, will constitute the complete interpretation of the equation $V = 0$. The analysis is essentially independent of the number of logical symbols involved in the function V, and the object of the proposition will, therefore, in all instances, be attained by the following Rule:—

RULE.—*Develop the function V, and equate to 0 every constituent whose coefficient does not vanish. The interpretation of these results collectively will constitute the interpretation of the given equation.*

6. Let us take as an example the definition of "clean beasts," laid down in the Jewish law, viz., "Clean beasts are those which both divide the hoof and chew the cud," and let us assume

$$x = \text{clean beasts};$$
$$y = \text{beasts dividing the hoof};$$
$$z = \text{beasts chewing the cud}.$$

Then the given proposition will be represented by the equation

$$x = yz,$$

which we shall reduce to the form

$$x - yz = 0,$$

and seek that form of interpretation to which the present method leads. Fully developing the first member, we have

$$0\,xyz + xy\,(1-z) + x\,(1-y)\,z + x\,(1-y)\,(1-z)$$
$$-(1-x)yz + 0(1-x)y(1-z) + 0(1-x)(1-y)z + 0(1-x)(1-y)(1-z).$$

Whence the terms, whose coefficients do not vanish, give

$$xy(1-z) = 0, \quad xz(1-y) = 0, \quad x(1-y)(1-z) = 0, \quad (1-x)yz = 0.$$

These equations express a denial of the existence of certain classes of objects, viz.:

1st. Of beasts which are clean, and divide the hoof, but do not chew the cud.

2nd. Of beasts which are clean, and chew the cud, but do not divide the hoof.

3rd. Of beasts which are clean, and neither divide the hoof nor chew the cud.

4th. Of beasts which divide the hoof, and chew the cud, and are not clean.

Now all these several denials are really involved in the original proposition. And conversely, if these denials be granted, the original proposition will follow as a necessary consequence. They are, in fact, the separate elements of that proposition. Every primary proposition can thus be resolved into a series of denials of the existence of certain defined classes of things, and may, from that system of denials, be itself reconstructed. It might here be asked, how it is possible to make an assertive proposition out of a series of denials or negations? From what source is the positive element derived? I answer, that the mind assumes the existence of a universe not *à priori* as a fact independent of experience, but either *à posteriori* as a deduction from experience, or *hypothetically* as a foundation of the possibility of assertive reasoning. Thus from the Proposition, "There are no men who are not fallible," which is a negation or denial of the existence of "infallible men," it may be inferred either hypothetically, "All men (if men exist) are fallible," or absolutely, (experience having assured us of the existence of the race), "All men are fallible."

The form in which conclusions are exhibited by the method of this Proposition may be termed the form of "Single or Conjoint Denial."

FORM II.

7. As the previous form was derived from the development and interpretation of an equation whose second member is 0, the present form, which is supplementary to it, will be derived from the development and interpretation of an equation whose second member is 1. It is, however, readily suggested by the analysis of the previous Proposition.

Thus in the example last discussed we deduced from the equation

$$x - yz = 0$$

the conjoint denial of the existence of the classes represented by the constituents

$$xy(1-z), \quad xz(1-y), \quad x(1-y)(1-z), \quad (1-x)yz,$$

whose coefficients were not equal to 0. It follows hence that the remaining constituents represent classes which make up the universe. Hence we shall have

$$xyz + (1-x)y(1-z) + (1-x)(1-y)z + (1-x)(1-y)(1-z) = 1.$$

This is equivalent to the affirmation that all existing things belong to some one or other of the following classes, viz.:

1st. Clean beasts both dividing the hoof and chewing the cud.

2nd. Unclean beasts dividing the hoof, but not chewing the cud.

3rd. Unclean beasts chewing the cud, but not dividing the hoof.

4th. Things which are neither clean beasts, nor chewers of the cud, nor dividers of the hoof.

This form of conclusion may be termed the form of "Single or Disjunctive Affirmation,"—single when but one constituent appears in the final equation; disjunctive when, as above, more constituents than one are there found.

Any equation, $V = 0$, wherein V satisfies the law of duality, may also be made to yield this form of interpretation by reducing it to the form $1 - V = 1$, and developing the first member. The case, however, is really included in the next general form. Both the previous forms are of slight importance compared with the following one.

FORM III.

8. In the two preceding cases the functions to be developed were equated to 0 and to 1 respectively. In the present case I shall suppose the corresponding function equated to any logical symbol w. We are then to endeavour to interpret the equation $V = w$, V being a function of the logical symbols x, y, z, &c. In the first place, however, I deem it necessary to show how the equation $V = w$, or, as it will usually present itself, $w = V$, arises.

Let us resume the definition of "clean beasts," employed in the previous examples, viz., "Clean beasts are those which both divide the hoof and chew the cud," and suppose it required to determine the relation in which "beasts chewing the cud" stand to "clean beasts" and "beasts dividing the hoof." The equation expressing the given proposition is

$$x = yz,$$

and our object will be accomplished if we can determine z as an interpretable function of x and y.

Now treating x, y, z as symbols of quantity subject to a peculiar law, we may deduce from the above equation, by solution,

$$z = \frac{x}{y}.$$

But this equation is not at present in an interpretable form. If we can reduce it to such a form it will furnish the relation required.

On developing the second member of the above equation, we have

$$z = xy + \frac{1}{0}x(1-y) + 0(1-x)y + \frac{0}{0}(1-x)(1-y),$$

and it will be shown hereafter (Prop. 3) that this admits of the following interpretation:

"Beasts which chew the cud consist of all clean beasts (which also divide the hoof), together with an indefinite remainder (some, none, or all) of unclean beasts which do not divide the hoof."

9. Now the above is a particular example of a problem of the utmost generality in Logic, and which may thus be stated:—"Given any logical equation connecting the symbols x, y, z, w, required an interpretable expression for the relation of the class represented by w to the classes represented by the other symbols x, y, z, &c."

The solution of this problem consists in all cases in determining, from the equation *given*, the expression of the above symbol w, in terms of the other symbols, and rendering that expression interpretable by development. Now the equation given is always of the first degree with respect to each of the symbols involved. The required expression for w can therefore always be found. In fact, if we develop the given equation, whatever its form may be with respect to w, we obtain an equation of the form

$$Ew + E'(1-w) = 0, \qquad (1)$$

E and E' being functions of the remaining symbols. From the above we have

$$E' = (E' - E)w.$$

Therefore

$$w = \frac{E'}{E' - E} \qquad (2)$$

and expanding the second member by the rule of development, it will only remain to interpret the result in logic by the next proposition.

If the fraction $\frac{E'}{E' - E}$ has common factors in its numerator and denominator, we are not permitted to reject them, unless they are mere numerical constants. For the symbols x, y, &c., regarded as quantitative, may admit of such values 0 and 1 as to cause the common factors to become equal to 0, in which case the algebraic rule of reduction fails. This is the case contemplated in our remarks on the failure of the algebraic axiom of division (II. 14). To *express* the solution in the form (2), and without attempting to perform any unauthorized reductions, to interpret the result by the theorem of development, is a course strictly in accordance with the general principles of this treatise.

If the relation of the class expressed by $1 - w$ to the other classes, x, y, &c. is required, we deduce from (1), in like manner as above,

$$1 - w = \frac{E}{E - E'},$$

to the interpretation of which also the method of the following Proposition is applicable :

Proposition III.

10. *To determine the interpretation of any logical equation of the form $w = V$, in which w is a class symbol, and V a function of other class symbols quite unlimited in its form.*

Let the second member of the above equation be fully expanded. Each coefficient of the result will belong to some one of the four classes, which, with their respective interpretations, we proceed to discuss.

1st. Let the coefficient be 1. As this is the symbol of the universe, and as the product of any two class symbols represents those individuals which are found in both classes, any constituent which has unity for its coefficient must be interpreted without limitation, i. e. the whole of the class which it represents is implied.

2nd. Let the coefficient be 0. As in Logic, equally with Arithmetic, this is the symbol of Nothing, no part of the class represented by the constituent to which it is prefixed must be taken.

3rd. Let the coefficient be of the form $\frac{0}{0}$. Now, as in Arithmetic, the symbol $\frac{0}{0}$ represents an *indefinite number*, except when otherwise determined by some special circumstance, analogy would suggest that in the system of this work the same symbol should represent an *indefinite class*. That this is its true meaning will be made clear from the following example:

Let us take the Proposition, "Men not mortal do not exist;" represent this Proposition by symbols; and seek, in obedience to the laws to which those symbols have been proved to be subject, a reverse definition of "mortal beings," in terms of "men."

Now if we represent "men" by y, and "mortal beings" by x, the Proposition, "Men who are not mortals do not exist," will be expressed by the equation

$$y(1-x)=0,$$

from which we are to seek the value of x. Now the above equation gives

$$y-yx=0, \text{ or } yx=y.$$

Were this an ordinary algebraic equation, we should, in the next place, divide both sides of it by y. But it has been remarked in Chap. II. that the operation of division cannot be *performed* with the symbols with which we are now engaged. Our resource, then, is to *express* the operation, and develop the result by the method of the preceding chapter. We have, then, first,

$$x=\frac{y}{y},$$

and, expanding the second member as directed,

$$x=y+\frac{0}{0}(1-y).$$

This implies that mortals (x) consist of all men (y), together with such a remainder of beings which are not men ($1-y$), as will be indicated by the coefficient $\frac{0}{0}$. Now let us inquire what remainder of "not men" is implied by the premiss. It might happen that the remainder included all the beings who are not men, or it might include only some of them, and not others, or it might include none, and any one of these assumptions would be in perfect accordance with our premiss. In other words, whether those beings which are not men are *all*, or *some*, or *none*, of them *mortal*, the truth of the premiss which virtually asserts that all men are mortal, will be equally unaffected, and therefore the expression $\frac{0}{0}$ here indicates that *all*, *some*, or *none* of the class to whose expression it is affixed must be taken.

Although the above determination of the significance of the symbol $\frac{0}{0}$ is founded only upon the examination of a particular case, yet the principle involved in the demonstration is general, and there are no circumstances under which the symbol can present itself to which the same mode of analysis is inapplicable. We may properly term $\frac{0}{0}$ an *indefinite class symbol*, and may, if convenience should require, replace it by an uncompounded symbol v, subject to the fundamental law, $v(1-v)=0$.

4th. It may happen that the coefficient of a constituent in an expansion does not belong to any of the previous cases. To ascertain its true interpretation when this happens, it will be necessary to premise the following theorem:

11. THEOREM.—*If a function V, intended to represent any class or collection of objects, w, be expanded, and if the numerical coefficient, a, of any constituent in its development, do not satisfy the law.*

$$a(1-a)=0,$$

then the constituent in question must be made equal to 0.

To prove the theorem generally, let us represent the expansion given, under the form

$$w = a_1 t_1 + a_2 t_2 + a_3 t_3 + \&c., \tag{1}$$

in which t_1, t_2, t_3, &c. represent the constituents, and a_1, a_2, a_3, &c. the coefficients; let us also suppose that a_1 and a_2 do not satisfy the law

$$a_1(1-a_1)=0, \quad a_2(1-a_2)=0;$$

but that the other coefficients are subject to the law in question, so that we have

$$a_3^2 = a_3, \text{ \&c.}$$

Now multiply each side of the equation (1) by itself. The result will be

$$w = a_1^2 t_1 + a_2^2 t_2 + \text{\&c.} \qquad (2)$$

This is evident from the fact that it must represent the development of the equation

$$w = V^2,$$

but it may also be proved by actually squaring (1), and observing that we have

$$t_1^2 = t_1, \quad t_2^2 = t_2, \quad t_1 t_2 = 0, \text{ \&c.}$$

by the properties of constituents. Now subtracting (2) from (1), we have

$$(a_1 - a_1^2) t_1 + (a_2 - a_2^2) t_2 = 0.$$

Or, $$a_1 (1 - a_1) t_1 + a_2 (1 - a_2) t_2 = 0.$$

Multiply the last equation by t_1; then since $t_1 t_2 = 0$, we have

$$a_1 (1 - a_1) t_1 = 0, \text{ whence } t_1 = 0.$$

In like manner multiplying the same equation by t_2, we have

$$a_2 (1 - a_2) t_2 = 0, \text{ whence } t_2 = 0.$$

Thus it may be shown generally that any constituent whose coefficient is not subject to the same fundamental law as the symbols themselves must be separately equated to 0. The usual form under which such coefficients occur is $\frac{1}{0}$. This is the algebraic symbol of infinity. Now the nearer any number approaches to infinity (allowing such an expression), the more does it depart from the condition of satisfying the fundamental law above referred to.

The symbol $\frac{0}{0}$, whose interpretation was previously discussed, does not necessarily disobey the law we are here considering, for it admits of the numerical values 0 and 1 indifferently. Its actual interpretation, however, as an indefinite class symbol, cannot, I conceive, except upon the ground of analogy, be deduced from its arithmetical properties, but must be established experimentally.

12. We may now collect the results to which we have been led, into the following summary:

1st. The symbol 1, as the coefficient of a term in a development, indicates that the whole of the class which that constituent represents, is to be taken.

2nd. The coefficient 0 indicates that none of the class are to be taken.

3rd. The symbol $\frac{0}{0}$ indicates that a perfectly *indefinite* portion of the class, i. e. *some*, *none*, or *all* of its members are to be taken.

4th. Any other symbol as a coefficient indicates that the constituent to which it is prefixed must be equated to 0.

It follows hence that if the solution of a problem, obtained by development, be of the form

$$w = A + 0B + \frac{0}{0} C + \frac{1}{0} D,$$

that solution may be resolved into the two following equations, viz.,

$$w = A + vC, \tag{3}$$

$$D = 0, \tag{4}$$

v being an indefinite class symbol. The interpretation of (3) shows what elements enter, or may enter, into the composition of w, the class of things whose definition is required; and the interpretation of (4) shows what relations exist among the elements of the original problem, in perfect independence of w.

Such are the canons of interpretation. It may be added, that they are universal in their application, and that their use is always unembarrassed by exception or failure.

13. *Corollary.*—If V be an independently interpretable logical function, it will satisfy the symbolical law, $V(1 - V) = 0$.

By an independently interpretable logical function, I mean one which is interpretable, without presupposing any relation among the things represented by the symbols which it involves. Thus $x(1 - y)$ is independently interpretable, but $x - y$ is not so.

The latter function presupposes, as a condition of its interpretation, that the class represented by y is wholly contained in the class represented by x; the former function does not imply any such requirement.

Now if V is independently interpretable, and if w represent the collection of individuals which it contains, the equation $w = V$ will hold true without entailing as a consequence the vanishing of any of the constituents in the development of V; since such vanishing of constituents would imply relations among the classes of things denoted by the symbols in V. Hence the development of V will be of the form

$$a_1 t_1 + a_2 t_2 + \&c.$$

the coefficients a_1, a_2, &c. all satisfying the condition

$$a_1 (1 - a_1) = 0, \quad a_2 (1 - a_2) = 0, \&c.$$

Hence by the reasoning of Prop. 4, Chap. v. the function V will be subject to the law

$$V(1 - V) = 0.$$

This result, though evident *à priori* from the fact that V is supposed to represent a class or collection of things, is thus seen to follow also from the properties of the constituents of which it is composed. The condition $V(1 - V) = 0$ may be termed "the condition of interpretability of logical functions."

14. The general form of solutions, or logical conclusions developed in the last Proposition, may be designated as a "Relation between terms." I use, as before, the word "terms" to denote the parts of a proposition, whether simple or complex, which are connected by the copula "is" or "are." The classes of things represented by the individual symbols may be called the elements of the proposition.

15. Ex. 1.—Resuming the definition of "clean beasts," (VI.6), required a description of "unclean beasts."

Here, as before, x standing for "clean beasts," y for "beasts dividing the hoof," z for "beasts chewing the cud," we have

$$x = yz; \qquad (5)$$

whence

$$1 - x = 1 - yz;$$

and developing the second member,

$$1 - x = y(1 - z) + z(1 - y) + (1 - y)(1 - z);$$

which is interpretable into the following Proposition: *Unclean beasts are all which divide the hoof without chewing the cud, all which chew the cud without dividing the hoof, and all which neither divide the hoof nor chew the cud.*

Ex. 2.—The same definition being given, required a description of beasts which do not divide the hoof.

From the equation $x = yz$ we have

$$y = \frac{x}{z};$$

therefore,

$$1 - y = \frac{z - x}{z};$$

and developing the second member,

$$1 - y = 0\,xz + \frac{-1}{0}\,x(1 - z) + (1 - x)z + \frac{0}{0}(1 - x)(1 - z).$$

Here, according to the Rule, the term whose coefficients is $\frac{-1}{0}$, must be separately equated to 0, whence we have

$$1 - y = (1 - x)z + v(1 - x)(1 - z),$$
$$x(1 - z) = 0;$$

whereof the first equation gives by interpretation the Proposition: *Beasts which do not divide the hoof consist of all unclean beasts which chew the cud, and an indefinite remainder (some, none, or all) of unclean beasts which do not chew the cud.*

The second equation gives the Proposition: *There are no clean beasts which do not chew the cud.* This is one of the independent relations above referred to. We sought the direct relation of "Beasts not dividing the hoof," to "Clean beasts and beasts which chew the cud." It happens, however, that independently of any relation to beasts not dividing the hoof, there exists, in virtue of the premiss, a separate relation between clean beasts and beasts which chew the cud. This relation is also necessarily given by the process.

Ex. 3.—Let us take the following definition, viz.: "Responsible beings are all rational beings who are either free to act, or have voluntarily sacrificed their freedom," and apply to it the preceding analysis.

Let x stand for responsible beings.
y ,, rational beings.
z ,, those who are free to act,
w ,, those who have voluntarily sacrificed their freedom of action.

In the expression of this definition I shall assume, that the two alternatives which it presents, viz.: "Rational beings free to act," and "Rational beings whose freedom of action has been voluntarily sacrificed," are mutually exclusive, so that no individuals are found at once in both these divisions. This will permit us to interpret the proposition literally into the language of symbols, as follows:

$$x = yz + yw. \tag{6}$$

Let us first determine hence the relation of "rational beings" to responsible beings, beings free to act, and beings whose freedom of action has been voluntarily abjured. Perhaps this object will be better stated by saying, that we desire to express the relation among the elements of the premiss in such a form as will enable us to determine how far rationality may be inferred from responsibility, freedom of action, a voluntary sacrifice of freedom, and their contraries.

From (6) we have

$$y = \frac{x}{z + w},$$

and developing the second member, but rejecting terms whose coefficients are 0,

$$y = \frac{1}{2}xzw + xz(1 - w) + x(1 - z)w + \frac{1}{0}x(1 - z)(1 - w)$$
$$+ \frac{0}{0}(1 - x)(1 - z)(1 - w),$$

whence, equating to 0 the terms whose coefficients are $\frac{1}{2}$ and $\frac{1}{0}$, we have

$$y = xz(1 - w) + xw(1 - z) + v(1 - x)(1 - z)(1 - w); \tag{7}$$

$$xzw = 0; \tag{8}$$

$$x(1-z)(1-w)=0; \tag{9}$$

whence by interpretation—

DIRECT CONCLUSION.—*Rational beings are all responsible beings who are either free to act, not having voluntarily sacrificed their freedom, or not free to act, having voluntarily sacrificed their freedom, together with an indefinite remainder (some, none, or all) of beings not responsible, not free, and not having voluntarily sacrificed their freedom.*

FIRST INDEPENDENT RELATION.—*No responsible beings are at the same time free to act, and in the condition of having voluntarily sacrificed their freedom.*

SECOND.—*No responsible beings are not free to act, and at the same time in the condition of not having sacrificed their freedom.*

The independent relations above determined may, however, be put in another and more convenient form. Thus (8) gives

$$xw=\frac{0}{z}=0\,z+\frac{0}{0}(1-z), \text{ on development;}$$

or,

$$xw=v(1-z); \tag{10}$$

and in like manner (9) gives

$$x(1-w)=\frac{0}{1-z}=\frac{0}{0}z+0(1-z);$$

or,

$$x(1-w)=vz; \tag{11}$$

and (10) and (11) interpreted give the following Propositions:

1st. *Responsible beings who have voluntarily sacrificed their freedom are not free.*

2nd. *Responsible beings who have not voluntarily sacrificed their freedom are free.*

These, however, are merely different forms of the relations before determined.

16. In examining these results, the reader must bear in mind, that the sole province of a method of inference or analysis, is to determine those relations which are necessitated by the *connexion* of the terms in the original proposition. Accordingly, in estimating the completeness with which this object is effected, we have nothing whatever to do with those other relations which

may be suggested to our minds by the *meaning* of the terms employed, as distinct from their expressed connexion. Thus it seems obvious to remark, that "They who have voluntarily sacrificed their freedom are not free," this being a relation implied in the very meaning of the terms. And hence it might appear, that the first of the two independent relations assigned by the method is on the one hand needlessly limited, and on the other hand superfluous. However, if regard be had merely to the connexion of the terms in the original premiss, it will be seen that the relation in question is not liable to either of these charges. The solution, as expressed in the direct conclusion and the independent relations, conjointly, is perfectly complete, without being in any way superfluous.

If we wish to take into account the implicit relation above referred to, viz., "They who have voluntarily sacrificed their freedom are not free," we can do so by making this a distinct proposition, the proper expression of which would be

$$w = v(1 - z).$$

This equation we should have to employ together with that expressive of the original premiss. The mode in which such an examination must be conducted will appear when we enter upon the theory of systems of propositions in a future chapter. The sole difference of result to which the analysis leads is, that the first of the independent relations deduced above is superseded.

17. Ex. 4.—Assuming the same definition as in Example 2, let it be required to obtain a description of irrational persons.

We have

$$1 - y = 1 - \frac{x}{z + w}$$

$$= \frac{z + w - x}{z + w}$$

$$= \frac{1}{2} xzw + 0\, xz(1 - w) + 0\, x(1 - z)\, w - \frac{1}{0}\, x(1 - z)(1 - w)$$

$$+ (1-x)zw + (1-x)z(1-w) + (1-x)(1-z)w + \frac{0}{0}(1-x)(1-z)(1-w)$$

$$= (1-x)zw + (1-x)z(1-w) + (1-x)(1-z)w + v(1-x)(1-z)(1-w)$$

$$= (1-x)z + (1-x)(1-z)w + v(1-x)(1-z)(1-w),$$

with $\quad xzw = 0, \quad x(1 - z)(1 - w) = 0.$

The independent relations here given are the same as we before arrived at, as they evidently ought to be, since whatever relations prevail independently of the existence of a given class of objects y, prevail independently also of the existence of the contrary class $1 - y$.

The direct solution afforded by the first equation is:—*Irrational persons consist of all irresponsible beings who are either free to act, or have voluntarily sacrificed their liberty, and are not free to act; together with an indefinite remainder of irresponsible beings who have not sacrificed their liberty, and are not free to act.*

18. The propositions analyzed in this chapter have been of that species called definitions. I have discussed none of which the second or predicate term is particular, and of which the general type is $Y = vX$, Y and X being functions of the logical symbols x, y, z, &c., and v an indefinite class symbol. The analysis of such propositions is greatly facilitated (though the step is not an essential one) by the elimination of the symbol v, and this process depends upon the method of the next chapter. I postpone also the consideration of another important problem necessary to complete the theory of single propositions, but of which the analysis really depends upon the method of the reduction of systems of propositions to be developed in a future page of this work.

CHAPTER VII.

ON ELIMINATION.

1. IN the examples discussed in the last chapter, all the elements of the original premiss re-appeared in the conclusion, only in a different order, and with a different connexion. But it more usually happens in common reasoning, and especially when we have more than one premiss, that some of the elements are required not to appear in the conclusion. Such elements, or, as they are commonly called, "middle terms," may be considered as introduced into the original propositions only for the sake of that connexion which they assist to establish among the other elements, which are alone designed to enter into the expression of the conclusion.

2. Respecting such intermediate elements, or middle terms, some erroneous notions prevail. It is a general opinion, to which, however, the examples contained in the last chapter furnish a contradiction, that inference consists peculiarly in the elimination of such terms, and that the elementary type of this process is exhibited in the elimination of one middle term from two premises, so as to produce a single resulting conclusion into which that term does not enter. Hence it is commonly held, that *syllogism* is the basis, or else the common type, of all inference, which may thus, however complex its form and structure, be resolved into a series of syllogisms. The propriety of this view will be considered in a subsequent chapter. At present I wish to direct attention to an important, but hitherto unnoticed, point of difference between the system of Logic, as expressed by symbols, and that of common algebra, with reference to the subject of elimination. In the algebraic system we are able to eliminate one symbol from two equations, two symbols from three equations, and generally $n - 1$ symbols from n equations. There thus exists a definite connexion between the number of independent equations given,

and the number of symbols of quantity which it is possible to eliminate from them. But it is otherwise with the system of Logic. No fixed connexion there prevails between the number of equations given representing propositions or premises, and the number of typical symbols of which the elimination can be effected. From a single equation an indefinite number of such symbols may be eliminated. On the other hand, from an indefinite number of equations, a single class symbol only may be eliminated. We may affirm, that in this peculiar system, the problem of elimination is resolvable under all circumstances alike. This is a consequence of that remarkable law of duality to which the symbols of Logic are subject. To the equations furnished by the premises given, there is added another equation or system of equations drawn from the fundamental laws of thought itself, and supplying the necessary means for the solution of the problem in question. Of the many consequences which flow from the law of duality, this is perhaps the most deserving of attention.

3. As in Algebra it often happens, that the elimination of symbols from a given system of equations conducts to a mere identity in the form $0 = 0$, no independent relations connecting the symbols which remain; so in the system of Logic, a like result, admitting of a similar interpretation, may present itself. Such a circumstance does not detract from the generality of the principle before stated. The object of the method upon which we are about to enter is to eliminate any number of symbols from any number of logical equations, and to exhibit in the result the actual relations which remain. Now it may be, that no such residual relations exist. In such a case the truth of the method is shown by its leading us to a merely identical proposition.

4. The notation adopted in the following Propositions is similar to that of the last chapter. By $f(x)$ is meant any expression involving the logical symbol x, with or without other logical symbols. By $f(1)$ is meant what $f(x)$ becomes when x is therein changed into 1; by $f(0)$ what the same function becomes when x is changed into 0.

PROPOSITION I.

5. *If $f(x) = 0$ be any logical equation involving the class symbol x, with or without other class symbols, then will the equation*

$$f(1)\ f(0) = 0$$

be true, independently of the interpretation of x; and it will be the complete result of the elimination of x from the above equation.

In other words, the elimination of x from any given equation, $f(x)=0$, will be effected by successively changing in that equation x into 1, and x into 0, and multiplying the two resulting equations together.

Similarly the complete result of the elimination of any class symbols, x, y, &c., from any equation of the form $V=0$, will be obtained by completely expanding the first member of that equation in constituents of the given symbols, and multiplying together all the coefficients of those constituents, and equating the product to 0.

Developing the first member of the equation $f(x) = 0$, we have (V..10),

$$f(1)\ x + f(0)\ (1 - x) = 0;$$

or,

$$\{f(1) - f(0)\}\ x + f(0) = 0. \qquad (1)$$

$$\therefore x = \frac{f(0)}{f(0) - f(1)};$$

and

$$1 - x = -\frac{f(1)}{f(0) - f(1)}.$$

Substitute these expressions for x and $1 - x$ in the fundamental equation

$$x\ (1 - x) = 0,$$

and there results

$$-\frac{f(0)\ f(1)}{\{f(0) - f(1)\}^2} = 0;$$

or,

$$f(1)\ f(0) = 0, \qquad (2)$$

the form required.

6. It is seen in this process, that the elimination is really effected between the given equation $f(x) = 0$ and the universally true equation $x\ (1 - x) = 0$, expressing the fundamental law of logical symbols, *qua* logical. There exists, therefore, no need of more

than one premiss or equation, in order to render possible the elimination of a term, the necessary law of thought virtually supplying the other premiss or equation. And though the demonstration of this conclusion may be exhibited in other forms, yet the same element furnished by the mind itself will still be virtually present. Thus we might proceed as follows:

Multiply (1) by x, and we have

$$f(1)\,x = 0, \tag{3}$$

and let us seek by the forms of ordinary algebra to eliminate x from this equation and (1).

Now if we have two algebraic equations of the form

$$ax + b = 0,$$
$$a'x + b' = 0;$$

it is well known that the result of the elimination of x is

$$ab' - a'b = 0. \tag{4}$$

But comparing the above pair of equations with (1) and (3) respectively, we find

$$a = f(1) - f(0), \qquad b = f(0);$$
$$a' = f(1) \qquad\quad b' = 0;$$

which, substituted in (4), give

$$f(1)\,f(0) = 0,$$

as before. In this form of the demonstration, the fundamental equation $x(1 - x) = 0$, makes its appearance in the derivation of (3) from (1).

7. I shall add yet another form of the demonstration, partaking of a half logical character, and which may set the demonstration of this important theorem in a clearer light.

We have as before

$$f(1)\,x + f(0)\,(1 - x) = 0.$$

Multiply this equation first by x, and secondly by $1 - x$, we get

$$f(1)\,x = 0, \qquad f(0)\,(1 - x) = 0.$$

From these we have by solution and development,

$$f(1) = \frac{0}{x} = \frac{0}{0}(1-x), \text{ on development,}$$

$$f(0) = \frac{0}{1-x} = \frac{0}{0}x.$$

The direct interpretation of these equations is—

1st. Whatever individuals are included in the class represented by $f(1)$, are not x's.

2nd. Whatever individuals are included in the class represented by $f(0)$, are x's.

Whence by common logic, there are no individuals at once in the class $f(1)$ and in the class $f(0)$, i.e. there are no individuals in the class $f(1)f(0)$. Hence,

$$f(1)f(0) = 0. \qquad (5)$$

Or it would suffice to multiply together the developed equations, whence the result would immediately follow.

8. The theorem (5) furnishes us with the following Rule:

TO ELIMINATE ANY SYMBOL FROM A PROPOSED EQUATION.

RULE.—*The terms of the equation having been brought, by transposition if necessary, to the first side, give to the symbol successively the values* 1 *and* 0, *and multiply the resulting equations together.*

The first part of the Proposition is now proved.

9. Consider in the next place the general equation

$$f(x, y) = 0;$$

the first member of which represents any function of x, y, and other symbols.

By what has been shown, the result of the elimination of y from this equation will be

$$f(x, 1)f(x, 0) = 0;$$

for such is the form to which we are conducted by successively changing in the given equation y into 1, and y into 0, and multiplying the results together.

Again, if in the result obtained we change successively x into 1, and x into 0, and multiply the results together, we have

$$f(1, 1)f(1, 0)f(0, 1)f(0, 0) = 0; \qquad (6)$$

as the final result of elimination.

But the four factors of the first member of this equation are the four coefficients of the complete expansion of $f(x, y)$, the first member of the original equation; whence the second part of the Proposition is manifest.

EXAMPLES.

10. Ex. 1.—Having given the Proposition, "All men are mortal," and its symbolical expression, in the equation,

$$y = vx,$$

in which y represents "men," and x "mortals," it is required to eliminate the indefinite class symbol v, and to interpret the result.

Here bringing the terms to the first side, we have

$$y - vx = 0.$$

When $v = 1$ this becomes

$$y - x = 0;$$

and when $v = 0$ it becomes

$$y = 0;$$

and these two equations multiplied together, give

$$y - yx = 0,$$

or

$$y(1 - x) = 0,$$

it being observed that $y^2 = y$.

The above equation is the required result of elimination, and its interpretation is, *Men who are not mortal do not exist*,—an obvious conclusion.

If from the equation last obtained we seek a description of beings who are not mortal, we have

$$x = \frac{y}{y},$$

$$\therefore 1 - x = \frac{0}{y}.$$

Whence, by expansion, $1 - x = \frac{0}{0}(1 - y)$, which interpreted gives, *They who are not mortal are not men.* This is an example of what in the common logic is called conversion by contraposition, or negative conversion.*

Ex. 2.—Taking the Proposition, "No men are perfect," as represented by the equation

$$y = v(1 - x),$$

wherein y represents "men," and x "perfect beings," it is required to eliminate v, and find from the result a description both of *perfect beings* and of *imperfect beings*. We have

$$y - v(1 - x) = 0.$$

Whence, by the rule of elimination,

$$\{y - (1 - x)\} \times y = 0,$$

or

$$y - y(1 - x) = 0,$$

or

$$yx = 0;$$

which is interpreted by the Proposition, *Perfect men do not exist.* From the above equation we have

$$x = \frac{0}{y} = \frac{0}{0}(1 - y) \text{ by development;}$$

whence, by interpretation, *No perfect beings are men.* Similarly,

$$1 - x = 1 - \frac{0}{y} = \frac{y}{y} = y + \frac{0}{0}(1 - y),$$

which, on interpretation, gives, *Imperfect beings are all men with an indefinite remainder of beings, which are not men.*

11. It will generally be the most convenient course, in the treatment of propositions, to eliminate first the indefinite class symbol v, wherever it occurs in the corresponding equations. This will only modify their form, without impairing their significance. Let us apply this process to one of the examples of Chap. IV. For the Proposition, "No men are placed in exalted stations and free from envious regards," we found the expression

$$y = v(1 - xz),$$

and for the equivalent Proposition, "Men in exalted stations are not free from envious regards," the expression

$$yx = v(1 - z);$$

* Whately's Logic, Book II. chap. II. sec. 4.

and it was observed that these equations, v being an indefinite class symbol, were themselves equivalent. To prove this, it is only necessary to eliminate from each the symbol v. The first equation is

$$y - v(1 - xz) = 0,$$

whence, first making $v = 1$, and then $v = 0$, and multiplying the results, we have

$$(y - 1 + xz)\,y = 0,$$

or

$$yxz = 0.$$

Now the second of the given equations becomes on transposition

$$yx - v(1 - z) = 0;$$

whence

$$(yx - 1 + z)\,yx = 0,$$

or

$$yxz = 0,$$

as before. The reader will easily interpret the result.

12. Ex. 3.—As a subject for the general method of this chapter, we will resume Mr. Senior's definition of wealth, viz.: "Wealth consists of things transferable, limited in supply, and either productive of pleasure or preventive of pain." We shall consider this definition, agreeably to a former remark, as including all things which possess at once both the qualities expressed in the last part of the definition, upon which assumption we have, as our representative equation,

$$w = st\,\{pr + p(1 - r) + r(1 - p)\},$$

or

$$w = st\,\{p + r(1 - p)\},$$

wherein

w stands for wealth.
s ,, things limited in supply.
t ,, things transferable.
p ,, things productive of pleasure.
r ,, things preventive of pain.

From the above equation we can eliminate any symbols that we do not desire to take into account, and express the result by solution and development, according to any proposed arrangement of subject and predicate.

Let us first consider what the expression for w, wealth, would be if the element r, referring to prevention of pain, were eliminated. Now bringing the terms of the equation to the first side, we get

$$w - st(p + r - rp) = 0.$$

Making $r = 1$, the first member becomes $w - st$, and making $r = 0$ it becomes $w - stp$; whence we have by the Rule,

$$(w - st)(w - stp) = 0, \tag{7}$$

or

$$w - wstp - wst + stp = 0; \tag{8}$$

whence

$$w = \frac{stp}{st + stp - 1};$$

the development of the second member of which equation gives

$$w = stp + \frac{0}{0} st(1 - p). \tag{9}$$

Whence we have the conclusion,—*Wealth consists of all things limited in supply, transferable, and productive of pleasure, and an indefinite remainder of things limited in supply, transferable, and not productive of pleasure.* This is sufficiently obvious.

Let it be remarked that it is not necessary to perform the multiplication indicated in (7), and reduce that equation to the form (8), in order to determine the expression of w in terms of the other symbols. The process of development may in all cases be made to supersede that of multiplication. Thus if we develop (7) in terms of w, we find

$$(1 - st)(1 - stp)w + stp(1 - w) = 0,$$

whence

$$w = \frac{stp}{stp - (1 - st)(1 - stp)};$$

and this equation developed will give, as before,

$$w = stp + \frac{0}{0} st(1 - p).$$

13. Suppose next that we seek a description of things limited in supply, as dependent upon their relation to wealth, transferableness, and tendency to produce pleasure, omitting all reference to the prevention of pain.

From equation (8), which is the result of the elimination of r from the original equation, we have

$$w - s(wt + wtp - tp) = 0;$$

whence

$$s = \frac{w}{wt + wtp - tp}$$

$$= wtp + wt(1-p) + \frac{1}{0}w(1-t)p + \frac{1}{0}w(1-t)(1-p)$$

$$+ 0(1-w)tp + \frac{0}{0}(1-w)t(1-p) + \frac{0}{0}(1-w)(1-t)p$$

$$+ \frac{0}{0}(1-w)(1-t)(1-p).$$

We will first give the direct interpretation of the above solution, term by term; afterwards we shall offer some general remarks which it suggests; and, finally, show how the expression of the conclusion may be somewhat abbreviated.

First, then, the direct interpretation is, Things limited in supply consist of *All wealth transferable and productive of pleasure —all wealth transferable, and not productive of pleasure,—an indefinite amount of what is not wealth, but is either transferable, and not productive of pleasure, or intransferable and productive of pleasure, or neither transferable nor productive of pleasure.*

To which the terms whose coefficients are $\frac{1}{0}$ permit us to add the following independent relations, viz.:

1st. *Wealth that is intransferable, and productive of pleasure, does not exist.*

2ndly. *Wealth that is intransferable, and not productive of pleasure, does not exist.*

14. Respecting this solution I suppose the following remarks are likely to be made.

First, it may be said, that in the expression above obtained for "things limited in supply," the term "All wealth transferable," &c., is in part redundant; since all wealth is (as implied in the original proposition, and directly asserted in the *independent relations*) necessarily transferable.

I answer, that although in ordinary speech we should not

deem it necessary to add to "wealth" the epithet "transferable," if another part of our reasoning had led us to express the conclusion, that there is no wealth which is not transferable, yet it pertains to the perfection of this method that it in all cases fully defines the objects represented by each term of the conclusion, by stating the relation they bear to each quality or element of distinction that we have chosen to employ. This is necessary in order to keep the different parts of the solution really distinct and independent, and actually prevents redundancy. Suppose that the pair of terms we have been considering had not contained the word "transferable," and had unitedly been "All wealth," we could then logically resolve the single term "All wealth" into the two terms "All wealth transferable," and "All wealth intransferable." But the latter term is shown to disappear by the "independent relations." Hence it forms no part of the description required, and is therefore redundant. The remaining term agrees with the conclusion actually obtained.

Solutions in which there cannot, by logical divisions, be produced any superfluous or redundant terms, may be termed *pure solutions*. Such are all the solutions obtained by the method of development and elimination above explained. It is proper to notice, that if the common algebraic method of elimination were adopted in the cases in which that method is possible in the present system, we should not be able to depend upon the purity of the solutions obtained. Its want of generality would not be its only defect.

15. In the second place, it will be remarked, that the conclusion contains two terms, the aggregate significance of which would be more conveniently expressed by a single term. Instead of "All wealth productive of pleasure, and transferable," and "All wealth not productive of pleasure, and transferable," we might simply say, "All wealth transferable." This remark is quite just. But it must be noticed that whenever any such simplifications are possible, they are immediately suggested by the form of the equation we have to interpret; and if that equation be reduced to its simplest form, then the interpretation to which it conducts will be in its simplest form also. Thus in the original solution the terms wtp and $wt(1-p)$, which have unity for their

coefficient, give, on addition, wt; the terms $w(1-t)p$ and $w(1-t)(1-p)$, which have $\frac{1}{0}$ for their coefficient give $w(1-t)$; and the terms $(1-w)(1-t)p$ and $(1-w)(1-t)(1-p)$, which have $\frac{0}{0}$ for their coefficient, give $(1-w)(1-t)$. Whence the complete solution is

$$s = wt + \frac{0}{0}(1-w)(1-t) + \frac{0}{0}(1-w)t(1-p),$$

with the independent relation,

$$w(1-t) = 0, \text{ or } w = \frac{0}{0}t.$$

The interpretation would now stand thus:—

1st. *Things limited in supply consist of all wealth transferable, with an indefinite remainder of what is not wealth and not transferable, and of transferable articles which are not wealth, and are not productive of pleasure.*

2nd. *All wealth is transferable.*

This is the simplest form under which the general conclusion, with its attendant condition, can be put.

16. When it is required to eliminate two or more symbols from a proposed equation we can either employ (6) Prop. I., or eliminate them in succession, the order of the process being indifferent. From the equation

$$w = st(p + r - pr),$$

we have eliminated r, and found the result,

$$w - wst - wstp + stp = 0.$$

Suppose that it had been required to eliminate both r and t, then taking the above as the first step of the process, it remains to eliminate from the last equation t. Now when $t = 1$ the first member of that equation becomes

$$w - ws - wsp + sp,$$

and when $t = 0$ the same member becomes w. Whence we have

$$w(w - ws - wsp + sp) = 0,$$

or

$$w - ws = 0,$$

for the required result of elimination.

If from the last result we determine w, we have

$$w = \frac{0}{1-s} = \frac{0}{0}s,$$

whence "*All wealth is limited in supply.*" As p does not enter into the equation, it is evident that the above is true, irrespectively of any relation which the elements of the conclusion bear to the quality "productive of pleasure."

Resuming the original equation, let it be required to eliminate s and t. We have

$$w = st\,(p + r - pr).$$

Instead, however, of separately eliminating s and t according to the Rule, it will suffice to treat st as a single symbol, seeing that it satisfies the fundamental law of the symbols by the equation

$$st\,(1 - st) = 0.$$

Placing, therefore, the given equation under the form

$$w - st\,(p + r - pr) = 0;$$

and making st successively equal to 1 and to 0, and taking the product of the results, we have

$$(w - p - r + pr)\,w = 0,$$

or

$$w - wp - wr + wpr = 0,$$

for the result sought.

As a particular illustration, let it be required to deduce an expression for "things productive of pleasure" (p), in terms of "wealth" (w), and "things preventive of pain" (r).

We have, on solving the equation,

$$p = \frac{w\,(1-r)}{w\,(1-r)}$$

$$= \frac{0}{0}wr + w(1-r) + \frac{0}{0}(1-w)\,r + \frac{0}{0}(1-w)\,(1-r)$$

$$= w\,(1-r) + \frac{0}{0}wr + \frac{0}{0}(1-w).$$

Whence the following conclusion:—*Things productive of plea-*

sure are, all wealth not preventive of pain, an indefinite amount of wealth that is preventive of pain, and an indefinite amount of what is not wealth.

From the same equation we get

$$1 - p = 1 - \frac{w(1-r)}{w(1-r)} = \frac{0}{w(1-r)},$$

which developed, gives

$$w(1-p) = \frac{0}{0}wr + \frac{0}{0}(1-w).r + \frac{0}{0}(1-w).(1-r)$$

$$= \frac{0}{0}wr + \frac{0}{0}(1-w).$$

Whence, *Things not productive of pleasure are either wealth, preventive of pain, or what is not wealth.*

Equally easy would be the discussion of any similar case.

17. In the last example of elimination, we have eliminated the compound symbol *st* from the given equation, by treating it as a single symbol. The same method is applicable to any combination of symbols which satisfies the fundamental law of individual symbols. Thus the expression $p + r - pr$ will, on being multiplied by itself, reproduce itself, so that if we represent $p + r - pr$ by a single symbol as y, we shall have the fundamental law obeyed, the equation

$$y = y^2, \text{ or } y(1-y) = 0,$$

being satisfied. For the rule of elimination for symbols is founded upon the supposition that each individual symbol is subject to that law; and hence the elimination of any function or combination of such symbols from an equation, may be effected by a single operation, whenever that law is satisfied by the function.

Though the forms of interpretation adopted in this and the previous chapter show, perhaps better than any others, the direct significance of the symbols 1 and $\frac{0}{0}$, modes of expression more agreeable to those of common discourse may, with equal truth and propriety, be employed. Thus the equation (9) may be interpreted in the following manner: *Wealth is either limited in supply, transferable, and productive of pleasure, or limited in sup-*

ply, transferable, and not productive of pleasure. And reversely, *Whatever is limited in supply, transferable, and productive of pleasure, is wealth.* Reverse interpretations, similar to the above, are always furnished when the final development introduces terms having unity as a coefficient.

18. NOTE.—The fundamental equation $f(1)f(0) = 0$, expressing the result of the elimination of the symbol x from any equation $f(x) = 0$, admits of a remarkable interpretation.

It is to be remembered, that by the equation $f(x) = 0$ is implied some proposition in which the individuals represented by the class x, suppose "men," are referred to, together, it may be, with other individuals; and it is our object to ascertain whether there is implied in the proposition any relation among the other individuals, independently of those found in the class *men*. Now the equation $f(1) = 0$ expresses what the original proposition would become if *men* made up the universe, and the equation $f(0) = 0$ expresses what that original proposition would become if men ceased to exist, wherefore the equation $f(1)f(0) = 0$ expresses what in virtue of the original proposition would be equally true on either assumption, i. e. equally true whether "men" were "all things" or "nothing." Wherefore the theorem expresses that *what is equally true, whether a given class of objects embraces the whole universe or disappears from existence, is independent of that class altogether, and* vice versâ. Herein we see another example of the interpretation of formal results, immediately deduced from the mathematical laws of thought, into general axioms of philosophy.

CHAPTER VIII.

ON THE REDUCTION OF SYSTEMS OF PROPOSITIONS.

1. IN the preceding chapters we have determined sufficiently for the most essential purposes the theory of single primary propositions, or, to speak more accurately, of primary propositions expressed by a single equation. And we have established upon that theory an adequate method. We have shown how any element involved in the given system of equations may be eliminated, and the relation which connects the remaining elements deduced in any proposed form, whether of denial, of affirmation, or of the more usual relation of subject and predicate. It remains that we proceed to the consideration of systems of propositions, and institute with respect to them a similar series of investigations. We are to inquire whether it is possible from the equations by which a system of propositions is expressed to eliminate, *ad libitum*, any number of the symbols involved; to deduce by interpretation of the result the whole of the relations implied among the remaining symbols; and to determine in particular the expression of any single element, or of any interpretable combination of elements, in terms of the other elements, so as to present the conclusion in any admissible form that may be required. These questions will be answered by showing that it is possible to reduce any system of equations, or any of the equations involved in a system, to an equivalent single equation, to which the methods of the previous chapters may be immediately applied. It will be seen also, that in this reduction is involved an important extension of the theory of single propositions, which in the previous discussion of the subject we were compelled to forego. This circumstance is not peculiar in its nature. There are many special departments of science which cannot be completely surveyed from within, but require to be studied also from an external point of view, and to be regarded in connexion with

other and kindred subjects, in order that their full proportions may be understood.

This chapter will exhibit two distinct modes of reducing systems of equations to equivalent single equations. The first of these rests upon the employment of arbitrary constant multipliers. It is a method sufficiently simple in theory, but it has the inconvenience of rendering the subsequent processes of elimination and development, when they occur, somewhat tedious. It was, however, the method of reduction first discovered, and partly on this account, and partly on account of its simplicity, it has been thought proper to retain it. The second method does not require the introduction of arbitrary constants, and is in nearly all respects preferable to the preceding one. It will, therefore, generally be adopted in the subsequent investigations of this work.

2. We proceed to the consideration of the first method.

Proposition I.

Any system of logical equations may be reduced to a single equivalent equation, by multiplying each equation after the first by a distinct arbitrary constant quantity, and adding all the results, including the first equation, together.

By Prop. 2, Chap. VI., the interpretation of any single equation, $f(x, y \ldots) = 0$ is obtained by equating to 0 those constituents of the development of the first member, whose coefficients do not vanish. And hence, if there be given two equations, $f(x, y \ldots) = 0$, and $F(x, y \ldots) = 0$, their united import will be contained in the system of results formed by equating to 0 all those constituents which thus present themselves in both, or in either, of the given equations developed according to the Rule of Chap. VI. Thus let it be supposed, that we have the two equations

$$xy - 2x = 0, \qquad (1)$$

$$x - y = 0; \qquad (2)$$

The development of the first gives

$$-xy - 2x(1 - y) = 0;$$

whence,

$$xy = 0, \quad x(1 - y) = 0. \qquad (3)$$

The development of the second equation gives

$$x(1-y)-y(1-x)=0;$$

whence, $$x(1-y)=0, \quad y(1-x)=0. \tag{4}$$

The constituents whose coefficients do not vanish in both developments are xy, $x(1-y)$, and $(1-x)y$, and these would together give the system

$$xy=0, \quad x(1-y)=0, \quad (1-x)y=0; \tag{5}$$

which is equivalent to the two systems given by the developments separately, seeing that in those systems the equation $x(1-y)=0$ is repeated. Confining ourselves to the case of binary systems of equations, it remains then to determine a single equation, which on development shall yield the same constituents with coefficients which do not vanish, as the given equations produce.

Now if we represent by

$$V_1=0, \quad V_2=0,$$

the given equations, V_1 and V_2 being functions of the logical symbols x, y, z, &c.; then the single equation

$$V_1+cV_2=0, \tag{6}$$

c being an arbitrary constant quantity, will accomplish the required object. For let At represent any term in the full development V_1 wherein t is a constituent and A its numerical coefficient, and let Bt represent the corresponding term in the full development of V_2, then will the corresponding term in the development of (6) be

$$(A+cB)t.$$

The coefficient of t vanishes if A and B both vanish, but not otherwise. For if we assume that A and B do not both vanish, and at the same time make

$$A+cB=0, \tag{7}$$

the following cases alone can present themselves.

1st. That A vanishes and B does not vanish. In this case the above equation becomes

$$cB=0,$$

and requires that $c = 0$. But this contradicts the hypothesis that c is an *arbitrary* constant.

2nd. That B vanishes and A does not vanish. This assumption reduces (7) to

$$A = 0,$$

by which the assumption is itself violated.

3rd. That neither A nor B vanishes. The equation (7) then gives

$$c = \frac{-A}{B},$$

which is a definite value, and, therefore, conflicts with the hypothesis that c is arbitrary.

Hence the coefficient $A + cB$ vanishes when A and B both vanish, but not otherwise. Therefore, the same constituents will appear in the development of (6), with coefficients which do not vanish, as in the equations $V_1 = 0$, $V_2 = 0$, singly or together. And the equation $V_1 + cV_2 = 0$, will be equivalent to the system $V_1 = 0$, $V_2 = 0$.

By similar reasoning it appears, that the general system of equations

$$V_1 = 0, \quad V_2 = 0, \quad V_3 = 0, \text{ \&c.};$$

may be replaced by the single equation

$$V_1 + cV_2 + c'V_3 + \text{\&c.} = 0,$$

c, c', &c., being arbitrary constants. The equation thus formed may be treated in all respects as the ordinary logical equations of the previous chapters. The arbitrary constants c_1, c_2, &c., are not *logical* symbols. They do not satisfy the law,

$$c_1(1 - c_1) = 0, \quad c_2(1 - c_2) = 0.$$

But their introduction is justified by that general principle which has been stated in (II. 15) and (V. 6), and exemplified in nearly all our subsequent investigations, viz., that equations involving the symbols of Logic may be treated in all respects as if those symbols were symbols of quantity, subject to the special law $x(1 - x) = 0$, until in the final stage of solution they assume a form interpretable in that system of thought with which Logic is conversant.

3. The following example will serve to illustrate the above method.

Ex. 1.—Suppose that an analysis of the properties of a particular class of substances has led to the following general conclusions, viz.:

1st. That wherever the properties A and B are combined, either the property C, or the property D, is present also; but they are not jointly present.

2nd. That wherever the properties B and C are combined, the properties A and D are either both present with them, or both absent.

3rd. That wherever the properties A and B are both absent, the properties C and D are both absent also; and *vice versâ*, where the properties C and D are both absent, A and B are both absent also.

Let it then be required from the above to determine what may be concluded in any particular instance from the presence of the property A with respect to the presence or absence of the properties B and C, paying no regard to the property D.

Represent the property A by x;
" the property B by y;
" the property C by z;
" the property D by w.

Then the symbolical expression of the premises will be

$$xy = v\,\{w\,(1-z) + z\,(1-w)\};$$
$$yz = v\,\{xw + (1-x)\,(1-w)\};$$
$$(1-x)\,(1-y) = (1-z)\,(1-w).$$

From the first two of these equations, separately eliminating the indefinite class symbol v, we have

$$xy\,\{1 - w\,(1-z) - z\,(1-w)\} = 0;$$
$$yz\,\{1 - xw - (1-x)(1-w)\} = 0.$$

Now if we observe that by development

$$1 - w\,(1-z) - z\,(1-w) = wz + (1-w)\,(1-z),$$

and

$$1 - xw - (1-x)(1-w) = x\,(1-w) + w\,(1-x),$$

and in these expressions replace, for simplicity,

$$1 - x \text{ by } \bar{x}, \quad 1 - y \text{ by } \bar{y}, \text{ \&c.},$$

we shall have from the three last equations,

$$xy\,(wz + \bar{w}\bar{z}) = 0\,; \qquad (1)$$

$$yz\,(x\bar{w} + \bar{x}w) = 0\,; \qquad (2)$$

$$\bar{x}\bar{y} = \bar{v}\bar{z}\,; \qquad (3)$$

and from this system we must eliminate w.

Multiplying the second of the above equations by c, and the third by c', and adding the results to the first, we have

$$xy\,(wz + \bar{w}\bar{z}) + cyz\,(x\bar{w} + \bar{x}w) + c'\,(\bar{x}\bar{y} - \bar{w}\bar{x}) = 0.$$

When w is made equal to 1, and therefore $\bar{w}$ to 0, the first member of the above equation becomes

$$xyz + c\bar{x}yz + c'\bar{x}\bar{y}.$$

And when in the same member w is made 0 and $\bar{w} = 1$, it becomes

$$xy\bar{z} + cxyz + c'\bar{x}\bar{y} - c'\bar{z}.$$

Hence the result of the elimination of w may be expressed in the form

$$(xyz + c\bar{x}yz + c'\bar{x}\bar{y})\,(xy\bar{z} + cxyz + c'\bar{x}\bar{y} - c'\bar{z}) = 0\,; \qquad (4)$$

and from this equation x is to be determined.

Were we now to proceed as in former instances, we should multiply together the factors in the first member of the above equation; but it may be well to show that such a course is not at all necessary. Let us develop the first member of (4) with reference to x, the symbol whose expression is sought, we find

$$yz\,(y\bar{z} + cyz - c'\bar{z})\,x + (cyz + c'\bar{y})\,(c'\bar{y} - c'\bar{z})\,(1 - x) = 0\,;$$

or, $$cyzx + (cyz + c'\bar{y})\,(c'\bar{y} - c'\bar{z})\,(1 - x) = 0\,;$$

whence we find,

$$x = \frac{(cyz + c'\bar{y})\,(c'\bar{y} - c'\bar{z})}{(cyz + c'\bar{y})\,(c'\bar{y} - c'\bar{z}) - cyz}\,;$$

and developing the second member with respect to y and z,

$$x = 0\,yz + \frac{0}{0}\,y\bar{z} + \frac{c'^2}{c'^2}\,\bar{y}z + \frac{0}{0}\,\bar{y}\bar{z}\,;$$

or,

$$x = (1-y)\,z + \frac{0}{0}\,y\,(1-z) + \frac{0}{0}\,(1-y)\,(1-z)\,;$$

or,

$$x = (1-y)\,z + \frac{0}{0}\,(1-z)\,;$$

the interpretation of which is, *Wherever the property A is present, there either C is present and B absent, or C is absent.* And inversely, *Wherever the property C is present, and the property B absent, there the property A is present.*

These results may be much more readily obtained by the method next to be explained. It is, however, satisfactory to possess different modes, serving for mutual verification, of arriving at the same conclusion.

4. We proceed to the second method.

Proposition II.

If any equations, $V_1 = 0$, $V_2 = 0$, &c., are such that the developments of their first members consist only of constituents with positive coefficients, those equations may be combined together into a single equivalent equation by addition.

For, as before, let At represent any term in the development of the function V_1, Bt the corresponding term in the development of V_2, and so on. Then will the corresponding term in the development of the equation

$$V_1 + V_2 + \&c. = 0, \qquad (1)$$

formed by the addition of the several given equations, be

$$(A + B + \&c.)\,t.$$

But as by hypothesis the coefficients A, B, &c. are none of them negative, the aggregate coefficient $A + B$, &c. in the derived equation will only vanish when the separate coefficients A, B, &c. vanish together. Hence the same constituents will appear in the development of the equation (1) as in the several equations $V_1 = 0$, $V_2 = 0$, &c. of the original system taken collectively, and therefore the interpretation of the equation (1) will be equivalent to the collective interpretations of the several equations from which it is derived.

PROPOSITION III.

5. *If* $V_1 = 0$, $V_2 = 0$, *&c. represent any system of equations, the terms of which have by transposition been brought to the first side, then the combined interpretation of the system will be involved in the single equation,*

$$V_1^2 + V_2^2 + \&c. = 0,$$

formed by adding together the squares of the given equations.

For let any equation of the system, as $V_1 = 0$, produce on development an equation

$$a_1 t_1 + a_2 t_2 + \&c. = 0,$$

in which t_1, t_2, &c. are constituents, and a_1, a_2, &c. their corresponding coefficients. Then the equation $V_1^2 = 0$ will produce on development an equation

$$a_1^2 t_1 + a_2^2 t_2 + \&c. = 0,$$

as may be proved either from the law of the development or by squaring the function $a_1 t_1 + a_2 t_2$, &c. in subjection to the conditions

$$t_1^2 = t_1, \quad t_2^2 = t_2, \quad t_1 t_2 = 0,$$

assigned in Prop. 3, Chap. v. Hence the constituents which appear in the expansion of the equation $V_1^2 = 0$, are the same with those which appear in the expansion of the equation $V_1 = 0$, and they have positive coefficients. And the same remark applies to the equations $V_2 = 0$, &c. Whence, by the last Proposition, the equation

$$V_1^2 + V_2^2 + \&c. = 0$$

will be equivalent in interpretation to the system of equations

$$V_1 = 0, \quad V_2 = 0, \&c.$$

Corollary.—Any equation, $V = 0$, of which the first member already satisfies the condition

$$V^2 = V, \text{ or } V(1 - V) = 0,$$

does not need (as it would remain unaffected by) the process of squaring. Such equations are, indeed, immediately developable into a series of constituents, with coefficients equal to 1, Chap. v. Prop. 4.

PROPOSITION IV.

6. *Whenever the equations of a system have by the above process of squaring, or by any other process, been reduced to a form such that all the constituents exhibited in their development have positive coefficients, any derived equations obtained by elimination will possess the same character, and may be combined with the other equations by addition.*

Suppose that we have to eliminate a symbol x from any equation $V = 0$, which is such that none of the constituents, in the full development of its first member, have negative coefficients. That expansion may be written in the form

$$V_1 x + V_0 (1 - x) = 0,$$

V_1 and V_0 being each of the form

$$a_1 t_1 + a_2 t_2 \ldots + a_n t_n,$$

in which $t_1 t_2 \ldots t_n$ are constituents of the other symbols, and $a_1 a_2 \ldots a_n$ in each case positive or vanishing quantities. The result of elimination is

$$V_1 V_2 = 0;$$

and as the coefficients in V_1 and V_2 are none of them negative, there can be no negative coefficients in the product $V_1 V_2$. Hence the equation $V_1 V_2 = 0$ may be added to any other equation, the coefficients of whose constituents are positive, and the resulting equation will combine the full significance of those from which it was obtained.

PROPOSITION V.

7. *To deduce from the previous Propositions a practical rule or method for the reduction of systems of equations expressing propositions in Logic.*

We have by the previous investigations established the following points, viz.:

1st. That any equations which are of the form $V = 0$, V satisfying the fundamental law of duality $V(1 - V) = 0$, may be combined together by simple addition.

2ndly. That any other equations of the form $V = 0$ may be reduced, by the process of squaring, to a form in which the same principle of combination by mere addition is applicable.

It remains then only to determine what equations in the actual expression of propositions belong to the former, and what to the latter, class.

Now the general types of propositions have been set forth in the conclusion of Chap. IV. The division of propositions which they represent is as follows:

1st. Propositions, of which the subject is universal, and the predicate particular.

The symbolical type (IV. 15) is

$$X = vY,$$

X and Y satisfying the law of duality. Eliminating v, we have

$$X(1 - Y) = 0, \tag{1}$$

and this will be found also to satisfy the same law. No further reduction by the process of squaring is needed.

2nd. Propositions of which both terms are universal, and of which the symbolical type is

$$X = Y,$$

X and Y separately satisfying the law of duality. Writing the equation in the form $X - Y = 0$, and squaring, we have

$$X - 2XY + Y = 0,$$

or

$$X(1 - Y) + Y(1 - X) = 0. \tag{2}$$

The first member of this equation satisfies the law of duality, as is evident from its very form.

We may arrive at the same equation in a different manner. The equation

$$X = Y$$

is equivalent to the two equations

$$X = vY, \quad Y = vX,$$

(for to affirm that X's are identical with Y's is to affirm both that All X's are Y's, and that All Y's are X's). Now these equations give, on elimination of v,

$$X(1-Y)=0, \quad Y(1-X)=0,$$

which added, produce (2).

3rd. Propositions of which both terms are particular. The form of such propositions is

$$vX = vY,$$

but v is not quite arbitrary, and therefore must not be eliminated. For v is the representative of *some*, which, though it may include in its meaning *all*, does not include *none*. We must therefore transpose the second member to the first side, and square the resulting equation according to the rule.

The result will obviously be

$$vX(1-Y) + vY(1-X) = 0.$$

The above conclusions it may be convenient to embody in a Rule, which will serve for constant future direction.

8. RULE.—*The equations being so expressed as that the terms X and Y in the following typical forms obey the law of duality, change the equations*

$$X = vY \text{ into } X(1-Y) = 0,$$
$$X = Y \text{ into } X(1-Y) + Y(1-X) = 0.$$
$$vX = vY \text{ into } vX(1-Y) + vY(1-X) = 0.$$

Any equation which is given in the form $X=0$ will not need transformation, and any equation which presents itself in the form $X=1$ may be replaced by $1-X=0$, as appears from the second of the above transformations.

When the equations of the system have thus been reduced, any of them, as well as any equations derived from them by the process of elimination, may be combined by addition.

9. NOTE.—It has been seen in Chapter IV. that in literally translating the terms of a proposition, without attending to its real meaning, into the language of symbols, we may produce equations in which the terms X and Y do not obey the law of duality. The equation $w = st(p+r)$, given in (3) Prop. 3 of

the chapter referred to, is of this kind. Such equations, however, as it has been seen, have a meaning. Should it, for curiosity, or for any other motive, be determined to employ them, it will be best to reduce them by the Rule (VI. 5).

10. Ex. 2.—Let us take the following Propositions of Elementary Geometry:

1st. Similar figures consist of all whose corresponding angles are equal, and whose corresponding sides are proportional.

2nd. Triangles whose corresponding angles are equal have their corresponding sides proportional, and *vice versâ*.

To represent these premises, let us make

s = similar.
t = triangles.
q = having corresponding angles equal.
r = having corresponding sides proportional.

Then the premises are expressed by the following equations:

$$s = qr, \tag{1}$$

$$tq = tr. \tag{2}$$

Reducing by the Rule, or, which amounts to the same thing, bringing the terms of these equations to the first side, squaring each equation, and then adding, we have

$$s + qr - 2qrs + tq + tr - 2tqr = 0. \tag{3}$$

Let it be required to deduce a description of dissimilar figures formed out of the elements expressed by the terms, *triangles*, having corresponding angles equal, having corresponding sides proportional.

We have from (3),

$$s = \frac{tq + qr + rt - 2tqr}{2qr - 1},$$

$$\therefore 1 - s = \frac{qr - tq - rt + 2tqr - 1}{2qr - 1}. \tag{4}$$

And fully developing the second member, we find

$$\begin{aligned} 1 - s = 0tqr &+ 2tq(1-r) + 2tr(1-q) + t(1-q)(1-r) \\ &+ 0(1-t)qr + (1-t)q(1-r) + (1-t)r(1-q) \\ &+ (1-t)(1-q)(1-r). \end{aligned} \tag{5}$$

In the above development two of the terms have the coefficient 2, these must be equated to 0 by the Rule, then those terms whose coefficients are 0 being rejected, we have

$$1 - s = t(1-q)(1-r) + (1-t)q(1-r) + (1-t)r(1-q) + (1-t)(1-q)(1-r); \quad (6)$$

$$tq(1-r) = 0; \quad (7)$$

$$tr(1-q) = 0; \quad (8)$$

the direct interpretation of which is

1st. *Dissimilar figures consist of all triangles which have not their corresponding angles equal and sides proportional, and of all figures not being triangles which have either their angles equal, and sides not proportional, or their corresponding sides proportional, and angles not equal, or neither their corresponding angles equal nor corresponding sides proportional.*

2nd. *There are no triangles whose corresponding angles are equal, and sides not proportional.*

3rd. *There are no triangles whose corresponding sides are proportional and angles not equal.*

11. Such are the immediate interpretations of the final equation. It is seen, in accordance with the general theory, that in deducing a description of a particular class of objects, viz., dissimilar figures, in terms of certain other elements of the original premises, we obtain also the independent relations which exist among those elements in virtue of the same premises. And that this is not superfluous information, even as respects the immediate object of inquiry, may easily be shown. For example, the independent relations may always be made use of to reduce, if it be thought desirable, to a briefer form, the expression of that relation which is directly sought. Thus if we write (7) in the form

$$0 = tq(1-r),$$

and add it to (6), we get, since

$$t(1-q)(1-r) + tq(1-r) = t(1-r),$$

$$1 - s = t(1-r) + (1-t)q(1-r) + (1-t)r(1-q) + (1-t)(1-q)(1-r),$$

which, on interpretation, would give for the first term of the description of dissimilar figures, "Triangles whose corresponding sides are not proportional," instead of the fuller description originally obtained. A regard to convenience must always determine the propriety of such reduction.

12. A reduction which is always advantageous (VII. 15) consists in collecting the terms of the immediate description sought, as of the second member of (5) or (6), into as few groups as possible. Thus the third and fourth terms of the second member of (6) produce by addition the single term $(1-t)(1-q)$. If this reduction be combined with the last, we have

$$1-s=t(1-r)+(1-t)q(1-r)+(1-t)(1-q),$$

the interpretation of which is

Dissimilar figures consist of all triangles whose corresponding sides are not proportional, and all figures not being triangles which have either their corresponding angles unequal, or their corresponding angles equal, but sides not proportional.

The fulness of the general solution is therefore not a superfluity. While it gives us all the information that we seek, it provides us also with the means of expressing that information in the mode that is most advantageous.

13. Another observation, illustrative of a principle which has already been stated, remains to be made. Two of the terms in the full development of $1-s$ in (5) have 2 for their coefficients, instead of $\frac{1}{0}$. It will hereafter be shown that this circumstance indicates that the two premises were not independent. To verify this, let us resume the equations of the premises in their reduced forms, viz.,

$$s(1-qr)+qr(1-s)=0,$$

$$tq(1-r)+tr(1-q)=0.$$

Now if the first members of these equations have any common constituents, they will appear on multiplying the equations together. If we do this we obtain

$$stq(1-r)+str(1-q)=0.$$

Whence there will result

$$stq(1-r)=0, \quad str(1-q)=0,$$

these being equations which are deducible from either of the primitive ones. Their interpretations are—

Similar triangles which have their corresponding angles equal have their corresponding sides proportional.

Similar triangles which have their corresponding sides proportional have their corresponding angles equal.

And these conclusions are equally deducible from either premiss *singly*. In this respect, according to the definitions laid down, the premises are not independent.

14. Let us, in conclusion, resume the problem discussed in illustration of the first method of this chapter. and endeavour to ascertain, by the present method, what may be concluded from the presence of the property C, with reference to the properties A and B.

We found on eliminating the symbols v the following equations, viz.:

$$xy(wz+\bar{w}\bar{z})=0, \tag{1}$$

$$yz(x\bar{w}+\bar{x}w)=0, \tag{2}$$

$$\bar{x}\bar{y}=\bar{w}\bar{z}. \tag{3}$$

From these we are to eliminate w and determine z. Now (1) and (2) already satisfy the condition $V(1-V)=0$. The third equation gives, on bringing the terms to the first side, and squaring

$$\bar{x}\bar{y}(1-\bar{w}\bar{z})+\bar{w}\bar{z}(1-\bar{x}\bar{y})=0. \tag{4}$$

Adding (1) (2) and (4) together, we have

$$xy(wz+\bar{w}\bar{z})+yz(x\bar{w}+x)+\bar{x}\bar{y}(1-\bar{w}\bar{z})+\bar{w}\bar{z}(1-\bar{x}\bar{y})=0.$$

Eliminating w, we get

$$(xyz+yz\bar{x}+\bar{w}\bar{y})\{xy\bar{z}+yzx+\bar{x}\bar{y}z+\bar{z}(1-\bar{x}\bar{y})\}=0.$$

Now, on multiplying the terms in the second factor by those in the first successively, observing that

$$x\bar{x}=0, \quad y\bar{y}=0, \quad z\bar{z}=0,$$

nearly all disappear, and we have only left

$$xyz + x\bar{y}z = 0\,; \qquad (5)$$

whence

$$z = \frac{0}{xy + \bar{x}\bar{y}}$$

$$= 0xy + \frac{0}{0}\,x\bar{y} + \frac{0}{0}\,\bar{x}y + 0\bar{x}\bar{y}$$

$$= \frac{0}{0}\,x\bar{y} + \frac{0}{0}\,\bar{x}y,$$

furnishing the interpretation. *Wherever the property C is found, either the property A or the property B will be found with it, but not both of them together.*

From the equation (5) we may readily deduce the result arrived at in the previous investigation by the method of arbitrary constant multipliers, as well as any other proposed forms of the relation between x, y, and z; e. g. *If the property B is absent, either A and C will be jointly present, or C will be absent.* And conversely, *If A and C are jointly present, B will be absent.* The converse part of this conclusion is founded on the presence of a term xz with unity for its coefficient in the developed value of $\bar{y}$.

CHAPTER IX.

ON CERTAIN METHODS OF ABBREVIATION.

1. THOUGH the three fundamental methods of development, elimination, and reduction, established and illustrated in the previous chapters, are sufficient for all the practical ends of Logic, yet there are certain cases in which they admit, and especially the method of elimination, of being simplified in an important degree; and to these I wish to direct attention in the present chapter. I shall first demonstrate some propositions in which the principles of the above methods of abbreviation are contained, and I shall afterwards apply them to particular examples.

Let us designate as class terms any terms which satisfy the fundamental law $V(1-V)=0$. Such terms will individually be constituents; but, when occurring together, will not, as do the terms of a development, necessarily involve the same symbols in each. Thus $ax+bxy+cyz$ may be described as an expression consisting of three class terms, x, xy, and yz, multiplied by the coefficients a, b, c respectively. The principle applied in the two following Propositions, and which, in some instances, greatly abbreviates the process of elimination, is that of the *rejection of superfluous class terms;* those being regarded as superfluous which do not add to the constituents of the final result.

PROPOSITION I.

2. *From any equation,* $V=0$, *in which* V *consists of a series of class terms having positive coefficients, we are permitted to reject any term which contains another term as a factor, and to change every positive coefficient to unity.*

For the significance of this series of positive terms depends only upon the number and nature of the constituents of its final expansion, i. e. of its expansion with reference to all the symbols which it involves, and not at all upon the actual values of the coefficients (VI. 5). Now let x be any term of the series, and xy any other term having x as a factor. The expansion of x with reference to the symbols x and y will be

$$xy + x(1 - y),$$

and the expansion of the sum of the terms x and xy will be

$$2xy + x(1 - y).$$

But by what has been said, these expressions occurring in the first member of an equation, of which the second member is 0, and of which all the coefficients of the first member are positive, are equivalent; since there must exist simply the two constituents xy and $x(1 - y)$ in the final expansion, whence will simply arise the resulting equations

$$xy = 0, \quad x(1 - y) = 0.$$

And, therefore, the aggregate of terms $x + xy$ may be replaced by the single term x.

The same reasoning applies to all the cases contemplated in the Proposition. Thus, if the term x is repeated, the aggregate $2x$ may be replaced by x, because under the circumstances the equation $x = 0$ must appear in the final reduction.

Proposition II.

3. *Whenever in the process of elimination we have to multiply together two factors, each consisting solely of positive terms, satisfying the fundamental law of logical symbols, it is permitted to reject from both factors any common term, or from either factor any term which is divisible by a term in the other factor; provided always, that the rejected term be added to the product of the resulting factors.*

In the enunciation of this Proposition, the word "divisible" is a term of convenience, used in the algebraic sense, in which xy and $x(1 - y)$ are said to be divisible by x.

To render more clear the import of this Proposition, let it be supposed that the factors to be multiplied together are $x + y + z$ and $x + yw + t$. It is then asserted, that from these two factors we may reject the term x, and that from the second factor we may reject the term yw, provided that these terms be transferred to the final product. Thus, the resulting factors being $y + z$ and t, if to their product $yt + zt$ we add the terms x and yw, we have

$$x + yw + yt + zt,$$

as an expression equivalent to the product of the given factors $x + y + z$ and $x + yw + t$; equivalent namely in the process of elimination.

Let us consider, first, the case in which the two factors have a common term x, and let us represent the factors by the expressions $x + P$, $x + Q$, supposing P in the one case and Q in the other to be the sum of the positive terms additional to x.

Now,

$$(x + P)(x + Q) = x + xP + xQ + PQ. \qquad (1)$$

But the process of elimination consists in multiplying certain factors together, and equating the result to 0. Either then the second member of the above equation is to be equated to 0, or it is a factor of some expression which is to be equated to 0.

If the former alternative be taken, then, by the last Proposition, we are permitted to reject the terms xP and xQ, inasmuch as they are positive terms having another term x as a factor. The resulting expression is

$$x + PQ,$$

which is what we should obtain by rejecting x from both factors, and adding it to the product of the factors which remain.

Taking the second alternative, the only mode in which the second member of (1) can affect the final result of elimination must depend upon the number and nature of its constituents, both which elements are unaffected by the rejection of the terms xP and xQ. For that development of x includes all possible constituents of which x is a factor.

Consider finally the case in which one of the factors contains a term, as xy, divisible by a term, x, in the other factor.

Let $x + P$ and $xy + Q$ be the factors. Now

$$(x + P)(xy + Q) = xy + xQ + xyP + PQ.$$

But by the reasoning of the last Proposition, the term xyP may be rejected as containing another positive term xy as a factor, whence we have

$$xy + xQ + PQ$$
$$= xy + (x + P)\, Q.$$

But this expresses the rejection of the term xy from the second factor, and its transference to the final product. Wherefore the Proposition is manifest.

PROPOSITION III.

4. *If t be any symbol which is retained in the final result of the elimination of any other symbols from any system of equations, the result of such elimination may be expressed in the form*

$$Et + E'(1 - t) = 0,$$

in which E is formed by making in the proposed system $t = 1$, and eliminating the same other symbols; and E' by making in the proposed system $t = 0$, and eliminating the same other symbols.

For let $\phi(t) = 0$ represent the final result of elimination. Expanding this equation, we have

$$\phi(1)\, t + \phi(0)(1 - t) = 0.$$

Now by whatever process we deduce the function $\phi(t)$ from the proposed system of equations, by the same process should we deduce $\phi(1)$, if in those equations t were changed into 1; and by the same process should we deduce $\phi(0)$, if in the same equations t were changed into 0. Whence the truth of the proposition is manifest.

5. Of the three propositions last proved, it may be remarked, that though quite unessential to the strict development or application of the general theory, they yet accomplish important ends of a practical nature. By Prop. 1 we can simplify the results of addition; by Prop. 2 we can simplify those of multiplication; and by Prop. 3 we can break up any tedious process of elimination into two distinct processes, which will in general be of a much less complex character. This method will be very frequently adopted, when the final object of inquiry is the determination of the value of t, in terms of the other symbols which remain after the elimination is performed.

6. Ex. 1.—Aristotle, in the Nicomachean Ethics, Book II. Cap. 3, having determined that actions are virtuous, not as possessing in themselves a certain character, but as implying a cer-

tain condition of mind in him who performs them, viz., that he perform them knowingly, and with deliberate preference, and for their own sakes, and upon fixed principles of conduct, proceeds in the two following chapters to consider the question, whether virtue is to be referred to the genus of Passions, or Faculties, or Habits, together with some other connected points. He grounds his investigation upon the following premises, from which, also, he deduces the general doctrine and definition of moral virtue, of which the remainder of the treatise forms an exposition.

PREMISES.

1. Virtue is either a passion (πάθος), or a faculty (δύναμις), or a habit (ἕξις).

2. Passions are not things according to which we are praised or blamed, or in which we exercise deliberate preference.

3. Faculties are not things according to which we are praised or blamed, and which are accompanied by deliberate preference.

4. Virtue is something according to which we are praised or blamed, and which is accompanied by deliberate preference.

5. Whatever art or science makes its work to be in a good state avoids extremes, and keeps the mean in view relative to human nature (τὸ μέσον . . . πρὸς ἡμᾶς).

6. Virtue is more exact and excellent than any art or science.

This is an argument *à fortiori.* If science and true art shun defect and extravagance alike, much more does virtue pursue the undeviating line of moderation. If *they* cause their work to be in a good state, much more reason have to we to say that Virtue causeth her peculiar work to be "in a good state." Let the final premiss be thus interpreted. Let us also pretermit all reference to praise or blame, since the mention of these in the premises accompanies only the mention of deliberate preference, and this is an element which we purpose to retain. We may then assume as our representative symbols—

v = virtue.
p = passions.
f = faculties.
h = habits.
d = things accompanied by deliberate preference.

g = things causing their work to be in a good state.
m = things keeping the mean in view relative to human nature.

Using, then, q as an indefinite class symbol, our premises will be expressed by the following equations:

$$v = q\{p(1-f)(1-h) + f(1-p)(1-h) + h(1-p)(1-f)\}.$$
$$p = q(1-d).$$
$$f = q(1-d).$$
$$v = qd.$$
$$g = qm.$$
$$v = qg.$$

And separately eliminating from these the symbols q,

$$v\{1 - p(1-f)(1-h) - f(1-p)(1-h) - h(1-p)(1-f) = 0. \quad (1)$$
$$pd = 0. \quad (2)$$
$$fd = 0. \quad (3)$$
$$v(1-d) = 0. \quad (4)$$
$$g(1-m) = 0. \quad (5)$$
$$v(1-g) = 0. \quad (6)$$

We shall first eliminate from (2), (3), and (4) the symbol d, and then determine v in relation to p, f, and h. Now the addition of (2), (3), and (4) gives

$$(p+f)d + v(1-d) = 0.$$

From which, eliminating d in the ordinary way, we find

$$(p+f)v = 0. \quad (7)$$

Adding this to (1), and determining v, we find

$$v = \frac{0}{p+f+1-p(1-f)(1-h)-f(1-p)(1-h)-h(1-f)(1-p)}.$$

Whence by development,

$$v = \frac{0}{0}h(1-f)(1-p).$$

The interpretation of this equation is: *Virtue is a habit, and not a faculty or a passion.*

Next, we will eliminate f, p, and g from the original system of equations, and then determine v in relation to h, d, and m. We will in this case eliminate p and f together. On addition of (1), (2), and (3), we get

$$v\{1 - p(1-f)(1-h) - f(1-p)(1-h) - h(1-p)(1-f)\} + pd + fd = 0.$$

Developing this with reference to p and f, we have

$$(v + 2d)pf + (vh + d)p(1-f) + (vh + d)(1-p)f + v(1-h)(1-p)(1-f) = 0.$$

Whence the result of elimination will be

$$(v + 2d)(vh + d)(vh + d)v(1 - h) = 0.$$

Now $v + 2d = v + d + d$, which by Prop. I. is reducible to $v + d$. The product of this and the second factor is

$$(v + d)(vh + d),$$

which by Prop. II. reduces to

$$d + v(vh) \text{ or } vh + d.$$

In like manner, this result, multiplied by the third factor, gives simply $vh + d$. Lastly, this multiplied by the fourth factor, $v(1 - h)$, gives, as the final equation,

$$vd(1 - h) = 0. \qquad (8)$$

It remains to eliminate g from (5) and (6). The result is

$$v(1 - m) = 0. \qquad (9)$$

Finally, the equations (4), (8), and (9) give on addition

$$v(1 - d) + vd(1 - h) + v(1 - m) = 0,$$

from which we have

$$v = \frac{0}{1 - d + d(1 - h) + 1 - m}.$$

And the development of this result gives

$$v = \frac{0}{0}hdm,$$

f which the interpretation is,—*Virtue is a habit accompanied by deliberate preference, and keeping in view the mean relative to human nature.*

Properly speaking, this is not a definition, but a description of virtue. It is *all*, however, that can be correctly inferred from the premises. Aristotle specially connects with it the necessity of prudence, to determine the safe and middle line of action; and there is no doubt that the ancient theories of virtue generally partook more of an intellectual character than those (the theory of utility excepted) which have most prevailed in modern days. Virtue was regarded as consisting in the right state and habit of the whole mind, rather than in the single supremacy of conscience or the moral faculty. And to some extent those theories were undoubtedly right. For though unqualified obedience to the dictates of conscience is an essential element of virtuous conduct, yet the conformity of those dictates with those unchanging principles of rectitude (*αἰώνια δίκαια*) which are founded in, or which rather are themselves the foundation of the constitution of things, is another element. And generally this conformity, in any high degree at least, is inconsistent with a state of ignorance and mental hebetude. Reverting to the particular theory of Aristotle, it will probably appear to most that it is of too negative a character, and that the shunning of extremes does not afford a sufficient scope for the expenditure of the nobler energies of our being. Aristotle seems to have been imperfectly conscious of this defect of his system, when in the opening of his seventh book he spoke of an "heroic virtue"* rising above the measure of human nature.

7. I have already remarked (VIII. 1) that the theory of single equations or propositions comprehends questions which cannot be fully answered, except in connexion with the theory of systems of equations. This remark is exemplified when it is proposed to determine from a given single equation the relation, not of some single elementary class, but of some compound class, involving in its expression more than one element, in terms of the remaining elements. The following particular example, and the succeeding general problem, are of this nature.

---

* τὴν ὑπὲρ ἡμᾶς ἀρετὴν ἡρωϊκήν τινα καὶ θείαν.—NIC. ETH. Book vii.

Ex. 2.—Let us resume the symbolical expression of the definition of wealth employed in Chap. VII., viz.,

$$w = st\,\{p + r\,(1 - p)\},$$

wherein, as before,

w = wealth,
s = things limited in supply,
t = things transferable,
p = things productive of pleasure,
r = things preventive of pain;

and suppose it required to determine hence the relation of things transferable and productive of pleasure, to the other elements of the definition, viz., wealth, things limited in supply, and things preventive of pain.

The expression for things transferable and productive of pleasure is tp. Let us represent this by a new symbol y. We have then the equations

$$w = st\,\{p + r\,(1 - p)\},$$
$$y = tp,$$

from which, if we eliminate t and p, we may determine y as a function of w, s, and r. The result interpreted will give the relation sought.

Bringing the terms of these equations to the first side, we have

$$\begin{aligned} w - stp - str\,(1 - p) &= 0. \\ y - tp &= 0. \end{aligned} \qquad (3)$$

And adding the squares of these equations together,

$$w + stp + str\,(1 - p) - 2wstp - 2wstr\,(1 - p) + y + tp - 2ytp = 0. \quad (4)$$

Developing the first member with respect to t and p, in order to eliminate those symbols, we have

$$\begin{aligned} &(w + s - 2ws + 1 - y)\,tp + (w + sr - 2wsr + y)\,t\,(1 - p) \\ &\qquad + (w + y)\,(1 - t)\,p + (w + y)\,(1 - t)\,(1 - p); \quad (5) \end{aligned}$$

and the result of the elimination of t and p will be obtained by equating to 0 the product of the four coefficients of

$$tp,\ t(1 - p),\ (1 - t)\,p,\ \text{and}\ (1 - t)\,(1 - p).$$

Or, by Prop. 3, the result of the elimination of t and p from the above equation will be of the form

$$Ey + E'(1 - y),$$

wherein E is the result obtained by changing in the given equation y into 1, and then eliminating t and p; and E' the result obtained by changing in the same equation y into 0, and then eliminating t and p. And the mode in each case of eliminating t and p is to multiply together the coefficients of the four constituents tp, $t(1-p)$, &c.

If we make $y = 1$, the coefficients become—

1st. $w(1-s) + s(1-w)$.

2nd. $1 + w(1-sr) + s(1-w)r$, equivalent to 1 by Prop. I.

3rd and 4th. $1 + w$, equivalent to 1 by Prop. I.

Hence the value of E will be

$$w(1-s) + s(1-w).$$

Again, in (5) making $y = 0$, we have for the coefficients—

1st. $1 + w(1-s) + s(1-w)$, equivalent to 1.

2nd. $w(1-sr) + sr(1-w)$.

3rd and 4th. w.

The product of these coefficients gives

$$E' = w(1-sr).$$

The equation from which y is to be determined, therefore, is

$$\{w(1-s) + s(1-w)\} y + w(1-sr)(1-y) = 0,$$

$$\therefore y = \frac{w(1-sr)}{w(1-sr) - w(1-s) - s(1-w)};$$

and expanding the second member,

$$y = \frac{0}{0} wsr + ws(1-r) + \frac{1}{0} w(1-s)r + \frac{1}{0} w(1-s)(1-r)$$

$$+ 0(1-w)sr + 0(1-w)s(1-r) + \frac{0}{0}(1-w)(1-s)r$$

$$+ \frac{0}{0}(1-w)(1-s)(1-r);$$

whence reducing

$$y = ws(1-r) + \frac{0}{0}wsr + \frac{0}{0}(1-w)(1-s), \qquad (6)$$

$$\text{with } w(1-s) = 0. \qquad (7)$$

The interpretation of which is—

1st. *Things transferable and productive of pleasure consist of all wealth (limited in supply and) not preventive of pain, an indefinite amount of wealth (limited in supply and) preventive of pain, and an indefinite amount of what is not wealth and not limited in supply.*

2nd. *All wealth is limited in supply.*

I have in the above solution written in parentheses that part of the full description which is implied by the accompanying independent relation (7).

8. The following problem is of a more general nature, and will furnish an easy practical rule for problems such as the last.

General Problem.

Given any equation connecting the symbols $x, y \,..\, w, z \,..$

Required to determine the logical expression of any class expressed in any way by the symbols $x, y \,..$ in terms of the remaining symbols, w, z, &c.

Let us confine ourselves to the case in which there are but two symbols, x, y, and two symbols, w, z, a case sufficient to determine the general Rule.

Let $V = 0$ be the given equation, and let $\phi(x, y)$ represent the class whose expression is to be determined.

Assume $t = \phi(x, y)$, then, from the above two equations, x and y are to be eliminated.

Now the equation $V = 0$ may be expanded in the form

$$Axy + Bx(1-y) + C(1-x)y + D(1-x)(1-y) = 0, \quad (1)$$

A, B, C, and D being functions of the symbols w and z.

Again, as $\phi(x, y)$ represents a class or collection of things, it must consist of a constituent, or series of constituents, whose coefficients are 1.

Wherefore if the *full* development of $\phi(x, y)$ be represented in the form

$$axy + bx(1-y) + c(1-x)y + d(1-x)(1-y),$$

the coefficients a, b, c, d must each be 1 or 0.

Now reducing the equation $t = \phi(x, y)$ by transposition and squaring, to the form

$$t\{1-\phi(x, y)\} + \phi(x, y)(1-t) = 0;$$

and expanding with reference to x and y, we get

$$\begin{aligned}\{t(1-a) + a(1-t)\}\, xy + \{t(1-b) + b(1-t)\}\, x(1-y) \\ + \{t(1-c) + c(1-t)\}\,(1-x)y \\ + \{t(1-d) + d(1-t)\}\,(1-x)(1-y) = 0;\end{aligned}$$

whence, adding this to (1), we have

$$\begin{aligned}\{A + t(1-a) + a(1-t)\}\, xy \\ + \{B + t(1-b) + b(1-t)\}\, x(1-y) + \&c. = 0.\end{aligned}$$

Let the result of the elimination of x and y be of the form

$$Et + E'(1-t) = 0,$$

then E will, by what has been said, be the reduced product of what the coefficients of the above expansion become when $t = 1$, and E' the product of the same factors similarly reduced by the condition $t = 0$.

Hence E will be the reduced product

$$(A+1-a)(B+1-b)(C+1-c)(D+1-d).$$

Considering any factor of this expression, as $A + 1 - a$, we see that when $a = 1$ it becomes A, and when $a = 0$ it becomes $1 + A$, which reduces by Prop. I. to 1. Hence we may infer that E will be the product of the coefficients of those constituents in the development of V whose coefficients in the development of $\phi(x, y)$ are 1.

Moreover E' will be the reduced product

$$(A+a)(B+b)(C+c)(D+d).$$

Considering any one of these factors, as $A + a$, we see that this becomes A when $a = 0$, and reduces to 1 when $a = 1$; and so on for the others. Hence E' will be the product of the coefficients of those constituents in the development of y, whose coefficients in the development $\phi(x, y)$ are 0. Viewing these cases together, we may establish the following Rule:

9. *To deduce from a logical equation the relation of any class expressed by a given combination of the symbols x, y, &c, to the classes represented by any other symbols involved in the given equation.*

RULE.—*Expand the given equation with reference to the symbols x, y. Then form the equation*

$$Et + E'(1 - t) = 0,$$

in which E is the product of the coefficients of all those constituents in the above development, whose coefficients in the expression of the given class are 1, *and E' the product of the coefficients of those constituents of the development whose coefficients in the expression of the given class are* 0. *The value of t deduced from the above equation by solution and interpretation will be the expression required.*

NOTE.—*Although in the demonstration of this Rule V is supposed to consist solely of positive terms, it may easily be shown that this condition is unnecessary, and the Rule general, and that no preparation of the given equation is really required.*

10. Ex. 3.—The same definition of wealth being given as in Example 2, required an expression for *things transferable, but not productive of pleasure*, $t(1 - p)$, in terms of the other elements represented by w, s, and r.

The equation

$$w - stp - str(1 - p) = 0,$$

gives, when squared,

$$w + stp + str(1 - p) - 2wstp - 2wstr(1 - p) = 0;$$

and developing the first member with respect to t and p,

$$(w + s - 2ws)\,tp + (w + sr - 2wsr)\,t(1 - p) + w(1 - t)\,p \\ + w(1 - t)(1 - p) = 0.$$

The coefficients of which it is best to exhibit as in the following equation;

$$\{w(1 - s) + s(1 - w)\}\,tp + \{w(1 - sr) + sr(1 - w)\}\,t(1 - p) + w(1 - t)p \\ + w(1 - t)(1 - p) = 0.$$

Let the function $t(1-p)$ to be determined, be represented by z; then the full development of z in respect of t and p is

$$z = 0\,tp + t\,(1-p) + 0\,(1-t)\,p + 0\,(1-t)\,(1-p).$$

Hence, by the last problem, we have

$$Ez + E'\,(1-z) = 0;$$

where
$$E = w\,(1-sr) + sr\,(1-w);$$

$$E' = \{w\,(1-s) + s\,(1-w)\} \times w \times w = w\,(1-s);$$

$$\therefore\ \{w\,(1-sr) + sr\,(1-w)\}\,z + w\,(1-s)\,(1-z) = 0.$$

Hence,

$$z = \frac{w\,(1-s)}{2wsr - ws - sr}$$

$$= \frac{0}{0}\,wsr + 0\,ws\,(1-r) + \frac{1}{0}\,w\,(1-s)\,r + \frac{1}{0}\,w\,(1-s)\,(1-r),$$

$$+ 0\,(1-w)\,sr + \frac{0}{0}\,(1-w)\,s\,(1-r) + \frac{0}{0}\,(1-w)\,(1-s)\,r$$

$$+ \frac{0}{0}\,(1-w)\,(1-s)\,(1-r).$$

Or,
$$z = \frac{0}{0}\,wsr + \frac{0}{0}\,(1-w)\,s\,(1-r) + \frac{0}{0}\,(1-w)\,(1-s),$$

with
$$w\,(1-s) = 0.$$

Hence, *Things transferable and not productive of pleasure are either wealth (limited in supply and preventive of pain); or things which are not wealth, but limited in supply and not preventive of pain; or things which are not wealth, and are unlimited in supply.*

The following results, deduced in a similar manner, will be easily verified:

Things limited in supply and productive of pleasure which are not wealth,—are intransferable.

Wealth that is not productive of pleasure is transferable, limited in supply, and preventive of pain.

Things limited in supply which are either wealth, or are productive of pleasure, but not both,—are either transferable and preventive of pain, or intransferable.

11. From the domain of natural history a large number of curious examples might be selected. I do not, however, con-

ceive that such applications would possess any independent value. They would, for instance, throw no light upon the true principles of classification in the science of zoology. For the discovery of these, some basis of positive knowledge is requisite,—some acquaintance with organic structure, with teleological adaptation; and this is a species of knowledge which can only be derived from the use of external means of observation and analysis. Taking, however, any collection of propositions in natural history, a great number of logical problems present themselves, without regard to the system of classification adopted. Perhaps in forming such examples, it is better to avoid, as superfluous, the mention of that property of a class or species which is immediately suggested by its name, e. g. the ring-structure in the annelida, a class of animals including the earth-worm and the leech.

Ex. 4.—1. The annelida are soft-bodied, and either naked or enclosed in a tube.

2. The annelida consist of all invertebrate animals having red blood in a double system of circulating vessels.

Assume a = annelida;
s = soft-bodied animals;
n = naked;
t = enclosed in a tube;
i = invertebrate;
r = having red blood, &c.

Then the propositions given will be expressed by the equations

$$a = vs\,\{n(1-t) + t(1-n)\}; \qquad (1)$$

$$a = ir; \qquad (2)$$

to which we may add the implied condition,

$$nt = 0. \qquad (3)$$

On eliminating v, and reducing the system to a single equation, we have

$$a\,\{1 - sn(1-t) - st(1-n)\} + a(1-ir) + ir(1-a) + nt = 0. \qquad (4)$$

Suppose that we wish to obtain the relation in which soft-bodied animals enclosed in tubes are placed (by virtue of the premises) with respect to the following elements, viz., the possession of red blood, of an external covering, and of a vertebral column.

We must first eliminate a. The result is

$$ir\{1 - sn(1-t) - st(1-n)\} + nt = 0.$$

Then (IX. 9) developing with respect to s and t, and reducing the first coefficient by Prop. 1, we have

$$nst + ir(1-n)s(1-t) + (ir+n)(1-s)t + ir(1-s)(1-t) = 0. \quad (5)$$

Hence, if $st = w$, we find

$$nw + ir(1-n) \times (ir+n) \times ir(1-w) = 0;$$

or,

$$nw + ir(1-n)(1-w) = 0;$$

$$\therefore w = \frac{ir(1-n)}{ir(1-n) - n}$$

$$= 0\,irn + ir(1-n) + 0i(1-r)n + \frac{0}{0}i(1-r)(1-n)$$
$$+ 0(1-i)rn + \frac{0}{0}(1-i)r(1-n) + 0(1-i)(1-r)n$$
$$+ \frac{0}{0}(1-i)(1-r)(1-n);$$

or,

$$w = ir(1-n) + \frac{0}{0}i(1-r)(1-n) + \frac{0}{0}(1-i)(1-n).$$

Hence, *soft-bodied animals enclosed in tubes consist of all invertebrate animals having red blood and not naked, and an indefinite remainder of invertebrate animals not having red blood and not naked, and of vertebrate animals which are not naked.*

And in an exactly similar manner, the following reduced equations, the interpretation of which is left to the reader, have been deduced from the development (5).

$$s(1-t) = irn + \frac{0}{0}i(1-n) + \frac{0}{0}(1-i)$$

$$(1-s)\,t = \frac{0}{0}(1-i)r(1-n) + \frac{0}{0}(1-r)(1-n)$$

$$(1-s)(1-t) = \frac{0}{0}i(1-r) + \frac{0}{0}(1-i).$$

In none of the above examples has it been my object to exhibit in any special manner the power of the method. That, I conceive, can only be fully displayed in connexion with the mathematical theory of probabilities. I would, however, suggest to any who may be desirous of forming a correct opinion upon this point, that they examine by the rules of ordinary logic the following problem, *before* inspecting its solution; remembering at the same time, that whatever complexity it possesses might be multiplied indefinitely, with no other effect than to render its solution by the method of this work more operose, but not less certainly attainable.

Ex. 5. Let the observation of a class of natural productions be supposed to have led to the following general results.

1st, That in whichsoever of these productions the properties A and C are missing, the property E is found, together with one of the properties B and D, but not with both.

2nd, That wherever the properties A and D are found while E is missing, the properties B and C will either both be found, or both be missing.

3rd, That wherever the property A is found in conjunction with either B or E, or both of them, there either the property C or the property D will be found, but not both of them. And conversely, wherever the property C or D is found singly, there the property A will be found in conjunction with either B or E, or both of them.

Let it then be required to ascertain, first, what in any particular instance may be concluded from the ascertained presence of the property A, with reference to the properties B, C, and D; also whether any relations exist independently among the properties B, C, and D. Secondly, what may be concluded in like manner respecting the property B, and the properties A, C, and D.

It will be observed, that in each of the three data, the information conveyed respecting the properties A, B, C, and D, is complicated with another element, E, about which we desire to say nothing in our conclusion. It will hence be requisite to eliminate the symbol representing the property E from the system of equations, by which the given propositions will be expressed.

Let us represent the property A by x, B by y, C by z, D by w, E by v. The data are

$$\bar{x}z = qv\,(y\bar{w} + w\bar{y}); \qquad (1)$$

$$\bar{v}xw = q\,(yz + \bar{y}\,\bar{z}); \qquad (2)$$

$$xy + xv\bar{y} = w\bar{z} + z\bar{w}; \qquad (3)$$

$\bar{x}$ standing for $1 - x$, &c., and q being an indefinite class symbol. Eliminating q separately from the first and second equations, and adding the results to the third equation reduced by (5), Chap. VIII., we get

$$\bar{x}\,z\,(1 - vy\bar{w} - vw\bar{y}) + \bar{v}xw\,(y\bar{z} + z\bar{y}) + (xy + xv\bar{y})\,(wz + \bar{w}\bar{z})$$
$$+ (w\bar{z} + z\bar{w})\,(1 - xy - xv\bar{y}) = 0. \qquad (4)$$

From this equation v must be eliminated, and the value of x determined from the result. For effecting this object, it will be convenient to employ the method of Prop. 3 of the present chapter.

Let then the result of elimination be represented by the equation

$$Ex + E'\,(1 - x) = 0.$$

To find E make $x = 1$ in the first member of (4), we find

$$\bar{v}w\,(y\bar{z} + z\bar{y}) + (y + v\bar{y})\,(wz + \bar{w}\,\bar{z}) + (w\bar{z} + z\bar{w})\,\bar{v}\bar{y}.$$

Eliminating v, we have

$$(wz + \bar{w}\,\bar{z})\,\{w\,(y\bar{z} + z\bar{y}) + y\,(wz + \bar{w}\,\bar{z}) + \bar{y}\,(w\bar{z} + z\bar{w})\};$$

which, on actual multiplication, in accordance with the conditions $w\bar{w} = 0$, $z\bar{z} = 0$, &c., gives

$$E = wz + y\bar{w}\,\bar{z}.$$

Next, to find E' make $x = 0$ in (4), we have

$$z\,(1 - vy\bar{w} - v\bar{y}w) + w\bar{z} + z\bar{w}.$$

whence, eliminating v, and reducing the result by Propositions 1 and 2, we find

$$E' = w\bar{z} + z\bar{w} + \bar{y}\,\bar{w}\,\bar{z};$$

and, therefore, finally we have

$$(wz + y\bar{w}\,\bar{z})\,x + (w\bar{z} + z\bar{w} + \bar{y}\,\bar{w}\,\bar{z})\,\bar{x} = 0; \qquad (5)$$

from which

$$x = \frac{w\bar{z} + z\bar{w} + \bar{y}\bar{w}\bar{z}}{w\bar{z} + z\bar{w} + \bar{y}\bar{w}\bar{z} - wz - y\bar{w}\bar{z}};$$

wherefore, by development,

$$x = 0\,yzw + yz\bar{w} + y\bar{z}w + 0\,y\bar{z}\bar{w}$$
$$+\ 0\,\bar{y}zw + \bar{y}z\bar{w} + \bar{y}\bar{z}w + \bar{y}\bar{z}\bar{w};$$

or, collecting the terms in vertical columns,

$$x = z\bar{w} + \bar{z}w + \bar{y}\bar{z}\bar{w}; \qquad (6)$$

the interpretation of which is—

In whatever substances the property A is found, there will also be found either the property C or the property D, but not both, or else the properties B, C, and D, will all be wanting. And conversely, *where either the property C or the property D is found singly, or the properties B, C, and D, are together missing, there the property A will be found.*

It also appears that there is no independent relation among the properties B, C, and D.

Secondly, we are to find y. Now developing (5) with respect to that symbol,

$$(xwz + x\bar{w}\bar{z} + \bar{x}w\bar{z} + \bar{x}z\bar{w})\,y + (xwz + \bar{x}w\bar{z} + \bar{x}z\bar{w} + \bar{x}\bar{z}\bar{w})\,\bar{y} = 0;$$

whence, proceeding as before,

$$y = \bar{x}\bar{w}\bar{z} + \frac{0}{0}\,(\bar{x}wz + xw\bar{z} + xz\bar{w}), \qquad (7)$$

$$xzw = 0; \qquad (8)$$

$$\bar{x}\bar{z}w = 0; \qquad (9)$$

$$\bar{x}z\bar{w} = 0; \qquad (10)$$

From (10) reduced by solution to the form

$$\bar{x}z = \frac{0}{0}\,w;$$

we have the independent relation,—*If the property A is absent and C present, D is present.* Again, by addition and solution (8) and (9) give

$$xz + \bar{x}\bar{z} = \frac{0}{0}\,\bar{w}.$$

Whence we have for the general solution and the remaining independent relation:

1st. *If the property B be present in one of the productions, either the properties A, C, and D, are all absent, or some one alone of them is absent.* And conversely, *if they are all absent it may be concluded that the property A is present* (7).

2nd. *If A and C are both present or both absent, D will be absent, quite independently of the presence or absence of B* (8) and (9).

I have not attempted to verify these conclusions.

CHAPTER X.

OF THE CONDITIONS OF A PERFECT METHOD.

1. THE subject of Primary Propositions has been discussed at length, and we are about to enter upon the consideration of Secondary Propositions. The interval of transition between these two great divisions of the science of Logic may afford a fit occasion for us to pause, and while reviewing some of the past steps of our progress, to inquire what it is that in a subject like that with which we have been occupied constitutes perfection of method. I do not here speak of that perfection only which consists in power, but of that also which is founded in the conception of what is fit and beautiful. It is probable that a careful analysis of this question would conduct us to some such conclusion as the following, viz., that a perfect method should not only be an efficient one, as respects the accomplishment of the objects for which it is designed, but should in all its parts and processes manifest a certain unity and harmony. This conception would be most fully realized if even the very forms of the method were suggestive of the fundamental principles, and if possible of the *one* fundamental principle, upon which they are founded. In applying these considerations to the science of Reasoning, it may be well to extend our view beyond the mere analytical processes, and to inquire what is best as respects not only the mode or form of deduction, but also the system of data or premises from which the deduction is to be made.

2. As respects mere power, there is no doubt that the first of the methods developed in Chapter VIII. is, within its proper sphere, a perfect one. The introduction of arbitrary constants makes us independent of the forms of the premises, as well as of any conditions among the equations by which they are represented. But it seems to introduce a foreign element, and while it is a more laborious, it is also a less elegant form of solution than the second method of reduction demonstrated in the same

chapter. There are, however, conditions under which the latter method assumes a more perfect form than it otherwise bears. To make the one fundamental condition expressed by the equation

$$x(1-x)=0,$$

the universal type of form, would give a unity of character to both processes and results, which would not else be attainable. Were brevity or convenience the only valuable quality of a method, no advantage would flow from the adoption of such a principle. For to impose upon every step of a solution the character above described, would involve in some instances no slight labour of preliminary reduction. But it is still interesting to know that this can be done, and it is even of some importance to be acquainted with the conditions under which such a form of solution would spontaneously present itself. Some of these points will be considered in the present chapter.

PROPOSITION I.

3. *To reduce any equation among logical symbols to the form* $V=0$, *in which* V *satisfies the law of duality,*

$$V(1-V)=0.$$

It is shown in Chap. v. Prop. 4, that the above condition is satisfied whenever V is the sum of a series of constituents. And it is evident from Prop. 2, Chap. VI. that all equations are equivalent which, when reduced by transposition to the form $V=0$, produce, by development of the first member, the same series of constituents with coefficients which do not vanish; the particular numerical values of those coefficients being immaterial.

Hence the object of this Proposition may always be accomplished by bringing all the terms of an equation to the first side, fully expanding that member, and changing in the result all the coefficients which do not vanish into unity, except such as have already that value.

But as the development of functions containing many symbols conducts us to expressions inconvenient from their great length, it is desirable to show how, in the only cases which do practically offer themselves to our notice, this source of complexity may be avoided.

The great primary forms of equations have already been discussed in Chapter VIII. They are—

$$X = vY,$$
$$X = Y,$$
$$vX = vY.$$

Whenever the conditions $X(1 - X) = 0$, $Y(1 - Y) = 0$, are satisfied, we have seen that the two first of the above equations conduct us to the forms

$$X(1 - Y) = 0, \qquad (1)$$
$$X(1 - Y) + Y(1 - X) = 0; \qquad (2)$$

and under the same circumstances it may be shown that the last of them gives

$$v\{X(1 - Y) + Y(1 - X)\} = 0; \qquad (3)$$

all which results obviously satisfy, in their first members, the condition

$$V(1 - V) = 0.$$

Now as the above are the forms and conditions under which the equations of a logical system properly expressed do actually present themselves, it is always possible to reduce them by the above method into subjection to the law required. Though, however, the separate equations may thus satisfy the law, their equivalent sum (VIII. 4) may not do so, and it remains to show how upon it also the requisite condition may be imposed.

Let us then represent the equation formed by adding the several reduced equations of the system together, in the form

$$v + v' + v'', \text{ \&c.} = 0, \qquad (4)$$

this equation being singly equivalent to the system from which it was obtained. We suppose v, v', v'', &c. to be class terms (IX. 1) satisfying the conditions

$$v(1 - v) = 0, \quad v'(1 - v') = 0, \text{ \&c.}$$

Now the full interpretation of (4) would be found by developing the first member with respect to all the elementary symbols x, y, &c. which it contains, and equating to 0 all the constituents whose coefficients do not vanish; in other words, all the constituents which are found in either v, v', v'', &c. But those constituents consist of—1st, such as are found in v; 2nd, such as are not found in v, but are found in v'; 3rd, such as are neither found in v nor v', but are found in v'', and so on. Hence they will be such as are found in the expression

$$v + (1 - v)v' + (1 - v)(1 - v')v'' + \&c., \qquad (5)$$

an expression in which no constituents are repeated, and which obviously satisfies the law $V(1 - V) = 0$.

Thus if we had the expression

$$(1 - t) + v + (1 - z) + tzw,$$

in which the terms $1 - t$, $1 - z$ are bracketed to indicate that they are to be taken as single class terms, we should, in accordance with (5), reduce it to an expression satisfying the condition $V(1 - V) = 0$, by multiplying all the terms after the first by t, then all after the second by $1 - v$; lastly, the term which remains after the third by z; the result being

$$1 - t + tv + t(1 - v)(1 - z) + t(1 - v)zw. \qquad (6)$$

4. All logical equations then are reducible to the form $V = 0$, V satisfying the law of duality. But it would obviously be a higher degree of perfection if equations always presented themselves in such a form, without preparation of any kind, and not only exhibited this form in their original statement, but retained it unimpaired after those additions which are necessary in order to reduce systems of equations to single equivalent forms. That they do not spontaneously present this feature is not properly attributable to defect of method, but is a consequence of the fact that our premises are not always complete, and accurate, and independent. They are not complete when they involve material (as distinguished from formal) relations, which are not expressed. They are not accurate when they imply relations which are not intended. But setting aside these points, with which, in the present instance, we are less concerned, let it be considered in what sense they may fail of being independent.

5. A system of propositions may be termed independent, when it is not possible to deduce from any portion of the system a conclusion deducible from any other portion of it. Supposing the equations representing those propositions all reduced to the form

$$V = 0,$$

then the above condition implies that no constituent which can be made to appear in the development of a particular function V of the system, can be made to appear in the development of any other function V' of the same system. When this condition is not satisfied, the equations of the system are not independent. This may happen in various cases. Let all the equations satisfy in their first members the law of duality, then if there appears a positive term x in the expansion of one equation, and a term xy in that of another, the equations are not independent, for the term x is further developable into $xy + x(1 - y)$, and the equation

$$xy = 0$$

is thus involved in both the equations of the system. Again, let a term xy appear in one equation, and a term xz in another. Both these may be developed so as to give the common constituent xyz. And other cases may easily be imagined in which premises which appear at first sight to be quite independent are not really so. Whenever equations of the form $V = 0$ are thus not truly independent, though individually they may satisfy the law of duality,

$$V(1 - V) = 0,$$

the equivalent equation obtained by adding them together will not satisfy that condition, unless sufficient reductions by the method of the present chapter have been performed. When, on the other hand, the equations of a system both satisfy the above law, and are independent of each other, their sum will also satisfy the same law. I have dwelt upon these points at greater length than would otherwise have been necessary, because it appears to me to be important to endeavour to form to ourselves, and to keep before us in all our investigations, the pattern of an ideal perfection,—the object and the guide of future efforts. In the present class of inquiries the chief aim of improvement of method should be to facilitate, as far as is consistent with brevity, the transformation of equations, so as to make the fundamental condition above adverted to universal.

In connexion with this subject the following Propositions are deserving of attention.

Proposition II.

If the first member of any equation $V = 0$ *satisfy the condition* $V(1 - V) = 0$, *and if the expression of any symbol* t *of that equation be determined as a developed function of the other symbols, the coefficients of the expansion can only assume the forms* $1, 0, \frac{0}{0}, \frac{1}{0}$.

For if the equation be expanded with reference to t, we obtain as the result,

$$Et + E'(1 - t), \tag{1}$$

E and E' being what V becomes when t is successively changed therein into 1 and 0. Hence E and E' will themselves satisfy the conditions

$$E(1 - E) = 0, \quad E'(1 - E') = 0. \tag{2}$$

Now (1) gives

$$t = \frac{E'}{E' - E},$$

the second member of which is to be expanded as a function of the remaining symbols. It is evident that the only numerical values which E and E' can receive in the calculation of the coefficients will be 1 and 0. The following cases alone can therefore arise:

1st. $\quad E' = 1, \quad E = 1, \quad$ then $\dfrac{E'}{E' - E} = \dfrac{1}{0}$.

2nd. $\quad E' = 1, \quad E = 0, \quad$ then $\dfrac{E'}{E' - E} = 1$.

3rd. $\quad E' = 0, \quad E = 1, \quad$ then $\dfrac{E'}{E' - E} = 0$.

4th. $\quad E' = 0, \quad E = 0, \quad$ then $\dfrac{E'}{E' - E} = \dfrac{0}{0}$.

Whence the truth of the Proposition is manifest.

6. It may be remarked that the forms 1, 0, and $\frac{0}{0}$ appear in the solution of equations independently of any reference to the condition $V(1-V)=0$. But it is not so with the coefficient $\frac{1}{0}$. The terms to which this coefficient is attached when the above condition is satisfied may receive any other value except the three values 1, 0, and $\frac{0}{0}$, when that condition is not satisfied. It is permitted, and it would conduce to uniformity, to change any coefficient of a development not presenting itself in any of the four forms referred to in this Proposition into $\frac{1}{0}$, regarding this as the symbol proper to indicate that the coefficient to which it is attached should be equated to 0. This course I shall frequently adopt.

Proposition III.

7. *The result of the elimination of any symbols x, y, &c. from an equation $V=0$, of which the first member identically satisfies the law of duality,*

$$V(1-V)=0,$$

may be obtained by developing the given equation with reference to the other symbols, and equating to 0 the sum of those constituents whose coefficients in the expansion are equal to unity.

Suppose that the given equation $V=0$ involves but three symbols, x, y, and t, of which x and y are to be eliminated. Let the development of the equation, with respect to t, be

$$At+B(1-t)=0, \qquad (1)$$

A and B being free from the symbol t.

By Chap. IX. Prop. 3, the result of the elimination of x and y from the given equation will be of the form

$$Et+E'(1-t)=0, \qquad (2)$$

in which E is the result obtained by eliminating the symbols x and y from the equation $A=0$, E' the result obtained by eliminating from the equation $B=0$.

Now A and B must satisfy the condition

$$A(1-A)=0, \quad B(1-B)=0.$$

Hence A (confining ourselves for the present to this coefficient) will either be 0 or 1, or a constituent, or the sum of a part of the constituents which involve the symbols x and y. If $A = 0$ it is evident that $E = 0$; if A is a single constituent, or the sum of a part of the constituents involving x and y, E will be 0. For the *full* development of A, with respect to x and y, will contain terms with vanishing coefficients, and E is the product of *all* the coefficients. Hence when $A = 1$, E is equal to A, but in other cases E is equal to 0. Similarly, when $B = 1$, E' is equal to B, but in other cases E' vanishes. Hence the expression (2) will consist of that part, if any there be, of (1) in which the coefficients A, B are unity. And this reasoning is general. Suppose, for instance, that V involved the symbols x, y, z, t, and that it were required to eliminate x and y. Then if the development of V, with reference to z and t, were

$$zt + xz(1-t) + y(1-z)t + (1-z)(1-t),$$

the result sought would be

$$zt + (1-z)(1-t) = 0,$$

this being that portion of the development of which the coefficients are unity.

Hence, if from any system of equations we deduce a single equivalent equation $V = 0$, V satisfying the condition

$$V(1-V)=0,$$

the ordinary processes of elimination may be entirely dispensed with, and the single process of development made to supply their place.

8. It may be that there is no practical advantage in the method thus pointed out, but it possesses a theoretical unity and completeness which render it deserving of regard, and I shall accordingly devote a future chapter (XIV.) to its illustration. The progress of applied mathematics has presented other and signal examples of the reduction of systems of problems or equations to the dominion of some central but pervading law.

9. It is seen from what precedes that there is one class of propositions to which all the special appliances of the above methods of preparation are unnecessary. It is that which is characterized by the following conditions:

First, That the propositions are of the ordinary kind, implied by the use of the copula *is* or *are*, the predicates being particular.

Secondly, That the terms of the proposition are intelligible without the supposition of any understood relation among the elements which enter into the expression of those terms.

Thirdly, That the propositions are independent.

We may, if such speculation is not altogether vain, permit ourselves to conjecture that these are the conditions which would be obeyed in the employment of language as an instrument of expression and of thought, by unerring beings, declaring simply what they mean, without suppression on the one hand, and without repetition on the other. Considered both in their relation to the idea of a perfect language, and in their relation to the processes of an exact method, these conditions are equally worthy of the attention of the student.

CHAPTER XI.

OF SECONDARY PROPOSITIONS, AND OF THE PRINCIPLES OF THEIR SYMBOLICAL EXPRESSION.

1. THE doctrine has already been established in Chap. IV., that every logical proposition may be referred to one or the other of two great classes, viz., Primary Propositions and Secondary Propositions. The former of these classes has been discussed in the preceding chapters of this work, and we are now led to the consideration of Secondary Propositions, i. e. of Propositions concerning, or relating to, other propositions regarded as true or false. The investigation upon which we are entering will, in its general order and progress, resemble that which we have already conducted. The two inquiries differ as to the subjects of thought which they recognise, not as to the formal and scientific laws which they reveal, or the methods or processes which are founded upon those laws. Probability would in some measure favour the expectation of such a result. It consists with all that we know of the uniformity of Nature, and all that we believe of the immutable constancy of the Author of Nature, to suppose, that in the mind, which has been endowed with such high capabilities, not only for converse with surrounding scenes, but for the knowledge of itself, and for reflection upon the laws of its own constitution, there should exist a harmony and uniformity not less real than that which the study of the physical sciences makes known to us. Anticipations such as this are never to be made the primary rule of our inquiries, nor are they in any degree to divert us from those labours of patient research by which we ascertain what is the actual constitution of things within the particular province submitted to investigation. But when the grounds of resemblance have been properly and independently determined, it is not inconsistent, even with purely scientific ends, to make that resemblance a subject of meditation, to trace its extent, and to receive the intimations of truth, yet undiscovered, which it may

seem to us to convey. The necessity of a final appeal to fact is not thus set aside, nor is the use of analogy extended beyond its proper sphere,—the suggestion of relations which independent inquiry must either verify or cause to be rejected.

2. *Secondary Propositions are those which concern or relate to Propositions considered as true or false.* The relations of *things* we express by primary propositions. But we are able to make Propositions themselves also the subject of thought, and to express our judgments concerning them. The expression of any such judgment constitutes a secondary-proposition. There exists no proposition whatever of which a competent degree of knowledge would not enable us to make one or the other of these two assertions, viz., either that the proposition is true, or that it is false; and each of these assertions is a secondary proposition. "It is true that the sun shines;" "It is not true that the planets shine by their own light;" are examples of this kind. In the former example the Proposition "The sun shines," is asserted to be true. In the latter, the Proposition, "The planets shine by their own light," is asserted to be false. Secondary propositions also include all judgments by which we express a relation or dependence among propositions. To this class or division we may refer conditional propositions, as, "If the sun shine the day will be fair." Also most disjunctive propositions, as, "Either the sun will shine, or the enterprise will be postponed." In the former example we express the dependence of the truth of the Proposition, "The day will be fair," upon the truth of the Proposition, "The sun will shine." In the latter we express a relation between the two Propositions, "The sun will shine," "The enterprise will be postponed," implying that the truth of the one excludes the truth of the other. To the same class of secondary propositions we must also refer all those propositions which assert the simultaneous truth or falsehood of propositions, as, "It is not true both that 'the sun will shine' and that 'the journey will be postponed.'" The elements of distinction which we have noticed may even be blended together in the same secondary proposition. It may involve both the disjunctive element expressed by *either*, *or*, and the conditional element expressed by *if;* in addition to which, the connected propositions may themselves be of a compound

character. *If* "the sun shine," *and* "leisure permit," then *either* "the enterprise shall be commenced," *or* "some preliminary step shall be taken." In this example a number of propositions are connected together, not arbitrarily and unmeaningly, but in such a manner as to express a *definite* connexion between them,—a connexion having reference to their respective truth or falsehood. This combination, therefore, according to our definition, forms a Secondary Proposition.

The theory of Secondary Propositions is deserving of attentive study, as well on account of its varied applications, as for that close and harmonious analogy, already referred to, which it sustains with the theory of Primary Propositions. Upon each of these points I desire to offer a few further observations.

3. I would in the first place remark, that it is in the form of secondary propositions, at least as often as in that of primary propositions, that the reasonings of ordinary life are exhibited. The discourses, too, of the moralist and the metaphysician are perhaps less often concerning things and their qualities, than concerning principles and hypotheses, concerning truths and the mutual connexion and relation of truths. The conclusions which our narrow experience suggests in relation to the great questions of morals and society yet unsolved, manifest, in more ways than one, the limitations of their human origin; and though the existence of universal principles is not to be questioned, the partial formulæ which comprise our knowledge of their application are subject to conditions, and exceptions, and failure. Thus, in those departments of inquiry which, from the nature of their subject-matter, should be the most interesting of all, much of our actual knowledge is hypothetical. That there has been a strong tendency to the adoption of the same forms of thought in writers on speculative philosophy, will hereafter appear. Hence the introduction of a general method for the discussion of hypothetical and the other varieties of secondary propositions, will open to us a more interesting field of applications than we have before met with.

4. The discussion of the theory of Secondary Propositions is in the next place interesting, from the close and remarkable analogy which it bears with the theory of Primary Propositions. It

will appear, that the formal laws to which the operations of the mind are subject, are identical in expression in both cases. The mathematical processes which are founded on those laws are, therefore, identical also. Thus the methods which have been investigated in the former portion of this work will continue to be available in the new applications to which we are about to proceed. But while the laws and processes of the method remain unchanged, the rule of interpretation must be adapted to new conditions. Instead of classes of things, we shall have to substitute propositions, and for the relations of classes and individuals, we shall have to consider the connexions of propositions or of events. Still, between the two systems, however differing in purport and interpretation, there will be seen to exist a pervading harmonious relation, an analogy which, while it serves to facilitate the conquest of every yet remaining difficulty, is of itself an interesting subject of study, and a conclusive proof of that unity of character which marks the constitution of the human faculties.

Proposition I.

5. *To investigate the nature of the connexion of Secondary Propositions with the idea of Time.*

It is necessary, in entering upon this inquiry, to state clearly the nature of the analogy which connects Secondary with Primary Propositions.

Primary Propositions express relations among things, viewed as component parts of a universe within the limits of which, whether coextensive with the limits of the actual universe or not, the matter of our discourse is confined. The relations expressed are essentially *substantive*. Some, or all, or none, of the members of a given class, *are* also members of another class. The subjects to which primary propositions refer—the relations among those subjects which they express—are all of the above character.

But in treating of secondary propositions, we find ourselves concerned with another class both of subjects and relations. For the subjects with which we have to do are themselves propositions, so that the question may be asked,—Can we regard these subjects also as *things*, and refer them, by analogy with the previous case, to a universe of their own? Again, the relations among these subject propositions are relations of coexistent truth or falsehood, not of substantive equivalence. We do not say, when expressing the connexion of two distinct propositions, that the one *is* the other, but use some such forms of speech as the following, according to the meaning which we desire to convey: "*Either* the proposition X is true, *or* the proposition Y is true;" "If the proposition X is true, the proposition Y is true;" "The propositions X and Y are jointly true;" and so on.

Now, in considering any such relations as the above, we are not called upon to inquire into the whole extent of their possible meaning (for this might involve us in metaphysical questions of causation, which are beyond the proper limits of science); but it suffices to ascertain some meaning which they undoubtedly possess, and which is adequate for the purposes of logical deduction. Let us take, as an instance for examination, the conditional proposition, "If the proposition X is true, the proposition Y is true." An undoubted meaning of this proposition is, that the *time* in which the proposition X is true, is *time* in which the proposition Y is true. This indeed is only a relation of coexistence, and may or may not exhaust the meaning of the proposition, but it is a relation really involved in the statement of the proposition, and further, it suffices for all the purposes of logical inference.

The language of common life sanctions this view of the essential connexion of secondary propositions with the notion of time. Thus we limit the application of a primary proposition by the word "some," but that of a secondary proposition by the word "sometimes." To say, "Sometimes injustice triumphs," is equivalent to asserting that there are times in which the proposition "Injustice now triumphs," is a true proposition. There are indeed propositions, the truth of which is not thus limited to particular periods or conjunctures; propositions which are true throughout all time, and have received the appellation of "eternal truths." The distinction must be familiar to every reader of Plato and Aristotle, by the latter of whom, especially, it is employed to denote the contrast between the abstract verities of science, such as the propositions of geometry which are always true, and those contingent or phænomenal relations of things which are sometimes true and sometimes false. But the forms of language in which both kinds of propositions are expressed manifest a common dependence upon the idea of time; in the one case as limited to some finite duration, in the other as stretched out to eternity.

6. It may indeed be said, that in ordinary reasoning we are often quite unconscious of this notion of time involved in the very language we are using. But the remark, however just, only serves to show that we commonly reason by the aid of words and the forms of a well-constructed language, without attending to the ulterior grounds upon which those very forms have been established. The course of the present investigation will afford an illustration of the very same principle. I shall avail myself of the notion of time in order to determine the laws of the expression of secondary propositions, as well as the laws of combination of the symbols by which they are expressed. But when those laws and those forms are once determined, this notion of time (essential, as I believe it to be, to the above end) may practically be dispensed with. We may then pass from the forms of common language to the closely analogous forms of the symbolical instrument of thought here developed, and use its processes, and interpret its results, without any conscious recognition of the idea of time whatever.

Proposition II.

7. *To establish a system of notation for the expression of Secondary Propositions, and to show that the symbols which it involves are subject to the same laws of combination as the corresponding symbols employed in the expression of Primary Propositions.*

Let us employ the capital letters X, Y, Z, to denote the elementary propositions concerning which we desire to make some assertion touching their truth or falsehood, or among which we seek to express some relation in the form of a secondary proposition. And let us employ the corresponding small letters x, y, z, considered as expressive of mental operations, in the following sense, viz.: Let x represent an act of the mind by which we fix our regard upon that portion of time for which the proposition X is true; and let this meaning be understood when it is asserted that x *denotes* the time for which the proposition X is true. Let us further employ the connecting signs $+$, $-$, $=$, &c., in the following sense, viz.: Let $x + y$ denote the aggregate of those portions of time for which the propositions X and Y are respectively true, those times being entirely separated from each other. Similarly let $x - y$ denote that remainder of time which is left when we take away from the portion of time for which X is true, that (by supposition) included portion for which Y is true. Also, let $x = y$ denote that the time for which the proposition X is true, is identical with the time for which the proposition Y is true. We shall term x the *representative symbol* of the proposition X, &c.

From the above definitions it will follow, that we shall always have

$$x + y = y + x,$$

for either member will denote the same aggregate of time.

Let us further represent by xy the performance in succession of the two operations represented by y and x, i.e. the whole mental operation which consists of the following elements, viz., 1st, The mental selection of that portion of time for which the proposition Y is true. 2ndly, The mental selection, out of that portion of time, of such portion as it contains of the time in which the proposition X is true,—the result of these successive processes being the fixing of the mental regard upon the whole of that portion of time for which the propositions X and Y are both true.

From this definition it will follow, that we shall always have

$$xy = yx. \qquad (1)$$

For whether we select mentally, first that portion of time for which the proposition Y is true, then out of the result that contained portion for which X is true; or first, that portion of time for which the proposition X is true, then out of the result that contained portion of it for which the proposition Y is true; we shall arrive at the same final result, viz., that portion of time for which the propositions X and Y are both true.

By continuing this method of reasoning it may be established, that the laws of combination of the symbols x, y, z, &c., in the species of interpretation here assigned to them, are identical in expression with the laws of combination of the same symbols, in the interpretation assigned to them in the first part of this treatise. The reason of this final identity is apparent. For in both cases it is the same faculty, or the same combination of faculties, of which we study the operations; operations, the essential character of which is unaffected, whether we suppose them to be engaged upon that universe of things in which all existence is contained, or upon that whole of time in which all events are realized, and to some part, at least, of which all assertions, truths, and propositions, refer.

Thus, in addition to the laws above stated, we shall have by (4), Chap. II., the law whose expression is

$$x(y+z) = xy + xz; \qquad (2)$$

and more particularly the fundamental law of duality (2) Chap. II., whose expression is

$$x^2 = x, \text{ or, } x(1-x) = 0; \qquad (3)$$

a law, which while it serves to distinguish the system of thought in Logic from the system of thought in the science of quantity, gives to the processes of the former a completeness and a generality which they could not otherwise possess.

8. Again, as this law (3) (as well as the other laws) is satisfied by the symbols 0 and 1, we are led, as before, to inquire whether those symbols do not admit of interpretation in the present system of thought. The same course of reasoning which we before pursued shows that they do, and warrants us in the two following positions, viz.:

1st, That in the expression of secondary propositions, 0 represents *nothing* in reference to the element of time.

2nd, That in the same system 1 represents the universe, or whole of time, to which the discourse is supposed in any manner to relate.

As in primary propositions the universe of discourse is sometimes limited to a small portion of the actual universe of things, and is sometimes co-extensive with that universe; so in secondary propositions, the universe of discourse may be limited to a single day or to the passing moment, or it may comprise the whole duration of time. It may, in the most literal sense, be "eternal." Indeed, unless there is some limitation expressed or implied in the nature of the discourse, the proper interpretation of the symbol 1 in secondary propositions is "eternity;" even as its proper interpretation in the primary system is the actually existent universe.

9. Instead of appropriating the symbols x, y, z, to the representation of the truths of propositions, we might with equal propriety apply them to represent the occurrence of events. In fact, the occurrence of an event both implies, and is implied by, the truth of a proposition, viz., of the proposition which asserts the occurrence of the event. The one signification of the symbol x necessarily involves the other. It will greatly conduce to convenience to be able to employ our symbols in either of these really equivalent interpretations which the circumstances of a problem may suggest to us as most desirable; and of this liberty I shall avail myself whenever occasion requires. In problems of pure Logic I shall consider the symbols x, y, &c. as representing elementary propositions, among which relation is expressed in the premises. In the mathematical theory of probabilities, which, as before intimated (I. 12), rests upon a basis of Logic, and which it is designed to treat in a subsequent portion of this work, I shall employ the same symbols to denote the simple events, whose implied or required frequency of occurrence it counts among its elements.

Proposition III.

10. *To deduce general Rules for the expression of Secondary Propositions.*

In the various inquiries arising out of this Proposition, fulness of demonstration will be the less necessary, because of the exact analogy which they bear with similar inquiries already completed with reference to primary propositions. We shall first consider the expression of terms; secondly, that of the propositions by which they are connected.

As 1 denotes the whole duration of time, and x that portion of it for which the proposition X is true, $1 - x$ will denote that portion of time for which the proposition X is false.

Again, as xy denotes that portion of time for which the propositions X and Y are both true, we shall, by combining this and the previous observation, be led to the following interpretations, viz.:

The expression $x(1-y)$ will represent the time during which the proposition X is true, and the proposition Y false. The expression $(1-x)(1-y)$ will represent the time during which the propositions X and Y are simultaneously false.

The expression $x(1-y) + y(1-x)$ will express the time during which either X is true or Y true, but not both; for that time is the sum of the times in which they are singly and exclusively true. The expression $xy + (1-x)(1-y)$ will express the time during which X and Y are either both true or both false.

If another symbol z presents itself, the same principles remain applicable. Thus xyz denotes the time in which the propositions X, Y, and Z are simultaneously true; $(1-x)(1-y)(1-z)$ the time in which they are simultaneously false; and the sum of these expressions would denote the time in which they are either true or false together.

The general principles of interpretation involved in the above examples do not need any further illustrations or more explicit statement.

11. The laws of the expression of propositions may now be exhibited and studied in the distinct cases in which they present themselves. There is, however, one principle of fundamental importance to which I wish in the first place to direct attention. Although the principles of expression which have been laid down are perfectly general, and enable us to limit our assertions of the truth or falsehood of propositions to any particular portions of that whole of time (whether it be an unlimited eternity, or a period whose beginning and whose end are definitely fixed, or the passing moment) which constitutes the universe of our discourse, yet, in the actual procedure of human reasoning, such limitation is not commonly employed. When we assert that a proposition is true, we generally mean that it is true throughout the whole

duration of the time to which our discourse refers; and when different assertions of the unconditional truth or falsehood of propositions are jointly made as the premises of a logical demonstration, it is to the same universe of time that those assertions are referred, and not to particular and limited parts of it. In that necessary matter which is the object or field of the exact sciences every assertion of a truth may be the assertion of an "eternal truth." In reasoning upon transient phænomena (as of some social conjuncture) each assertion may be qualified by an immediate reference to the present time, "Now." But in both cases, unless there is a distinct expression to the contrary, it is to the same period of duration that each separate proposition relates. The cases which then arise for our consideration are the following:

1st. *To express the Proposition, "The proposition X is true."*

We are here required to express that within those limits of time to which the matter of our discourse is confined the proposition X is true. Now the time for which the proposition X is true is denoted by x, and the extent of time to which our discourse refers is represented by 1. Hence we have

$$x = 1 \qquad (4)$$

as the expression required.

2nd. *To express the Proposition, "The proposition X is false."*

We are here to express that within the limits of time to which our discourse relates, the proposition X is false; or that within those limits there is no portion of time for which it is true. Now the portion of time for which it is true is x. Hence the required equation will be

$$x = 0. \qquad (5)$$

This result might also be obtained by equating to the whole duration of time 1, the expression for the time during which the proposition X is false, viz., $1 - x$. This gives

$$1 - x = 1,$$

whence

$$x = 0.$$

3rd. *To express the disjunctive Proposition, "Either the pro-*

position X is true or the proposition Y is true;" it being thereby implied that the said propositions are mutually exclusive, that is to say, that one only of them is true.

The time for which either the proposition X is true or the proposition Y is true, but not both, is represented by the expression $x(1-y)+y(1-x)$. Hence we have

$$x(1-y)+y(1-x)=1, \qquad (6)$$

for the equation required.

If in the above Proposition the particles *either*, *or*, are supposed not to possess an absolutely disjunctive power, so that the possibility of the simultaneous truth of the propositions X and Y is not excluded, we must add to the first member of the above equations the term xy. We shall thus have

$$xy+x(1-y)+(1-x)y=1,$$

or

$$x+(1-x)y=1. \qquad (7)$$

4th. *To express the conditional Proposition, "If the proposition Y is true, the proposition X is true."*

Since whenever the proposition Y is true, the proposition X is true, it is necessary and sufficient here to express, that the time in which the proposition Y is true is time in which the proposition X is true; that is to say, that it is some indefinite portion of the whole time in which the proposition X is true. Now the time in which the proposition Y is true is y, and the whole time in which the proposition X is true is x. Let v be a symbol of time indefinite, then will vx represent an indefinite portion of the whole time x. Accordingly, we shall have

$$y=vx$$

as the expression of the proposition given.

12. When v is thus regarded as a symbol of time indefinite, vx may be understood to represent the whole, or an indefinite part, or no part, of the whole time x; for any one of these meanings may be realized by a particular determination of the arbitrary symbol v. Thus, if v be determined to represent a time in which the whole time x is included, vx will represent the whole time x. If v be determined to represent a time, some part of which is in-

cluded in the time x, but which does not fill up the measure of that time, vx will represent a part of the time x. If, lastly, v is determined to represent a time, of which no part is common with any part of the time x, vx will assume the value 0, and will be equivalent to "no time," or "never."

Now it is to be observed that the proposition, "If Y is true, X is true," contains no assertion of the truth of either of the propositions X and Y. It may equally consist with the supposition that the truth of the proposition Y is a condition indispensable to the truth of the proposition X, in which case we shall have $v = 1$; or with the supposition that although Y expresses a condition which, when realized, assures us of the truth of X, yet X may be true without implying the fulfilment of that condition, in which case v denotes a time, some part of which is contained in the whole time x; or, lastly, with the supposition that the proposition Y is not true at all, in which case v represents some time, no part of which is common with any part of the time x. All these cases are involved in the general supposition that v is a symbol of time indefinite.

5th. *To express a proposition in which the conditional and the disjunctive characters both exist.*

The general form of a conditional proposition is, "If Y is true, X is true," and its expression is, by the last section, $y = vx$. We may properly, in analogy with the usage which has been established in primary propositions, designate Y and X as the *terms* of the conditional proposition into which they enter; and we may further adopt the language of the ordinary Logic, which designates the term Y, to which the particle *if* is attached, the "antecedent" of the proposition, and the term X the "consequent."

Now instead of the terms, as in the above case, being simple propositions, let each or either of them be a disjunctive proposition involving different terms connected by the particles *either*, *or*, as in the following illustrative examples, in which X, Y, Z, &c. denote simple propositions.

1st. If either X is true or Y is true, then Z is true.

2nd. If X is true, then either Y is true or Z true.

3rd. If either X is true or Y is true, then either Z and W are both true, or they are both false.

It is evident that in the above cases the relation of the antecedent to the consequent is not affected by the circumstance that one of those terms or both are of a disjunctive character. Accordingly it is only necessary to obtain, in conformity with the principles already established, the proper expressions for the antecedent and the consequent, to affect the latter with the indefinite symbol v, and to equate the results. Thus for the propositions above stated we shall have the respective equations,

1st. $$x(1-y)+(1-x)y=vz.$$

2nd. $$x=v\{y(1-z)+z(1-y)\}.$$

3rd. $$x(1-y)+y(1-x)=v\{zw+(1-z)(1-w)\}.$$

The rule here exemplified is of general application.

Cases in which the disjunctive and the conditional elements enter in a manner different from the above into the expression of a compound proposition, are conceivable, but I am not aware that they are ever presented to us by the natural exigencies of human reason, and I shall therefore refrain from any discussion of them. No serious difficulty will arise from this omission, as the general principles which have formed the basis of the above applications are perfectly general, and a slight effort of thought will adapt them to any imaginable case.

13. In the laws of expression above stated those of interpretation are implicitly involved. The equation

$$x=1$$

must be understood to express that the proposition X is true; the equation

$$x=0,$$

that the proposition X is false. The equation

$$xy=1$$

will express that the propositions X and Y are both true together; and the equation

$$xy=0$$

that they are not both together true.

In like manner the equations

$$x(1-y)+y(1-x)=1,$$
$$x(1-y)+y(1-x)=0,$$

will respectively assert the truth and the falsehood of the disjunctive Proposition, "Either X is true or Y is true." The equations

$$y=vx$$
$$y=v(1-x)$$

will respectively express the Propositions, "If the proposition Y is true, the proposition X is true." "If the proposition Y is true, the proposition X is false."

Examples will frequently present themselves, in the succeeding chapters of this work, of a case in which some terms of a particular member of an equation are affected by the indefinite symbol v, and others not so affected. The following instance will serve for illustration. Suppose that we have

$$y=xz+vx(1-z).$$

Here it is implied that the time for which the proposition Y is true consists of all the time for which X and Z are together true, together with an indefinite portion of the time for which X is true and Z false. From this it may be seen, 1st, That if Y is true, either X and Z are together true, or X is true and Z false; 2ndly, If X and Z are together true, Y is true. The latter of these may be called the reverse interpretation, and it consists in taking the antecedent out of the second member, and the consequent from the first member of the equation. The existence of a term in the second member, whose coefficient is unity, renders this latter mode of interpretation possible. The general principle which it involves may be thus stated:

14. PRINCIPLE.—*Any constituent term or terms in a particular member of an equation which have for their coefficient unity, may be taken as the antecedent of a proposition, of which all the terms in the other member form the consequent.*

Thus the equation

$$y=xz+vx(1-z)+(1-x)(1-z)$$

would have the following interpretations:

Direct Interpretation.—*If the proposition Y is true, then either X and Z are true, or X is true and Z false, or X and Z are both false.*

Reverse Interpretation.—*If either X and Z are true, or X and Z are false, Y is true.*

The aggregate of these partial interpretations will express the whole significance of the equation given.

15. We may here call attention again to the remark, that although the idea of time appears to be an essential element in the theory of the interpretation of secondary propositions, it may practically be neglected as soon as the laws of expression and of interpretation are definitely established. The forms to which those laws give rise seem, indeed, to correspond with the forms of a perfect language. Let us imagine any known or existing language freed from idioms and divested of superfluity, and let us express in that language any given proposition in a manner the most simple and literal,—the most in accordance with those principles of pure and universal thought upon which all languages are founded, of which all bear the manifestation, but from which all have more or less departed. The transition from such a language to the notation of analysis would consist of no more than the substitution of one set of signs for another, without essential change either of form or character. For the elements, whether things or propositions, among which relation is expressed, we should substitute letters; for the disjunctive conjunction we should write +; for the connecting copula or sign of relation, we should write =. This analogy I need not pursue. Its reality and completeness will be made more apparent from the study of those forms of expression which will present themselves in subsequent applications of the present theory, viewed in more immediate comparison with that imperfect yet noble instrument of thought—the English language.

16. Upon the general analogy between the theory of Primary and that of Secondary Propositions, I am desirous of adding a few remarks before dismissing the subject of the present chapter.

We might undoubtedly have established the theory of Primary Propositions upon the simple notion of space, in the same way as that of secondary propositions has been established upon the notion of time. Perhaps, had this been done, the analogy which we are contemplating would have been in somewhat closer accordance with the view of those who regard space and time as merely "forms of the human understanding," conditions of knowledge imposed by the very constitution of the mind upon all that is submitted to its apprehension. But this view, while on the one hand it is incapable of demonstration, on the other hand ties us down to the recognition of "place," τὸ ποῦ, as an essential category of existence. The question, indeed, whether it is so or not, lies, I apprehend, beyond the reach of our faculties; but it may be, and I conceive has been, established, that the formal processes of reasoning in primary propositions do not require, as an essential condition, the manifestation in space of the things about which we reason; that they would remain applicable, with equal strictness of demonstration, to forms of existence, if such there be, which lie beyond the realm of sensible extension. It is a fact, perhaps, in some degree analogous to this, that we are able in many known examples in geometry and dynamics, to exhibit the formal analysis of problems founded upon some intellectual conception of space different from that which is presented to us by the senses, or which can be realized by the imagination.* I conceive, therefore, that the idea of space is not

---

* Space is presented to us in perception, as possessing the three dimensions of length, breadth, and depth. But in a large class of problems relating to the properties of curved surfaces, the rotations of solid bodies around axes, the vibrations of elastic media, &c., this limitation appears in the analytical investigation to be of an arbitrary character, and if attention were paid to the processes of solution alone, no reason could be discovered why space should not exist in four or in any greater number of dimensions. The intellectual procedure in the imaginary world thus suggested can be apprehended by the clearest light of analogy.

The existence of space in three dimensions, and the views thereupon of the religious and philosophical mind of antiquity, are thus set forth by Aristotle:—*Μεγέθος δὲ τὸ μὲν ἐφ' ἓν, γραμμή, τὸ δ' ἐπὶ δύο ἐπίπεδον, τὸ δ' ἐπὶ τρία σῶμα· Καὶ παρὰ ταῦτα οὐκ ἔστιν ἄλλο μέγεθος, διὰ τὸ τριά πάντα εἶναι καὶ τὸ τρὶς πάντῃ. Κάθαπερ γάρ φασι καὶ οἱ Πυθαγόρειοι, τὸ πᾶν καὶ τὰ πάντα τοῖς τρισὶν ὥρισται. Τελευτὴ γὰρ καὶ μέσον καὶ ἀρχὴ τὸν ἀριθμὸν ἔχει τὸν τοῦ παντός· ταῦτα δὲ τὸν τῆς τριάδος. Διὸ παρὰ τῆς φύσεως εἰληφότες ὥσπερ νόμους ἐκείνης, καὶ πρὸς τὰς ἁγιστείας χρώμεθα τῶν θεῶν τῷ ἀριθμῷ τούτῳ.*—*De Cælo*, 1.

essential to the development of a theory of primary propositions, but am disposed, though desiring to speak with diffidence upon a question of such extreme difficulty, to think that the idea of time is essential to the establishment of a theory of secondary propositions. There seem to be grounds for thinking, that without any change in those faculties which are concerned in *reasoning*, the manifestation of space to the human mind might have been different from what it is, but not (at least the same) grounds for supposing that the manifestation of time could have been otherwise than we perceive it to be. Dismissing, however, these speculations as possibly not altogether free from presumption, let it be affirmed that the real ground upon which the symbol 1 represents in primary propositions the universe of things, and not the space they occupy, is, that the sign of identity = connecting the members of the corresponding equations, implies that the things which they represent are identical, not simply that they are found in the same portion of space. Let it in like manner be affirmed, that the reason why the symbol 1 in secondary propositions represents, not the universe of events, but the eternity in whose successive moments and periods they are evolved, is, that the same sign of identity connecting the logical members of the corresponding equations implies, not that the events which those members represent are identical, but that the times of their occurrence are the same. These reasons appear to me to be decisive of the immediate question of interpretation. In a former treatise on this subject (Mathematical Analysis of Logic, p. 49), following the theory of Wallis respecting the Reduction of Hypothetical Propositions, I was led to interpret the symbol 1 in secondary propositions as the universe of "cases" or "conjunctures of circumstances;" but this view involves the necessity of a definition of what is meant by a "case," or "conjuncture of circumstances;" and it is certain, that whatever is involved in the term beyond the notion of time is alien to the objects, and restrictive of the processes, of formal Logic.

CHAPTER XII.

OF THE METHODS AND PROCESSES TO BE ADOPTED IN THE TREATMENT OF SECONDARY PROPOSITIONS.

1. IT has appeared from previous researches (XI. 7) that the laws of combination of the literal symbols of Logic are the same, whether those symbols are employed in the expression of primary or in that of secondary propositions, the sole existing difference between the two cases being a difference of interpretation. It has also been established (V. 6), that whenever distinct systems of thought and interpretation are connected with the same system of formal laws, i. e., of laws relating to the combination and use of symbols, the attendant processes, intermediate between the expression of the primary conditions of a problem and the interpretation of its symbolical solution, are the same in both. Hence, as between the systems of thought manifested in the two forms of primary and of secondary propositions, this community of formal law exists, the processes which have been established and illustrated in our discussion of the former class of propositions will, without any modification, be applicable to the latter.

2. Thus the laws of the two fundamental processes of elimination and development are the same in the system of secondary as in the system of primary propositions. Again, it has been seen (Chap. VI. Prop. 2) how, in primary propositions, the interpretation of any proposed equation devoid of fractional forms may be effected by developing it into a series of constituents, and equating to 0 every constituent whose coefficient does not vanish. To the equations of secondary propositions the same method is applicable, and the interpreted result to which it finally conducts us is, as in the former case (VI. 6), a system of co-existent denials. But while in the former case the force of those denials is expended upon the existence of certain classes of things, in the latter it relates to the truth of certain combinations of the ele-

mentary propositions involved in the *terms* of the given premises. And as in primary propositions it was seen that the system of denials admitted of conversion into various other forms of propositions (VI. 7), &c., such conversion will be found to be possible here also, the sole difference consisting not in the forms of the equations, but in the nature of their interpretation.

3. Moreover, as in primary propositions, we can find the expression of any element entering into a system of equations, in terms of the remaining elements (VI. 10), or of any selected number of the remaining elements, and interpret that expression into a logical inference, the same object can be accomplished by the same means, difference of interpretation alone excepted, in the system of secondary propositions. The elimination of those elements which we desire to banish from the final solution, the reduction of the system to a single equation, the algebraic solution and the mode of its development into an interpretable form, differ in no respect from the corresponding steps in the discussion of primary propositions.

To remove, however, any possible difficulty, it may be desirable to collect under a general Rule the different cases which present themselves in the treatment of secondary propositions.

RULE.—*Express symbolically the given propositions* (XI. 11).

Eliminate separately from each equation in which it is found the indefinite symbol v (VII. 5).

Eliminate the remaining symbols which it is desired to banish from the final solution: always before elimination reducing to a single equation those equations in which the symbol or symbols to be eliminated are found (VIII. 7). *Collect the resulting equations into a single equation* $V = 0$.

Then proceed according to the particular form in which it is desired to express the final relation, as—

1st. *If in the form of a denial, or system of denials, develop the function* V, *and equate to* 0 *all those constituents whose coefficients do not vanish.*

2ndly. *If in the form of a disjunctive proposition, equate to* 1 *the sum of those constituents whose coffiecients vanish.*

3rdly. *If in the form of a conditional proposition having a sim-*

ple element, as x or $1 - x$, for its antecedent, determine the algebraic expression of that element, and develop that expression.

4thly. *If in the form of a conditional proposition having a compound expression, as xy, $xy + (1-x)(1-y)$, &c., for its antecedent, equate that expression to a new symbol t, and determine t as a developed function of the symbols which are to appear in the consequent, either by ordinary methods or by the special method* (IX. 9).

5thly. *Interpret the results by* (XI. 13, 14).

If it only be desired to ascertain whether a particular elementary proposition x is true or false, we must eliminate all the symbols but x; then the equation $x = 1$ will indicate that the proposition is true, $x = 0$ that it is false, $0 = 0$ that the premises are insufficient to determine whether it is true or false.

4. Ex. 1.—The following prediction is made the subject of a curious discussion in Cicero's fragmentary treatise, De Fato:—"Si quis (Fabius) natus est oriente Canicula, is in mari non morietur." I shall apply to it the method of this chapter. Let y represent the proposition, "Fabius was born at the rising of the dogstar;" x the proposition, "Fabius will die in the sea." In saying that x *represents* the proposition, "Fabius, &c.," it is only meant that x is a symbol so appropriated (XI. 7) to the above proposition, that the equation $x = 1$ declares, and the equation $x = 0$ denies, the truth of that proposition. The equation we have to discuss will be

$$y = v(1 - x). \qquad (1)$$

And, first, let it be required to reduce the given proposition to a negation or system of negations (XII. 3). We have, on transposition,

$$y - v(1 - x) = 0.$$

Eliminating v,

$$y\{y - (1 - x)\} = 0,$$

or,

$$y - y(1 - x) = 0,$$

or,

$$yx = 0. \qquad (2)$$

The interpretation of this result is:—"It is not true that Fabius was born at the rising of the dogstar, and will die in the sea."

Cicero terms this form of proposition, "Conjunctio ex repugnantibus;" and he remarks that Chrysippus thought in this way to evade the difficulty which he imagined to exist in contingent assertions respecting the future: "Hoc loco Chrysippus æstuans falli sperat Chaldæos cæterosque divinos, neque eos usuros esse conjunctionibus ut ita sua percepta pronuntient: Si quis natus est oriente Caniculâ is in mari non morietur; sed potius ita dicant: Non et natus est quis oriente Caniculâ, et in mari morietur. O licentiam jocularem! . . . . Multa genera sunt enuntiandi, nec ullum distortius quam hoc quo Chrysippus sperat Chaldæos contentos Stoicorum causâ fore."—*Cic. De Fato*, 7, 8.

5. To reduce the given proposition to a disjunctive form.

The constituents not entering into the first member of (2) are

$$x(1-y),\ (1-x)y,\ (1-x)(1-y).$$

Whence we have

$$y(1-x)+x(1-y)+(1-x)(1-y)=1. \qquad (3)$$

The interpretation of which is :—*Either Fabius was born at the rising of the dogstar, and will not perish in the sea; or he was not born at the rising of the dogstar, and will perish in the sea; or he was not born at the rising of the dogstar, and will not perish in the sea.*

In cases like the above, however, in which there exist constituents differing from each other only by a single factor, it is, as we have seen (VII. 15), most convenient to collect such constituents into a single term. If we thus connect the first and third terms of (3), we have

$$(1-y)x+1-x=1;$$

and if we similarly connect the second and third, we have

$$y(1-x)+1-y=1.$$

These forms of the equation severally give the interpretations—

Either Fabius was not born under the dogstar, and will die in the sea, or he will not die in the sea.

Either Fabius was born under the dogstar, and will not die in the sea, or he was not born under the dogstar.

It is evident that these interpretations are strictly equivalent to the former one.

Let us ascertain, in the form of a conditional proposition, the consequences which flow from the hypothesis, that "Fabius will perish in the sea."

In the equation (2), which expresses the result of the elimination of v from the original equation, we must seek to determine x as a function of y.

We have

$$x = \frac{0}{y} = 0\,y + \frac{0}{0}(1-y) \text{ on expansion,}$$

or,

$$x = \frac{0}{0}(1-y);$$

the interpretation of which is,—*If Fabius shall die in the sea, he was not born at the rising of the dogstar.*

These examples serve in some measure to illustrate the connexion which has been established in the previous sections between primary and secondary propositions, a connexion of which the two distinguishing features are identity of process and analogy of interpretation.

6. Ex. 2.—There is a remarkable argument in the second book of the Republic of Plato, the design of which is to prove the immutability of the Divine Nature. It is a very fine example both of the careful induction from familiar instances by which Plato arrives at general principles, and of the clear and connected logic by which he deduces from them the particular inferences which it is his object to establish. The argument is contained in the following dialogue:

"Must not that which departs from its proper form be changed either by itself or by another thing? Necessarily so. Are not things which are in the best state least changed and disturbed, as the body by meats and drinks, and labours, and every species of plant by heats and winds, and such like affections? Is not the healthiest and strongest the least changed? Assuredly. And does not any trouble from without least disturb and change that soul which is strongest and wisest? And as to all made vessels, and furnitures, and garments, according to the same principle, are not those which are well wrought, and in a good condition, least changed by time and other accidents? Even so. And whatever is in a right state, either by nature or by art, or by both these, admits of the smallest change from any other thing. So it seems. But God and things divine are in every sense in the best state. Assuredly. In this way, then, God should least of all bear many forms? Least, indeed, of all. Again, should He transform and change Himself? Manifestly He must do so, if He is changed at all. Changes He then Himself to that which is more good and fair, or to that which is worse and baser? Necessarily to the worse, if he be changed. For never shall we say that God is indigent of beauty or of virtue. You speak most rightly, said I, and the matter being so, seems it to you, O Adimantus, that God or man *willingly* makes himself in any sense worse? Impossible, said he. Impossible, then, it is, said I, that a god should wish to change himself; but ever being fairest and best, each of them ever remains absolutely in the same form."

The premises of the above argument are the following:

1st. If the Deity suffers change, He is changed either by Himself or by another.

2nd. If He is in the best state, He is not changed by another.

3rd. The Deity is in the best state.

4th. If the Deity is changed by Himself, He is changed to a worse state.

5th. If He acts willingly, He is not changed to a worse state.

6th. The Deity acts willingly.

Let us express the elements of these premises as follows:

Let x represent the proposition, "The Deity suffers change."
y, He is changed by Himself.
z, He is changed by another.
s, He is in the best state.
t, He is changed to a worse state.
w, He acts willingly.

Then the premises expressed in symbolical language yield, after elimination of the indefinite class symbols v, the following equations:

$$xyz + x(1-y)(1-z) = 0, \quad (1)$$
$$sz = 0, \quad (2)$$
$$s = 1, \quad (3)$$
$$y(1-t) = 0, \quad (4)$$
$$wt = 0, \quad (5)$$
$$w = 1. \quad (6)$$

Retaining x, I shall eliminate in succession z, s, y, t, and w (this being the order in which those symbols occur in the above system), and interpret the successive results.

Eliminating z from (1) and (2), we get

$$xs(1-y) = 0. \quad (7)$$

Eliminating s from (3) and (7),

$$x(1-y) = 0. \quad (8)$$

Eliminating y from (4) and (8),

$$x(1-t) = 0. \quad (9)$$

Eliminating t from (5) and (9),

$$xw = 0. \quad (10)$$

Eliminating w from (6) and (10),

$$x = 0. \quad (11)$$

These equations, beginning with (8), give the following results:

From (8) we have $x = \frac{0}{0}y$, therefore, *If the Deity suffers change, He is changed by Himself.*

From (9), $x = \frac{0}{0}t$, *If the Deity suffers change, He is changed to a worse state.*

From (10), $x = \frac{0}{0}(1-w)$. *If the Deity suffers change, He does not act willingly.*

From (11), *The Deity does not suffer change.* This is Plato's result.

Now I have before remarked, that the order of elimination is indifferent. Let us in the present case seek to verify this fact by eliminating the same symbols in a reverse order, beginning with w. The resulting equations are,

$$t = 0, \quad y = 0, \quad x(1-z) = 0, \quad z = 0, \quad x = 0;$$

yielding the following interpretations:

God is not changed to a worse state.
He is not changed by Himself.
If He suffers change, He is changed by another.
He is not changed by another.
He is not changed.

We thus reach by a different route the same conclusion.

Though as an exhibition of the *power* of the method, the above examples are of slight value, they serve as well as more complicated instances would do, to illustrate its nature and character.

7. It may be remarked, as a final instance of analogy between the system of primary and that of secondary propositions, that in the latter system also the fundamental equation,

$$x(1-x) = 0,$$

admits of interpretation. It expresses the axiom, *A proposition cannot at the same time be true and false.* Let this be compared with the corresponding interpretation (III. 15). Solved under the form

$$x = \frac{0}{1-x} = \frac{0}{0}x,$$

by development, it furnishes the respective axioms: "A thing is what it is:" "If a proposition is true, it is true:" forms of what has been termed "The principle of identity." Upon the nature and the value of these axioms the most opposite opinions have been entertained. Some have regarded them as the very pith and marrow of philosophy. Locke devoted to them a chapter, headed, "On Trifling Propositions."* In both these views there seems to have been a mixture of truth and error. Regarded as supplanting experience, or as furnishing materials for the vain and wordy janglings of the schools, such propositions are worse than trifling. Viewed, on the other hand, as intimately allied with the very laws and conditions of thought, they rise into at least a speculative importance.

---

* Essay on the Human Understanding, Book IV. Chap. viii.

CHAPTER XIII.

ANALYSIS OF A PORTION OF DR. SAMUEL CLARKE'S "DEMONSTRATION OF THE BEING AND ATTRIBUTES OF GOD," AND OF A PORTION OF THE "ETHICA ORDINE GEOMETRICO DEMONSTRATA" OF SPINOZA.

1. THE general order which, in the investigations of the following chapter, I design to pursue, is the following. I shall examine what are the actual premises involved in the demonstrations of some of the general propositions of the above treatises, whether those premises be expressed or implied. By the actual premises I mean whatever propositions are assumed in the course of the argument, without being proved, and are employed as parts of the foundation upon which the final conclusion is built. The premises thus determined, I shall express in the language of symbols, and I shall then deduce from them by the methods developed in the previous chapters of this work, the most important inferences which they involve, in addition to the particular inferences actually drawn by the authors. I shall in some instances modify the premises by the omission of some fact or principle which is contained in them, or by the addition or substitution of some new proposition, and shall determine how by such change the ultimate conclusions are affected. In the pursuit of these objects it will not devolve upon me to inquire, except incidentally, how far the metaphysical principles laid down in these celebrated productions are worthy of confidence, but only to ascertain what conclusions may justly be drawn from given premises; and in doing this, to exemplify the perfect liberty which we possess as concerns both the choice and the order of the elements of the final or concluding propositions, viz., as to determining what elementary propositions are true or false, and what are true or false under given restrictions, or in given combinations.

2. The chief practical difficulty of this inquiry will consist,

not in the application of the method to the premises once determined, but in ascertaining what the premises are. In what are regarded as the most rigorous examples of reasoning applied to metaphysical questions, it will occasionally be found that different trains of thought are blended together; that particular but essential parts of the demonstration are given parenthetically, or out of the main course of the argument; that the meaning of a premiss may be in some degree ambiguous; and, not unfrequently, that arguments, viewed by the strict laws of formal reasoning, are incorrect or inconclusive. The difficulty of determining and distinctly exhibiting the true premises of a demonstration may, in such cases, be very considerable. But it is a difficulty which must be overcome by all who would ascertain whether a particular conclusion is proved or not, whatever form they may be prepared or disposed to give to the ulterior process of reasoning. It is a difficulty, therefore, which is not peculiar to the method of this work, though it manifests itself more distinctly in connexion with this method than with any other. So intimate, indeed, is this connexion, that it is impossible, employing the method of this treatise, to form even a conjecture as to the validity of a conclusion, without a distinct apprehension and exact statement of all the premises upon which it rests. In the more usual course of procedure, nothing is, however, more common than to examine some of the steps of a train of argument, and thence to form a vague general impression of the scope of the whole, without any such preliminary and thorough analysis of the premises which it involves.

The necessity of a rigorous determination of the real premises of a demonstration ought not to be regarded as an evil; especially as, when that task is accomplished, every source of doubt or ambiguity is removed. In employing the method of this treatise, the order in which premises are arranged, the mode of connexion which they exhibit, with every similar circumstance, may be esteemed a matter of indifference, and the process of inference is conducted with a precision which might almost be termed mechanical.

3. The "Demonstration of the Being and Attributes of God," consists of a series of propositions or theorems, each

of them proved by means of premises resolvable, for the most part, into two distinct classes, viz., facts of observation, such as the existence of a material world, the phænomenon of motion, &c., and hypothetical principles, the authority and universality of which are supposed to be recognised *à priori*. It is, of course, upon the truth of the latter, assuming the correctness of the reasoning, that the validity of the demonstration really depends. But whatever may be thought of its claims in this respect, it is unquestionable that, as an intellectual performance, its merits are very high. Though the trains of argument of which it consists are not in general very clearly arranged, they are almost always specimens of correct Logic, and they exhibit a subtlety of apprehension and a force of reasoning which have seldom been equalled, never perhaps surpassed. We see in them the consummation of those intellectual efforts which were awakened in the realm of metaphysical inquiry, at a period when the dominion of hypothetical principles was less questioned than it now is, and when the rigorous demonstrations of the newly risen school of mathematical physics seemed to have furnished a model for their direction. They appear to me for this reason (not to mention the dignity of the subject of which they treat) to be deserving of high consideration; and I do not deem it a vain or superfluous task to expend upon some of them a careful analysis.

4. The Ethics of Benedict Spinoza is a treatise, the object of which is to prove the identity of God and the universe, and to establish, upon this doctrine, a system of morals and of philosophy. The analysis of its main argument is extremely difficult, owing not to the complexity of the separate propositions which it involves, but to the use of vague definitions, and of axioms which, through a like defect of clearness, it is perplexing to determine whether we ought to accept or to reject. While the reasoning of Dr. Samuel Clarke is in part verbal, that of Spinoza is so in a much greater degree; and perhaps this is the reason why, to some minds, it has appeared to possess a formal cogency, to which in reality it possesses no just claim. These points will, however, be considered in the proper place.

CLARKE'S DEMONSTRATION.

PROPOSITION I.

5. "*Something has existed from eternity.*"

The proof is as follows:—

"For since something now is, 'tis manifest that something always was. Otherwise the things that now are must have risen out of nothing, absolutely and without cause. Which is a plain contradiction in terms. For to say a thing is produced, and yet that there is no cause at all of that production, is to say that something is effected when it is effected by nothing, that is, at the same time when it is not effected at all. Whatever exists has a cause of its existence, either in the necessity of its own nature, and thus it must have been of itself eternal: or in the will of some other being, and then that other being must, at least in the order of nature and causality, have existed before it."

Let us now proceed to analyze the above demonstration. Its first sentence is resolvable into the following propositions:

1st. Something is.

2nd. If something is, either something always was, or the things that now are must have risen out of nothing.

The next portion of the demonstration consists of a proof that the second of the above alternatives, viz., "The things that now are have risen out of nothing," is impossible, and it may formally be resolved as follows:

3rd. If the things that now are have risen out of nothing, something has been effected, and at the same time that something has been effected by nothing.

4th. If that something has been effected by nothing, it has not been effected at all.

The second portion of this argument appears to be a mere assumption of the point to be proved, or an attempt to make that point clearer by a different verbal statement.

The third and last portion of the demonstration contains a distinct proof of the truth of either the original proposition to be proved, viz., "Something always was," or the point proved in the second part of the demonstration, viz., the untenable nature of the hypothesis, that "the things that now are have risen out of nothing." It is resolvable as follows :—

5th. If something is, either it exists by the necessity of its own nature, or it exists by the will of another being.

6th. If it exists by the necessity of its own nature, something always was.

7th. If it exists by the will of another being, then the proposition, that the things which exist have arisen out of nothing, is false.

The last proposition is not expressed in the same form in the text of Dr. Clarke; but his expressed conclusion of the prior existence of another Being is clearly meant as equivalent to a denial of the proposition that the things which now are have risen out of nothing.

It appears, therefore, that the demonstration consists of two distinct trains of argument: one of those trains comprising what I have designated as the *first* and *second* parts of the demonstration; the other comprising the *first* and *third* parts. Let us consider the latter train.

The premises are :—

1st. Something is.

2nd. If something is, either something always was, or the things that now are have risen out of nothing.

3rd. If something is, either it exists in the necessity of its own nature, or it exists by the will of another being.

4th. If it exists in the necessity of its own nature, something always was.

5th. If it exists by the will of another being, then the hypothesis, that the things which now are have risen out of nothing, is false.

We must now express symbolically the above proposition.

Let x = Something is.

y = Something always was.

z = The things which now are have risen from nothing.

p = It exists in the necessity of its own nature (i. e. the *something* spoken of above).

q = It exists by the will of another Being.

It must be understood, that by the expression, Let x = "Something is," is meant no more than that x is the representative symbol of that proposition (XI. 7), the equations $x = 1$, $x = 0$, respectively declaring its truth and its falsehood.

The equations of the premises are:—

1st. $x = 1$;
2nd. $x = v\{y(1-z) + z(1-y)\}$;
3rd. $x = v\{p(1-q) + q(1-p)\}$;
4th. $p = vy$;
5th. $q = v(1-z)$;

and on eliminating the several indefinite symbols v, we have

$$1 - x = 0; \quad (1)$$

$$x\{yz + (1-y)(1-z)\} = 0; \quad (2)$$

$$x\{pq + (1-p)(1-q)\} = 0; \quad (3)$$

$$p(1-y) = 0; \quad (4)$$

$$qz = 0. \quad (5)$$

6. First, I shall examine whether any conclusions are deducible from the above, concerning the truth or falsity of the single propositions represented by the symbols y, z, p, q, viz., of the propositions, "Something always was;" "The things which now are have risen from nothing;" "The something which is exists by the necessity of its own nature;" "The something which is exists by the will of another being."

For this purpose we must separately eliminate all the symbols but y, all these but z, &c. The resulting equation will determine whether any such separate relations exist.

To eliminate x from (1), (2), and (3), it is only necessary to substitute in (2) and (3) the value of x derived from (1). We find as the results,

$$yz + (1-y)(1-z) = 0. \quad (6)$$

$$pq + (1-p)(1-q) = 0. \quad (7)$$

To eliminate p we have from (4) and (7), by addition,

$$p(1-y) + pq + (1-p)(1-q) = 0; \quad (8)$$

whence we find,

$$(1-y)(1-q) = 0. \quad (9)$$

To eliminate q from (5) and (9), we have

$$qz + (1 - y)(1 - q) = 0;$$

whence we find

$$z(1 - y) = 0. \tag{10}$$

There now remain but the two equations (6) and (10), which, on addition, give

$$yz + 1 - y = 0.$$

Eliminating from this equation z, we have

$$1 - y = 0, \text{ or, } y = 1. \tag{11}$$

Eliminating from the same equation y, we have

$$z = 0. \tag{12}$$

The interpretation of (11) is

Something always was.

The interpretation of (12) is

The things which are have not risen from nothing.

Next resuming the system (6), (7), with the two equations (4), (5), let us determine the two equations involving p and q respectively.

To eliminate y we have from (4) and (6),

$$p(1 - y) + yz + (1 - y)(1 - z) = (0);$$

whence $$(p + 1 - z)z = 0, \text{ or, } pz = 0. \tag{13}$$

To eliminate z from (5) and (13), we have

$$qz + pz = 0;$$

whence we get,

$$0 = 0.$$

There remains then but the equation (7), from which eliminating q, we have $0 = 0$ for the final equation, in p.

Hence there is no conclusion derivable from the premises affirming the simple truth or falsehood of the proposition, "The something which is exists in the necessity of its own nature." And as, on eliminating p, there is the same result, $0 = 0$, for the ultimate equation in q, it also follows, that *there is no conclusion deducible from the premises as to the simple truth or falsehood of the proposition, "The something which is exists by the will of another Being."*

Of relations connecting more than one of the propositions represented by the elementary symbols, it is needless to consider any but that which is denoted by the equation (7) connecting p and q, inasmuch as the propositions represented by the remaining symbols are absolutely true or false independently of any connexion of the kind here spoken of. The interpretation of (7), placed under the form

$$p(1-q)+q(1-p)=1, \text{ is,}$$

The something which is, either exists in the necessity of its own nature, or by the will of another being.

I have exhibited the details of the above analysis with a, perhaps, needless fulness and prolixity, because in the examples which will follow, I propose rather to indicate the steps by which results are obtained, than to incur the danger of a wearisome frequency of repetition. The conclusions which have resulted from the above application of the method are easily verified by ordinary reasoning.

The reader will have no difficulty in applying the method to the other train of premises involved in Dr. Clarke's first Proposition, and deducing from them the two first of the conclusions to which the above analysis has led.

Proposition II.

7. *Some one unchangeable and independent Being has existed from eternity.*

The premises from which the above proposition is proved are the following:

1st. Something has always existed.

2nd. If something has always existed, either there has existed some one unchangeable and independent being, or the whole of existing things has been comprehended in a succession of changeable and dependent beings.

3rd. If the universe has consisted of a succession of changeable and dependent beings, either that series has had a cause from without, or it has had a cause from within.

4th. It has not had a cause from without (because it includes, by hypothesis, all things that exist).

5th. It has not had a cause from within (because no part is necessary, and if no part is necessary, the whole cannot be necessary).

Omitting, merely for brevity, the subsidiary proofs contained in the parentheses of the fourth and fifth premiss, we may represent the premises as follows:

Let x = Something has always existed.

y = There has existed some one unchangeable and independent being.

z = There has existed a succession of changeable and dependent beings.

p = That series has had a cause from without.

q = That series has had a cause from within.

Then we have the following system of equations, viz.:

1st. $x = 1$;

2nd. $x = v\{y(1-z) + z(1-y)\}$;

3rd. $z = v\{p(1-q) + (1-p)q\}$;

4th. $p = 0$;

5th. $q = 0$:

which, on the separate elimination of the indefinite symbols v, gives

$$1 - x = 0; \quad (1)$$

$$x\{yz + (1-y)(1-z)\} = 0; \quad (2)$$

$$z\{pq + (1-p)(1-q)\} = 0; \quad (3)$$

$$p = 0; \quad (4)$$

$$q = 0. \quad (5)$$

The elimination from the above system of x, p, q, and y, conducts to the equation

$$z = 0.$$

And the elimination of x, p, q, and z, conducts in a similar manner to the equation

$$y = 1.$$

Of which equations the respective interpretations are:

1st. *The whole of existing things has not been comprehended in a succession of changeable and dependent beings.*

2nd. *There has existed some one unchangeable and independent being.*

The latter of these is the proposition which Dr. Clarke proves. As, by the above analysis, all the propositions represented by the literal symbols x, y, z, p, q, are determined as absolutely true or false, it is needless to inquire into the existence of any further relations connecting those propositions together.

Another proof is given of Prop. II., which for brevity I pass over. It may be observed, that the "impossibility of infinite succession," the proof of which forms a part of Clarke's argument, has commonly been assumed as a fundamental principle of metaphysics, and extended to other questions than that of causation. Aristotle applies it to establish the necessity of first principles of demonstration;* the necessity of an end (the good), in human actions, &c.† There is, perhaps, no principle more frequently referred to in his writings. By the schoolmen it was similarly applied to prove the impossibility of an infinite subordination of genera and species, and hence the necessary existence of universals. Apparently the impossibility of our forming a definite and complete conception of an infinite series, i. e. of comprehending it as a *whole*, has been confounded with a logical inconsistency, or contradiction in the idea itself.

8. The analysis of the following argument depends upon the theory of Primary Propositions.

PROPOSITION III.

That unchangeable and independent Being must be self-existent.

The premises are:—

1. Every being must either have come into existence out of nothing, or it must have been produced by some external cause, or it must be self-existent.

2. No being has come into existence out of nothing.

3. The unchangeable and independent Being has not been produced by an external cause.

For the symbolical expression of the above, let us assume,

---

* Metaphysics, III. 4; Anal. Post. I. 19, *et seq.*

† Nic. Ethics, Book I. Cap. II.

x = Beings which have arisen out of nothing.
y = Beings which have been produced by an external cause.
z = Beings which are self-existent.
w = The unchangeable and independent Being.

Then we have

$$x(1-y)(1-z)+y(1-x)(1-z)+z(1-x)(1-y)=1, \quad (1)$$

$$x=0, \quad (2)$$

$$w=v(1-y), \quad (3)$$

from the last of which eliminating v,

$$wy=0. \quad (4)$$

Whenever, as above, the value of a symbol is given as 0 or 1, it is best eliminated by simple substitution. Thus the elimination of x gives

$$y(1-z)+z(1-y)=1; \quad (5)$$

or,
$$yz+(1-y)(1-z)=0. \quad (6)$$

Now adding (4) and (6), and eliminating y, we get

$$w(1-z)=0,$$

$$\therefore\ w=vz;$$

the interpretation of which is,—*The unchangeable and independent being is necessarily self-existing.*

Of (5), in its actual form, the interpretation is,—*Every being has either been produced by an external cause, or it is self-existent.*

9. In Dr. Samuel Clarke's observations on the above proposition occurs a remarkable argument, designed to prove that the material world is not the self-existent being above spoken of. The passage to which I refer is the following:

"If matter be supposed to exist necessarily, then in that necessary existence there is either included the power of gravitation, or not. If not, then in a world merely material, and in which no intelligent being presides, there never could have been any motion; because motion, as has been already shown, and is now granted in the question, is not necessary of itself. But if the power of gravitation be included in the pretended necessary existence of matter: then, it following necessarily that there must be a vacuum (as the incomparable Sir Isaac Newton has abundantly demonstrated that there must, if gravitation be an universal quality or affection of matter), it follows likewise, that matter is not a necessary being. For if a vacuum actually be, then it is plainly more than possible for matter not to be."—(pp. 25, 26).

It will, upon attentive examination, be found that the actual premises involved in the above demonstration are the following:

1st. If matter is a necessary being, either the property of gravitation is necessarily present, or it is necessarily absent.

2nd. If gravitation is necessarily absent, and the world is not subject to any presiding Intelligence, motion does not exist.

3rd. If the property of gravitation is necessarily present, the existence of a vacuum is necessary.

4th. If the existence of a vacuum is necessary, matter is not a necessary being.

5th. If matter is a necessary being, the world is not subject to a presiding Intelligence.

6th. Motion exists.

Of the above premises the first four are expressed in the demonstration; the fifth is implied in the connexion of its first and second sentences; and the sixth expresses a fact, which the author does not appear to have thought it necessary to state, but which is obviously a part of the ground of his reasoning. Let us represent the elementary propositions in the following manner:

Let x = Matter is a necessary being.
y = Gravitation is necessarily present.
t = Gravitation is necessarily absent.
z = The world is merely material, and not subject to any presiding Intelligence.
w = Motion exists.
v = A vacuum is necessary.

Then the system of premises will be represented by the following equations, in which q is employed as the symbol of time indefinite:

$$\begin{aligned}
x &= q\{y(1-t)+(1-y)t\}.\\
tz &= q(1-w).\\
y &= qv.\\
v &= q(1-x).\\
x &= qz.\\
w &= 1.
\end{aligned}$$

From which, if we eliminate the symbols q, we have the following system, viz.:

$$x\{yt+(1-y)(1-t)\}=0. \quad (1)$$
$$tzw=0. \quad (2)$$
$$y(1-v)=0. \quad (3)$$
$$vx=0. \quad (4)$$
$$x(1-z)=0. \quad (5)$$
$$1-w=0. \quad (6)$$

Now if from these equations we eliminate w, v, z, y, and t, we obtain the equation

$$x=0,$$

which expresses the proposition, *Matter is not a necessary being.* This is Dr. Clarke's conclusion. If we endeavour to eliminate any other set of five symbols (except the set v, z, y, t, and x, which would give $w=1$), we obtain a result of the form $0=0$. It hence appears that *there are no other conclusions expressive of the absolute truth or falsehood of any of the elementary propositions designated by single symbols.*

Of conclusions expressed by equations involving two symbols, there exists but the following, viz.:—*If the world is merely material, and not subject to a presiding Intelligence, gravitation is not necessarily absent.* This conclusion is expressed by the equation

$$tz=0, \text{ whence } z=q(1-t).$$

If in the above analysis we suppress the concluding premiss, expressing the fact of the existence of motion, and leave the hypothetical principles which are embodied in the remaining premises untouched, some remarkable conclusions follow. To these I shall direct attention in the following chapter.

10. Of the remainder of Dr. Clarke's argument I shall briefly state the substance and connexion, dwelling only on certain por-

tions of it which are of a more complex character than the others, and afford better illustrations of the method of this work.

In Prop. IV. it is shown that the substance or essence of the self-existent being is incomprehensible. The tenor of the reasoning employed is, that we are ignorant of the essential nature of all other things,—much more, then, of the essence of the self-existent being.

In Prop. V. it is contended that "though the substance or essence of the self-existent being is itself absolutely incomprehensible to us, yet many of the essential attributes of his nature are strictly demonstrable, as well as his existence."

In Prop. VI. it is argued that "the self-existent being must of necessity be infinite and omnipresent;" and it is contended that his infinity must be "an infinity of fulness as well as of immensity." The ground upon which the demonstration proceeds is, that an absolute necessity of existence must be independent of time, place, and circumstance, free from limitation, and therefore excluding all imperfection. And hence it is inferred that the self-existent being must be "a most simple, unchangeable, incorruptible being, without parts, figure, motion, or any other such properties as we find in matter."

The premises actually employed may be exhibited as follows:

1. If a finite being is self-existent, it is a contradiction to suppose it not to exist.

2. A finite being may, without contradiction, be absent from one place.

3. That which may without contradiction be absent from one place may without contradiction be absent from all places.

4. That which may without contradiction be absent from all places may without contradiction be supposed not to exist.

Let us assume

x = Finite beings.

y = Things self-existent.

z = Things which it is a contradiction to suppose not to exist.

w = Things which may be absent without contradiction from one place.

t = Things which without contradiction may be absent from every place.

We have on expressing the above, and eliminating the indefinite symbols,

$$xy(1-z)=0. \quad (1)$$
$$x(1-w)=0. \quad (2)$$
$$w(1-t)=0. \quad (3)$$
$$tz=0. \quad (4)$$

Eliminating in succession t, w, and z, we get

$$xy=0,$$
$$\therefore y=\frac{0}{0}(1-x);$$

the interpretation of which is,—*Whatever is self-existent is infinite.*

In Prop. VII. it is argued that the self-existent being must of necessity be One. The order of the proof is, that the self-existent being is "necessarily existent," that "necessity absolute in itself is simple and uniform, and without any possible difference or variety," that all "variety or difference of existence" implies dependence; and hence that "whatever exists necessarily is the one simple essence of the self-existent being."

The conclusion is also made to flow from the following premises:—

1. If there are two or more necessary and independent beings, either of them may be supposed to exist alone.

2. If either may be supposed to exist alone, it is not a contradiction to suppose the other not to exist.

3. If it is not a contradiction to suppose this, there are not two necessary and independent beings.

Let us represent the elementary propositions as follows:—

x = there exist two necessary independent beings.
y = either may be supposed to exist alone.
z = it is not a contradiction to suppose the other not to exist.

We have then, on proceeding as before,

$$x(1-y)=0. \quad (1)$$
$$y(1-z)=0. \quad (2)$$
$$zx=0. \quad (3)$$

Eliminating y and z, we have

$$x = 0.$$

Whence, *There do not exist two necessary and independent beings.*

11. To the premises upon which the two previous propositions rest, it is well known that Bishop Butler, who at the time of the publication of the "Demonstration," was a student in a nonconformist academy, made objection in some celebrated letters, which, together with Dr. Clarke's replies to them, are usually appended to editions of the work. The real question at issue is the validity of the principle, that "whatsoever is absolutely necessary at all is absolutely necessary in every part of space, and in every point of duration,"—a principle assumed in Dr. Clarke's reasoning, and explicitly stated in his reply to Butler's first letter. In his second communication Butler says: "I do not conceive that the idea of ubiquity is contained in the idea of self-existence, or *directly follows from it*, any otherwise than as whatever exists must exist *somewhere*." That is to say, necessary existence implies existence in some part of space, but not in every part. It does not appear that Dr. Clarke was ever able to dispose effectually of this objection. The whole of the correspondence is extremely curious and interesting. The objections of Butler are precisely those which would occur to an acute mind impressed with the conviction, that upon the sifting of first principles, rather than upon any mechanical dexterity of reasoning, the successful investigation of truth mainly depends. And the replies of Dr. Clarke, although they cannot be admitted as satisfactory, evince, in a remarkable degree, that peculiar intellectual power which is manifest in the work from which the discussion arose.

12. In Prop. VIII. it is argued that the self-existent and original cause of all things must be an Intelligent Being.

The main argument adduced in support of this proposition is, that as the cause is more excellent than the effect, the self-existent being, as the cause and original of all things, must contain in itself the perfections of all things; and that Intelligence is one of the perfections manifested in a part of the creation. It is further argued that this perfection is not a modification of figure, divisibility, or any of the known properties of matter; for these are not perfections, but *limitations*. To this is added the *à posteriori* argument from the manifestation of design in the frame of the universe.

There is appended, however, a distinct argument for the existence of an intelligent self-existent being, founded upon the phænomenal existence of motion in the universe. I shall briefly exhibit this proof, and shall apply to it the method of the present treatise.

The argument, omitting unimportant explanations, is as follows:—

"'Tis evident there is some such a thing as motion in the world; which either began at some time or other, or was eternal. If it began in time, then the question is granted that the first cause is an intelligent being. . . . On the contrary, if motion was eternal, either it was eternally caused by some eternal intelligent being, or it must of itself be necessary and self-existent, or else, without any necessity in its own nature, and without any external necessary cause, it must have existed from eternity by an endless successive communication. If motion was eternally caused by some eternal intelligent being, this also is granting the question as to the present dispute. If it was of itself necessary and self-existent, then it follows that it must be a contradiction in terms to suppose any matter to be at rest. And yet, at the same time, because the determination of this self-existent motion must be every way at once, the effect of it would be nothing else but a perpetual rest. . . . But if it be said that motion, without any necessity in its own nature, and without any external necessary cause, has existed from eternity merely by an endless successive communication, as Spinoza inconsistently enough seems to assert, this I have before shown (in the proof of the second general proposition of this discourse) to be a plain contradiction. It remains, therefore, that motion must of necessity be originally caused by something that is intelligent."

The premises of the above argument may be thus disposed:

1. If motion began in time, the first cause is an intelligent being.

2. If motion has existed from eternity, either it has been eternally caused by some eternal intelligent being, or it is self-existent, or it must have existed by endless successive communication.

3. If motion has been eternally caused by an eternal intelligent being, the first cause is an intelligent being.

4. If it is self-existent, matter is at rest and not at rest.

5. That motion has existed by endless successive communication, and that at the same time it is not self-existent, and has not been eternally caused by some eternal intelligent being, is false.

To express these propositions, let us assume—

x = Motion began in time (and therefore)

$1 - x$ = Motion has existed from eternity.

y = The first cause is an intelligent being.

p = Motion has been eternally caused by some eternal intelligent being.

q = Motion is self-existent.

r = Motion has existed by endless successive communication.

s = Matter is at rest.

The equations of the premises then are—

$$x = vy.$$

$$1 - x = v\{p(1-q)(1-r) + q(1-p)(1-r) + r(1-p)(1-q)\}.$$

$$p = vy.$$

$$q = vs(1-s) = 0.$$

$$r(1-q)(1-p) = 0.$$

Since, by the fourth equation, $q = 0$, we obtain, on substituting for q its value in the remaining equations, the system

$$x = vy, \qquad 1 - x = v\{p(1-r) + r(1-p)\},$$

$$p = vy, \qquad r(1-p) = 0,$$

from which eliminating the indefinite symbols v, we have the final reduced system,

$$x(1-y) = 0, \quad (1)$$

$$(1-x)\{pr + (1-p)(1-r)\} = 0, \quad (2)$$

$$p(1-y) = 0. \quad (3)$$

$$r(1-p) = 0. \quad (4)$$

We shall first seek the value of y, the symbol involved in Dr. Clarke's conclusion. First, eliminating x from (1) and (2), we have

$$(1-y)\{pr+(1-p)(1-r)\}=0. \tag{5}$$

Next, to eliminate r from (4) and (5), we have

$$r(1-p)+(1-y)\{pr+(1-p)(1-r)\}=0,$$

$$\therefore \{1-p+(1-y)p\}\times(1-y)(1-p)=0;$$

whence

$$(1-y)(1-p)=0. \tag{6}$$

Lastly, eliminating p from (3) and (6), we have

$$1-y=0,$$
$$\therefore\ y=1,$$

which expresses the required conclusion, *The first cause is an intelligent being.*

Let us now examine what other conclusions are deducible from the premises.

If we substitute the value just found for y in the equations (1), (2), (3), (4), they are reduced to the following pair of equations, viz.,

$$(1-x)\{pr+(1-p)(1-r)\}=0, \quad r(1-p)=0. \tag{7}$$

Eliminating from these equations x, we have

$$r(1-p)=0, \text{ whence } r=vp,$$

which expresses the conclusion, *If motion has existed by endless successive communication, it has been eternally caused by an eternal intelligent being.*

Again eliminating, from the given pair, r, we have

$$(1-x)(1-p)=0,$$

or,

$$1-x=vp,$$

which expresses the conclusion, *If motion has existed from eternity, it has been eternally caused by some eternal intelligent being.*

Lastly, from the same original pair eliminating p, we get

$$(1-x)r=0,$$

which, solved in the form

$$1-x=v(1-r),$$

gives the conclusion, *If motion has existed from eternity, it has not existed by an endless successive communication.*

Solved under the form

$$r = vx,$$

the above equation leads to the equivalent conclusion, *If motion exists by an endless successive communication, it began in time.*

13. Now it will appear to the reader that the first and last of the above four conclusions are inconsistent with each other. The two consequences drawn from the hypothesis that motion exists by an endless successive communication, viz., 1st, that it has been eternally caused by an eternal intelligent being; 2ndly, that it began in time,—are plainly at variance. Nevertheless, they are both rigorous deductions from the original premises. The opposition between them is not of a *logical,* but of what is technically termed a *material,* character. This opposition might, however, have been formally stated in the premises. We might have added to them a formal proposition, asserting that "whatever is *eternally* caused by an eternal intelligent being, does not begin in time." Had this been done, no such opposition as now appears in our conclusions could have presented itself. Formal logic can only take account of relations which are formally expressed (VI. 16); and it may thus, in particular instances, become necessary to express, in a formal manner, some connexion among the premises which, without actual statement, is involved in the very meaning of the language employed.

To illustrate what has been said, let us add to the equations (2) and (4) the equation

$$px = 0,$$

which expresses the condition above adverted to. We have

$$(1 - x) \{pr + (1 - p) (1 - r)\} + r (1 - p) + px = 0. \qquad (8)$$

Eliminating p from this, we find simply

$$r = 0,$$

which expresses the proposition, *Motion does not exist by an endless successive communication.* If now we substitute for r its value in (8), we have

$$(1 - x) (1 - p) + px = 0, \text{ or, } 1 - x = p;$$

whence we have the interpretation, *If motion has existed from eternity, it has been eternally caused by an eternal intelligent being;* together with the converse of that proposition.

In Prop. IX. it is argued, that "the self-existent and original cause of all things is not a necessary agent, but a being endued with liberty and choice." The proof is based mainly upon his possession of intelligence, and upon the existence of final causes, implying design and choice. To the objection that the supreme cause operates by necessity for the production of what is best, it is replied, that this is a necessity of fitness and wisdom, and not of nature.

14. In Prop. x. it is argued, that "the self-existent being, the supreme cause of all things, must of necessity have infinite power." The ground of the demonstration is, that as "all the powers of all things are derived from him, nothing can make any difficulty or resistance to the execution of his will." It is defined that the infinite power of the self-existent being does not extend to the "making of a thing which implies a contradiction," or the doing of that "which would imply imperfection (whether natural or moral) in the being to whom such power is ascribed," but that it does extend to the creation of matter, and of an immaterial, cogitative substance, endued with a power of beginning motion, and with a liberty of will or choice. Upon this doctrine of liberty it is contended that we are able to give a satisfactory answer to "that ancient and great question, πόθεν τὸ κακὸν, what is the cause and original of evil?" The argument on this head I shall briefly exhibit.

"All that we call evil is either an evil of imperfection, as the want of certain faculties or excellencies which other creatures have; or natural evil, as pain, death, and the like; or moral evil, as all kinds of vice. The first of these is not properly an evil; for every power, faculty, or perfection, which any creature enjoys, being the free gift of God, . . it is plain the want of any certain faculty or perfection in any kind of creatures, which never belonged to their natures is no more an evil to them, than their never having been created or brought into being at all could properly have been called an evil. The second kind of evil, which we call natural evil, is either a necessary consequence of the former, as death to a creature on whose nature immortality was never conferred; and then it is no more properly an evil than the former. Or else it is counterpoised on the whole with as great or greater good, as the afflictions and sufferings of good men, and then also it is not properly an evil; or else, lastly, it is a punishment, and then it is a necessary consequence of the third and last kind of evil, viz., moral evil. And this arises wholly from the abuse of liberty which God gave to His creatures for other purposes, and which it was reasonable and fit to give them for the perfection and order of the whole creation. Only they, contrary to God's intention and command, have abused what was necessary to the perfection of the whole, to the corruption and depravation of themselves. And thus all sorts of evils have entered into the world without any diminution to the infinite goodness of the Creator and Governor thereof."—p. 112.

The main premises of the above argument may be thus stated:

1st. All reputed evil is either evil of imperfection, or natural evil, or moral evil.

2nd. Evil of imperfection is not absolute evil.

3rd. Natural evil is either a consequence of evil of imperfection, or it is compensated with greater good, or it is a consequence of moral evil.

4th. That which is either a consequence of evil of imperfection, or is compensated with greater good, is not absolute evil.

5th. All absolute evils are included in reputed evils.

To express these premises let us assume—

w = reputed evil.
x = evil of imperfection.
y = natural evil.
z = moral evil.
p = consequence of evil of imperfection.
q = compensated with greater good.
r = consequence of moral evil.
t = absolute evil.

Then, regarding the premises as Primary Propositions, of which all the predicates are particular, and the conjunctions *either*, *or*, as absolutely disjunctive, we have the following equations:

$$w = v\{x(1-y)(1-q) + y(1-x)(1-z) + z(1-x)(1-y)\}$$

$$x = v(1-t).$$

$$y = v\{p(1-q)(1-r) + q(1-p)(1-r) + r(1-p)(1-q)\}$$

$$p(1-q) + q(1-p) = v(1-t).$$

$$t = vw.$$

From which, if we separately eliminate the symbol v, we have

$$w\{1 - x(1-y)(1-z) - y(1-x)(1-z) - z(1-x)(1-y)\} = 0, \quad (1)$$

$$xt = 0, \quad (2)$$

$$y\{1 - p(1-q)(1-r) - q(1-p)(1-r) - r(1-p)(1-q)\} = 0, \quad (3)$$

$$\{p(1-q) + q(1-p)\}t = 0, \quad (4)$$

$$t(1-w) = 0. \quad (5)$$

Let it be required, first, to find what conclusion the premises warrant us in forming respecting absolute evils, as concerns their dependence upon moral evils, and the consequences of moral evils.

For this purpose we must determine t in terms of z and r. The symbols w, x, y, p, q must therefore be eliminated. The process is easy, as any set of the equations is reducible to a single equation by addition.

Eliminating w from (1) and (5), we have

$$t\{1 - x(1-y)(1-z) - y(1-x)(1-z) - z(1-x)(1-y)\} = 0. \quad (6)$$

The elimination of p from (3) and (4) gives

$$yqr + yqt + yt(1-r)(1-q) = 0. \quad (7)$$

The elimination of q from this gives

$$yt(1-r) = 0. \quad (8)$$

The elimination of x between (2) and (6) gives

$$t\{yz + (1-y)(1-z)\} = 0. \quad (9)$$

The elimination of y from (8) and (9) gives

$$t(1-z)(1-r) = 0.$$

This is the only relation existing between the elements t, z, and r.

We hence get

$$t = \frac{0}{(1-z)(1-r)}$$

$$= \frac{0}{0} zr + \frac{0}{0} z(1-r) + \frac{0}{0}(1-z)r + 0(1-z)(1-r)$$

$$= \frac{0}{0} z + \frac{0}{0}(1-z)r;$$

the interpretation of which is, *Absolute evil is either moral evil, or it is, if not moral evil, a consequence of moral evil.*

Any of the results obtained in the process of the above solution furnish us with interpretations. Thus from (8) we might deduce

$$t = \frac{0}{y(1-r)} = \frac{0}{0} yr + \frac{0}{0}(1-y)r + \frac{0}{0}(1-y)(1-r)$$

$$= \frac{0}{0} yr + \frac{0}{0}(1-y);$$

whence, *Absolute evils are either natural evils, which are the consequences of moral evils, or they are not natural evils at all.*

A variety of other conclusions may be deduced from the given equations in reply to questions which may be arbitrarily proposed. Of such I shall give a few examples, without exhibiting the intermediate processes of solution.

Quest. 1.—Can any relation be deduced from the premises connecting the following elements, viz.: absolute evils, consequences of evils of imperfection, evils compensated with greater good?

Ans.—*No relation exists.* If we eliminate all the symbols but z, p, q, the result is $0 = 0$.

Quest. 2.—Is any relation implied between absolute evils, evils of imperfection, and consequences of evils of imperfection.

Ans.—The final relation between x, t, and p is

$$xt + pt = 0;$$

whence

$$t = \frac{0}{p+x} = \frac{0}{0}(1-p)(1-x).$$

Therefore, *Absolute evils are neither evils of imperfection, nor consequences of evils of imperfection.*

Quest. 3.—Required the relation of natural evils to evils of imperfection and evils compensated with greater good.

We find

$$pqy = 0,$$

$$\therefore y = \frac{0}{pq} = \frac{0}{0} p (1 - q) + \frac{0}{0} (1 - p).$$

Therefore, *Natural evils are either consequences of evils of imperfection which are not compensated with greater good, or they are not consequences of evils of imperfection at all.*

Quest. 4.—In what relation do those natural evils which are not moral evils stand to absolute evils and the consequences of moral evils?

If $y (1 - z) = s$, we find, after elimination,

$$ts (1 - r) = 0;$$

$$\therefore s = \frac{0}{t (1 - r)} = \frac{0}{0} tr + \frac{0}{0} (1 - t).$$

Therefore, *Natural evils, which are not moral evils, are either absolute evils, which are the consequences of moral evils, or they are not absolute evils at all.*

The following conclusions have been deduced in a similar manner. The subject of each conclusion will show of what particular things a description was required, and the predicate will show what elements it was designed to involve:—

Absolute evils, which are not consequences of moral evils, are moral and not natural evils.

Absolute evils which are not moral evils are natural evils, which are the consequences of moral evils.

Natural evils which are not consequences of moral evils are not absolute evils.

Lastly, let us seek a description of evils which are not absolute, expressed in terms of natural and moral evils.

We obtain as the final equation,

$$1 - t = yz + \frac{0}{0} y (1 - z) + \frac{0}{0} (1 - y) z + (1 - y) (1 - z).$$

The direct interpretation of this equation is a necessary truth, but the reverse interpretation is remarkable. *Evils which are both natural and moral, and evils which are neither natural nor moral, are not absolute evils.*

This conclusion, though it may not express a truth, is certainly involved in the given premises, as *formally* stated.

15. Let us take from the same argument a somewhat fuller system of premises, and let us in those premises suppose that the particles, *either*, *or*, are not absolutely disjunctive, so that in the meaning of the expression, "either evil of imperfection, or natural evil, or moral evil," we include whatever possesses one or more of these qualities.

Let the premises be—

1. All evil (w) is either evil of imperfection (x), or natural evil (y), or moral evil (z).
2. Evil of imperfection (x) is not absolute evil (t).
3. Natural evil (y) is either a consequence of evil of imperfection (p), or it is compensated with greater good (q), or it is a consequence of moral evil (r).
4. Whatever is a consequence of evil of imperfection (p) is not absolute evil (t).
5. Whatever is compensated with greater good (q) is not absolute evil (t).
6. Moral evil (z) is a consequence of the abuse of liberty (u).
7. That which is a consequence of moral evil (r) is a consequence of the abuse of liberty (u).
8. Absolute evils are included in reputed evils.

The premises expressed in the usual way give, after the elimination of the indefinite symbols v, the following equations:

$$w(1-x)(1-y)(1-z)=0, \quad (1)$$

$$xt=0, \quad (2)$$

$$y(1-p)(1-q)(1-r)=0, \quad (3)$$

$$pt=0, \quad (4)$$

$$qt=0, \quad (5)$$

$$z(1-u)=0, \quad (6)$$

$$r(1-u)=0, \quad (7)$$

$$t(1-w)=0. \quad (8)$$

Each of these equations satisfies the condition $V(1-V)=0$.

The following results are easily deduced—

Natural evil is either absolute evil, which is a consequence of moral evil, or it is not absolute evil at all.

All evils are either absolute evils, which are consequences of the abuse of liberty, or they are not absolute evils.

Natural evils are either evils of imperfection, which are not absolute evils, or they are not evils of imperfection at all.

Absolute evils are either natural evils, which are consequences of the abuse of liberty, or they are not natural evils, and at the same time not evils of imperfection.

Consequences of the abuse of liberty include all natural evils which are absolute evils, and are not evils of imperfection, with an indefinite remainder of natural evils which are not absolute, and of evils which are not natural.

16. These examples will suffice for illustration. The reader can easily supply others if they are needed. We proceed now to examine the most essential portions of the demonstration of Spinoza.

DEFINITIONS.

1. By a *cause of itself (causa sui)*, I understand that of which the essence involves existence, or that of which the nature cannot be conceived except as existing.

2. That thing is said to be finite or bounded in its own kind (*in suo genere finita*) which may be bounded by another thing of the same kind; e. g. Body is said to be finite, because we can always conceive of another body greater than a given one. So thought is bounded by other thought. But body is not bounded by thought, nor thought by body.

3. By substance, I understand that which is in itself (*in se*), and is conceived by itself (*per se concipitur*), i. e., that whose conception does not require to be formed from the conception of another thing.

4. By attribute, I understand that which the intellect perceives in substance, as constituting its very essence.

5. By mode, I understand the affections of substance, or that which is in another thing, by which thing also it is conceived.

6. By God, I understand the Being absolutely infinite, that

is the substance consisting of infinite attributes, each of which expresses an eternal and infinite essence.

Explanation.—I say absolutely infinite, not infinite in its own kind. For to whatever is only infinite in its own kind we may deny the possession of (some) infinite attributes. But when a thing is absolutely infinite, whatsoever expresses essence and involves no negation belongs to its essence.

7. That thing is termed *free*, which exists by the sole necessity of its own nature, and is determined to action by itself alone; *necessary*, or rather constrained, which is determined by another thing to existence and action, in a certain and determinate manner.

8. By eternity, I understand existence itself, in so far as it is conceived necessarily to follow from the sole definition of the eternal thing.

Explanation.—For such existence, as an eternal truth, is conceived as the essence of the thing, and therefore cannot be explained by mere duration or time, though the latter should be conceived as without beginning and without end.

AXIOMS.

1. All things which exist are either in themselves (*in se*) or in another thing.

2. That which cannot be conceived by another thing ought to be conceived by itself.

3. From a given determinate cause the effect necessarily follows, and, contrariwise, if no determinate cause be granted, it is impossible that an effect should follow.

4. The knowledge of the effect depends upon, and involves, the knowledge of the cause.

5. Things which have nothing in common cannot be understood by means of each other; or the conception of the one does not involve the conception of the other.

6. A true idea ought to agree with its own object. (*Idea vera debet cum suo ideato convenire.*)

7. Whatever can be conceived as non-existing does not involve existence in its essence.

Other definitions are implied, and other axioms are virtually assumed, in some of the demonstrations. Thus, in Prop. I., "Substance is prior in nature to its affections," the proof of which consists in a mere reference to Defs. 3 and 5, there seems to be an assumption of the following axiom, viz., "That by which a thing is conceived is prior in nature to the thing conceived." Again, in the demonstration of Prop. V. the converse of this axiom is assumed to be true. Many other examples of the same kind occur. It is impossible, therefore, by the mere processes of Logic, to deduce the whole of the conclusions of the first book of the Ethics from the axioms and definitions which are prefixed to it, and which are given above. In the brief analysis which will follow, I shall endeavour to present in their proper order what appear to me to be the real premises, whether formally stated or implied, and shall show in what manner they involve the conclusions to which Spinoza was led.

17. I conceive, then, that in the course of his demonstration, Spinoza effects several parallel divisions of the universe of possible existence, as,

1st. Into things which are in themselves, x, and things which are in some other thing, x'; whence, as these classes of thing together make up the universe, we have

$$x + x' = 1; \quad \text{(Ax. I.)}$$

or,

$$x = 1 - x'.$$

2nd. Into things which are conceived by themselves, y, and things which are conceived through some other thing, y'; whence

$$y = 1 - y'. \quad \text{(Ax. II.)}$$

3rd. Into substance, z, and modes, z'; whence

$$z = 1 - z'. \quad \text{(Def. III. V.)}$$

4th. Into things free, f, and things necessary, f'; whence

$$f = 1 - f'. \quad \text{(Def. VII.)}$$

5th. Into things which are causes and self-existent, e, and things caused by some other thing, e'; whence

$$e = 1 - e'. \quad \text{(Def. I. Ax. VII.)}$$

And his reasoning proceeds upon the expressed or assumed principle, that these divisions are not only parallel, but equivalent. Thus in Def. III., Substance is made equivalent with that which is conceived by itself; whence

$$z = y.$$

Again, Ax. IV., as it is actually applied by Spinoza, establishes the identity of cause with that by which a thing is conceived; whence

$$y = e.$$

Again, in Def. VII., things free are identified with things self-existent; whence

$$f = e.$$

Lastly, in Def. V., mode is made identical with that which is in another thing; whence $z' = x'$, and therefore,

$$z = x.$$

All these results may be collected together into the following series of equations, viz.:

$$x = y = z = f = e = 1 - x' = 1 - y' = 1 - f' = 1 - z' = 1 - e'.$$

And any two members of this series connected together by the sign of equality express a conclusion, whether drawn by Spinoza or not, which is a legitimate consequence of his system. Thus the equation

$$z = 1 - e',$$

expresses the sixth proposition of his system, viz., One substance cannot be produced by another. Similarly the equation

$$z = e,$$

expresses his seventh proposition, viz., "It pertains to the nature of substance to exist." This train of deduction it is unnecessary to pursue. Spinoza applies it chiefly to the deduction according to his views of the properties of the Divine Nature, having first endeavoured to prove that the only substance is God. In the steps of this process, there appear to me to exist some fallacies, dependent chiefly upon the ambiguous use of words, to which it will be necessary here to direct attention.

18. In Prop. v. it is endeavoured to show, that "There cannot exist two or more substances of the same nature or attribute." The proof is virtually as follows: If there are more substances than one, they are distinguished either by attributes or modes; if by attributes, then there is only one substance of the same attribute; if by modes, then, laying aside these as non-essential, there remains no *real* ground of distinction. Hence there exists but one substance of the same attribute. The assumptions here involved are inconsistent with those which are found in other parts of the treatise. Thus substance, Def. IV., is apprehended by the intellect through the means of attribute. By Def. VI. it may have many attributes. One substance may, therefore, *conceivably* be distinguished from another by a difference in some of its attributes, while others remain the same.

In Prop. VIII. it is attempted to show that, All substance is necessarily infinite. The proof is as follows. There exists but one substance, of one attribute, Prop. v.; and it pertains to its nature to exist, Prop. VII. It will, therefore, be of its nature to exist either as finite or infinite. But not as finite, for, by Def. II. it would require to be bounded by another substance of the same nature, which also ought to exist *necessarily*, Prop. VII. Therefore, there would be two substances of the same attribute, which is absurd, Prop. v. Substance, therefore, is infinite.

In this demonstration the word "finite" is confounded with the expression, "Finite in its own kind," Def. II. It is thus assumed that nothing can be finite, unless it is bounded by another thing of the same kind. This is not consistent with the ordinary meaning of the term. Spinoza's use of the term finite tends to make space the only form of substance, and all existing things but affections of space, and this, I think, is really one of the ultimate foundations of his system.

The first scholium applied to the above Proposition is remarkable. I give it in the original words: "Quum finitum esse revera sit ex parte negatio, et infinitum absoluta affirmatio existentiæ alicujus naturæ, sequitur ergo ex sola Prop. VII. omnem substantiam debere esse infinitam." Now this is in reality an assertion of the principle affirmed by Clarke, and controverted by

Butler (XIII. 11), that necessary existence implies existence in every part of space. Probably this principle will be found to lie at the basis of every attempt to demonstrate, *à priori*, the existence of an Infinite Being.

From the general properties of substance above stated, and the definition of God as the substance consisting of infinite attributes, the peculiar doctrines of Spinoza relating to the Divine Nature necessarily follow. As substance is self-existent, free, causal in its very nature, the thing in which other things are, and by which they are conceived; the same properties are also asserted of the Deity. He is self-existent, Prop. XI.; indivisible, Prop. XIII.; the only substance, Prop. XIV.; the Being in which all things are, and by which all things are conceived, Prop. XV.; free, Prop. XVII.; the immanent cause of all things, Prop. XVIII. The proof that God is the only substance is drawn from Def. VI., which is interpreted into a declaration that "God is the Being absolutely infinite, of whom no attribute which expresses the essence of substance can be denied." Every conceivable attribute being thus assigned by definition to Him, and it being determined in Prop. V. that there cannot exist two substances of the same attribute, it follows that God is the only substance.

Though the "Ethics" of Spinoza, like a large portion of his other writings, is presented in the geometrical form, it does not afford a good praxis for the symbolical method of this work. Of course every train of reasoning admits, when its ultimate premises are truly determined, of being treated by that method; but in the present instance, such treatment scarcely differs, except in the use of letters for words, from the processes employed in the original demonstrations. Reasoning which consists so largely of a play upon terms defined as equivalent, is not often met with; and it is rather on account of the interest attaching to the subject, than of the merits of the demonstrations, highly as by some they are esteemed, that I have devoted a few pages here to their exposition.

19. It is not possible, I think, to rise from the perusal of the arguments of Clarke and Spinoza without a deep conviction of the futility of all endeavours to establish, entirely *à priori*, the existence

of an Infinite Being, His attributes, and His relation to the universe. The fundamental principle of all such speculations, viz., that whatever we can clearly conceive, must exist, fails to accomplish its end, even when its truth is admitted. For how shall the finite comprehend the infinite? Yet must the possibility of such conception be granted, and in something more than the sense of a mere withdrawal of the limits of phænomenal existence, before any solid ground can be established for the knowledge, *à priori*, of things infinite and eternal. Spinoza's affirmation of the reality of such knowledge is plain and explicit: "Mens humana adæquatum habet cognitionem æternæ et infinitæ essentiæ Deï" (Prop. XLVII., Part 2nd). Let this be compared with Prop. XXXIV., Part 2nd: "Omnis idea quae in nobis est absoluta sive adæquata et perfecta, vera est;" and with Axiom VI., Part 1st, "Idea vera debet cum suo ideato convenire." Moreover, this species of knowledge is made the essential constituent of all other knowledge: "De natura rationis est res sub quadam æternitatis specie percipere" (Prop. XLIV., Cor. II., Part 2nd). Were it said, that there is a tendency in the human mind to rise in contemplation from the particular towards the universal, from the finite towards the infinite, from the transient towards the eternal; and that this tendency suggests to us, with high probability, the existence of more than sense perceives or understanding comprehends; the statement might be accepted as true for at least a a large number of minds. There is, however, a class of speculations, the character of which must be explained in part by reference to other causes,—impatience of probable or limited knowledge, so often all that we can really attain to; a desire for absolute certainty where intimations sufficient to mark out before us the path of duty, but not to satisfy the demands of the speculative intellect, have alone been granted to us; perhaps, too, dissatisfaction with the present scene of things. With the undue predominance of these motives, the more sober procedure of analogy and probable induction falls into neglect. Yet the latter is, beyond all question, the course most adapted to our present condition. To infer the existence of an intelligent cause from the teeming evidences of surrounding design, to rise to the conception of a moral Governor of the world, from the study of the constitution and the moral provisions of our own nature;—these, though but the feeble steps of an understanding limited in its faculties and its materials of knowledge, are of more avail than the ambitious attempt to arrive at a certainty unattainable on the ground of natural religion. And as these were the most ancient, so are they still the most solid foundations, Revelation being set apart, of the belief that the course of this world is not abandoned to chance and inexorable fate.

CHAPTER XIV.

EXAMPLE OF THE ANALYSIS OF A SYSTEM OF EQUATIONS BY THE METHOD OF REDUCTION TO A SINGLE EQUIVALENT EQUATION $V = 0$, WHEREIN V SATISFIES THE CONDITION $V(1 - V) = 0$.

1. LET us take the remarkable system of premises employed in the previous Chapter, to prove that "Matter is not a necessary being;" and suppressing the 6th premiss, viz., Motion exists,—examine some of the consequences which flow from the remaining premises. This is in reality to accept as true Dr. Clarke's hypothetical principles; but to suppose ourselves ignonorant of the fact of the existence of motion. Instances may occur in which such a selection of a portion of the premises of an argument may lead to interesting consequences, though it is with other views that the present example has been resumed. The premises actually employed will be—

1. If matter is a necessary being, either the property of gravitation is necessarily present, or it is necessarily absent.

2. If gravitation is necessarily absent, and the world is not subject to any presiding intelligence, motion does not exist.

3. If gravitation is necessarily present, a vacuum is necessary.

4. If a vacuum is necessary, matter is not a necessary being.

5. If matter is a necessary being, the world is not subject to a presiding intelligence.

If, as before, we represent the elementary propositions by the following notation, viz.:

x = Matter is a necessary being.
y = Gravitation is necessarily present.
w = Motion exists.
t = Gravitation is necessarily absent.
z = The world is merely material, and not subject to a presiding intelligence.
v = A vacuum is necessary.

We shall on expression of the premises and elimination of the indefinite class symbols (q), obtain the following system of equations:

$$xyt + x\bar{y}\bar{t} = 0,$$
$$tzw = 0,$$
$$y\bar{v} = 0,$$
$$vx = 0,$$
$$x\bar{z} = 0;$$

in which for brevity $\bar{y}$ stands for $1 - y$, $\bar{t}$ for $1 - t$, and so on; whence, also, $1 - \bar{t} = t$, $1 - \bar{y} = y$, &c.

As the first members of these equations involve only positive terms, we can form a single equation by adding them together (VIII. Prop. 2), viz.:

$$xyt + x\bar{y}\bar{t} + y\bar{v} + vx + x\bar{z} + tzw = 0,$$

and it remains to reduce the first member so as to cause it to satisfy the condition $V(1 - V) = 0$.

For this purpose we will first obtain its development with reference to the symbols x and y. The result is—

$$(t + \bar{v} + v + \bar{z} + tzw)\, xy + (\bar{t} + v + \bar{z} + tzw)\, x\bar{y}$$
$$+ (\bar{v} + tzw)\, \bar{x}y + tzw\bar{x}\bar{y} = 0.$$

And our object will be accomplished by reducing the four coefficients of the development to equivalent forms, themselves satisfying the condition required.

Now the first coefficient is, since $v + \bar{v} = 1$,

$$1 + t + \bar{z} + tzw,$$

which reduces to unity (IX. Prop. 1).

The second coefficient is

$$\bar{t} + v + \bar{z} + tzw;$$

and its reduced form (X. 3) is

$$\bar{t} + tv + t\bar{v}\bar{z} + t\bar{v}zw.$$

The third coefficient, $\bar{v} + tzw$, reduces by the same method to $\bar{v} + tzwv$; and the last coefficient tzw needs no reduction. Hence the development becomes

$$xy + (\bar{t} + tv + t\bar{v}\bar{z} + t\bar{v}zw)\,x\bar{y} + (\bar{v} + tzwv)\,\bar{x}y + tzw\bar{x}\bar{y} = 0; \quad (1)$$

and this is the form of reduction sought.

2. Now according to the principle asserted in Prop. III., Chap. X., the whole relation connecting any particular set of the symbols in the above equation may be deduced by developing that equation with reference to the particular symbols in question, and retaining in the result only those constituents whose coefficients are unity. Thus, if x and y are the symbols chosen, we are immediately conducted to the equation

$$xy = 0,$$

whence we have

$$y = \frac{0}{0}(1 - x),$$

with the interpretation, *If gravitation is necessarily present, matter is not a necessary being.*

Let us next seek the relation between x and w. Developing (1) with respect to those symbols, we get

$$(y + \bar{t}\bar{y} + tv\bar{y} + t\bar{v}\bar{z}\bar{y} + t\bar{v}z\bar{y})\,xw + (y + \bar{t}\bar{y} + tv\bar{y} + t\bar{v}\bar{z}\bar{y})\,x\bar{w}$$
$$+ (vy + tzvy + tz\bar{y})\,\bar{x}w + \bar{v}y\bar{x}\bar{w} = 0.$$

The coefficient of xw, and it alone, reduces to unity. For $t\bar{v}\bar{z}\bar{y} + t\bar{v}z\bar{y} = t\bar{v}\bar{y}$, and $tv\bar{y} + t\bar{v}\bar{y} = t\bar{y}$, and $\bar{t}\bar{y} + t\bar{y} = \bar{y}$, and lastly, $y + \bar{y} = 1$. This is always the mode in which such reductions take place. Hence we get

$$xw = 0,$$

$$\therefore w = \frac{0}{0}(1 - x),$$

of which the interpretation is, *If motion exists, matter is not a necessary being.*

If, in like manner, we develop (1) with respect to x and z, we get the equation

$$x\bar{z} = 0,$$

$$\therefore x = \frac{0}{0}z,$$

with the interpretation, *If matter is a necessary being, the world is merely material, and without a presiding intelligence.*

This, indeed, is only the fifth premiss reproduced, but it shows that there is no other relation connecting the two elements which it involves.

If we seek the whole relation connecting the elements x, w, and y, we find, on developing (1) with reference to those symbols, and proceeding as before,

$$xy + xw\bar{y} = 0.$$

Suppose it required to determine hence the consequences of the hypothesis, "Motion does not exist," relatively to the questions of the necessity of matter, and the necessary presence of gravitation. We find

$$w = \frac{-xy}{x\bar{y}},$$

$$\therefore 1 - w = \frac{x}{x\bar{y}} = \frac{1}{0}xy + x\bar{y} + \frac{0}{0}\bar{x};$$

or,

$$1 - w = x\bar{y} + \frac{0}{0}\bar{x}, \text{ with } xy = 0.$$

The direct interpretation of the first equation is, *If motion does not exist, either matter is a necessary being, and gravitation is not necessarily present, or matter is not a necessary being.*

The reverse interpretation is, *If matter is a necessary being, and gravitation not necessary, motion does not exist.*

In exactly the same mode, if we sought the full relation between x, z, and w, we should find

$$xzw + x\bar{z} = 0.$$

From this we may deduce

$$z = x\bar{w} + \frac{0}{0}\bar{x}, \text{ with } xw = 0.$$

Therefore, *If the world is merely material, and not subject to any presiding intelligence, either matter is a necessary being, and motion does not exist, or matter is not a necessary being.*

Also, *reversely, If matter is a necessary being, and there is no such thing as motion, the world is merely material.*

3. We might, of course, extend the same method to the de-

termination of the consequences of any complex hypothesis u, such as, "The world is merely material, and without any presiding intelligence (z), but motion exists" (w), with reference to any other elements of doubt or speculation involved in the original premises, such as, "Matter is a necessary being" (x), "Gravitation is a necessary quality of matter," (y). We should, for this purpose, connect with the general equation (1) a new equation,

$$u = wz,$$

reduce the system thus formed to a single equation, $V = 0$, in which V satisfies the condition $V(1 - V) = 0$, and proceed as above to determine the relation between u, x, and y, and finally u as a developed function of x and y. But it is very much better to adopt the methods of Chapters VIII. and IX. I shall here simply indicate a few results, with the leading steps of their deduction, and leave their verification to the reader's choice.

In the problem last mentioned we find, as the relation connecting x, y, w, and z,

$$xw + x\bar{w}y + x\bar{w}\bar{y}\bar{z} = 0.$$

And if we write $u = xy$, and then eliminate the symbols x and y by the general problem, Chap. IX., we find

$$xu + xy\bar{u} = 0,$$

whence

$$u = \frac{1}{0}xy + 0x\bar{y} + \frac{0}{0}\bar{x};$$

wherefore

$$wz = \frac{0}{0}\bar{x} \text{ with } xy = 0.$$

Hence, *If the world is merely material, and without a presiding intelligence, and at the same time motion exists, matter is not a necessary being.*

Now it has before been shown that *if motion exists, matter is not a necessary being*, so that the above conclusion tells us even less than we had before ascertained to be (inferentially) true. Nevertheless, that conclusion is the proper and complete answer to the question which was proposed, which was, to determine simply the consequences of a certain complex hypothesis.

4. It would thus be easy, even from the limited system of premises before us, to deduce a great variety of additional inferences, involving, in the conditions which are given, any proposed combinations of the elementary propositions. If the condition is one which is inconsistent with the premises, the fact will be indicated by the form of the solution. The value which the method will assign to the combination of symbols expressive of the proposed condition will be 0. If, on the other hand, the fulfilment of the condition in question imposes no restriction upon the propositions among which relation is sought, so that every combination of those propositions is equally possible,—the fact will also be indicated by the form of the solution. Examples of each of these cases are subjoined.

If in the ordinary way we seek the consequences which would flow from the condition that *matter is a necessary being*, and at the same time that *motion exists*, as affecting the Propositions, *The world is merely material, and without a presiding intelligence*, and, *Gravitation is necessarily present*, we shall obtain the equation

$$xw = 0,$$

which indicates that the condition proposed is inconsistent with the premises, and therefore cannot be fulfilled.

If we seek the consequences which would flow from the condition that *Matter is not a necessary being*, and at the same time that *Motion does exist*, with reference to the same elements as above, viz., *the absence of a presiding intelligence*, and the *necessity of gravitation*,—we obtain the following result,

$$(1 - x)\,w = \frac{0}{0}\,yz + \frac{0}{0}\,y\,(1 - z) + \frac{0}{0}\,(1 - y)\,z + \frac{0}{0}\,(1 - y)\,(1 - z),$$

which might literally be interpreted as follows:

If matter is not a necessary being, and motion exists, then either the world is merely material and without a presiding intelligence, and gravitation is necessary, or one of these two results follows without the other, or they both fail of being true. Wherefore of the four possible combinations, of which some one is true of necessity, and of which of necessity one only can be true, it is affirmed that any one may be true. Such a result is a truism—a mere *necessary* truth. Still it contains the only answer which can be given to the question proposed.

I do not deem it necessary to vindicate against the charge of laborious trifling these applications. It may be requisite to enter with some fulness into details useless in themselves, in order to establish confidence in general principles and methods. When this end shall have been accomplished in the subject of the present inquiry, let all that has contributed to its attainment, but has afterwards been found superfluous, be forgotten.

CHAPTER XV.

THE ARISTOTELIAN LOGIC AND ITS MODERN EXTENSIONS, EXAMINED BY THE METHOD OF THIS TREATISE.

1. THE logical system of Aristotle, modified in its details, but unchanged in its essential features, occupies so important a place in academical education, that some account of its nature, and some brief discussion of the leading problems which it presents, seem to be called for in the present work. It is, I trust, in no narrow or harshly critical spirit that I approach this task. My object, indeed, is not to institute any direct comparison between the time-honoured system of the schools and that of the present treatise; but, setting truth above all other considerations, to endeavour to exhibit the real nature of the ancient doctrine, and to remove one or two prevailing misapprehensions respecting its extent and sufficiency.

That which may be regarded as essential in the spirit and procedure of the Aristotelian, and of all cognate systems of Logic, is the attempted classification of the allowable forms of inference, and the distinct reference of those forms, collectively or individually, to some general principle of an axiomatic nature, such as the "dictum of Aristotle:" Whatsoever is affirmed or denied of the genus may in the same sense be affirmed or denied of any species included under that genus. Concerning such general principles it may, I think, be observed, that they either state directly, but in an abstract form, the argument which they are supposed to elucidate, and, so stating that argument, affirm its validity; or involve in their expression technical terms which, after definition, conduct us again to the same point, viz., the abstract statement of the supposed allowable forms of inference. The idea of classification is thus a pervading element in those systems. Furthermore, they exhibit Logic as resolvable into two great branches, the one of which is occupied with the treatment of categorical, the other with that of hypothetical or conditional propositions. The distinction is nearly identical with that of primary and secondary propositions in the present work. The discussion of the theory of categorical propositions is, in all the ordinary treatises of Logic, much more full and elaborate than that of hypothetical propositions, and is occupied partly with ancient scholastic distinctions, partly with the canons of deductive inference. To the latter application only is it necessary to direct attention here.

2. Categorical propositions are classed under the four following heads, viz.:

		TYPE.
1st.	Universal affirmative Propositions:	All Y's are X's.
2nd.	Universal negative "	No Y's are X's.
3rd.	Particular affirmative "	Some Y's are X's.
4th.	Particular negative "	Some Y's are not X's.

To these forms, four others have recently been added, so as to constitute in the whole eight forms (see the next article) susceptible, however, of reduction to six, and subject to relations which have been discussed with great fulness and ability by Professor De Morgan, in his Formal Logic. A scheme somewhat different from the above has been given to the world by Sir W. Hamilton, and is made the basis of a method of syllogistic inference, which is spoken of with very high respect by authorities on the subject of Logic.*

The processes of Formal Logic, in relation to the above system of propositions, are described as of two kinds, viz., "Conversion" and "Syllogism." By Conversion is meant the expression of any proposition of the above kind in an equivalent form, but with a reversed order of terms. By Syllogism is meant the deduction from two such propositions having a common term, whether subject or predicate, of some third proposition inferentially involved in the two, and forming the "conclusion." It is maintained by most writers on Logic, that these processes, and according to some, the single process of Syllogism, furnish the universal types of reasoning, and that it is the business of the mind, in any train of demonstration, to conform itself, whether

* Thomson's Outlines of the Laws of Thought, p. 177.

consciously or unconsciously, to the particular models of the processes which have been classified in the writings of logicians.

3. The course which I design to pursue is to show how these processes of Syllogism and Conversion may be conducted in the most general manner upon the principles of the present treatise, and, viewing them thus in relation to a system of Logic, the foundations of which, it is conceived, have been laid in the ultimate laws of thought, to seek to determine their true place and essential character.

The expressions of the eight fundamental types of proposition in the language of symbols are as follows:

1. All Y's are X's,	$y = vx.$	
2. No Y's are X's,	$y = v(1-x).$	
3. Some Y's are X's,	$vy = vx.$	
4. Some Y's are not-X's,	$vy = v(1-x).$	
5. All not-Y's are X's,	$1-y = vx.$	(1)
6. No not-Y's are X's,	$1-y = v(1-x).$	
7. Some not-Y's are X's,	$v(1-y) = vx.$	
8. Some not-Y's are not-X's,	$v(1-y) = v(1-x).$	

In referring to these forms, it will be convenient to apply, in a sense shortly to be explained, the epithets of logical quantity, "universal" and "particular," and of quality, "affirmative" and "negative," to the terms of propositions, and not to the propositions themselves. We shall thus consider the term "All Y's," as universal-affirmative; the term "Y's," or "Some Y's," as particular-affirmative; the term "All not-Y's," as universal-negative; the term "Some not-Y's," as particular-negative. The expression "No Y's," is not properly a *term* of a proposition, for the true meaning of the proposition, "No Y's are X's," is "All Y's are not-X's." The subject of that proposition is, therefore, universal-affirmative, the predicate particular-negative. That there is a real distinction between the conceptions of "men" and "not men" is manifest. This distinction is all that I contemplate when applying as above the designations of affirmative and negative, without, however, insisting upon the etymological propriety of the application to the terms of propositions. The designations positive and privative would have been more appropriate, but the former term is already employed in a fixed sense in other parts of this work.

4. From the symbolical forms above given the laws of conversion immediately follow. Thus from the equation

$$y = vx,$$

representing the proposition, "All Y's are X's," we deduce, on eliminating v,

$$y(1 - x) = 0,$$

which gives by solution with reference to $1 - x$,

$$1 - x = \frac{0}{0}(1 - y);$$

the interpretation of which is,

All not-X's are not-Y's.

This is an example of what is called "negative conversion." In like manner, the equation

$$y = v(1 - x),$$

representing the proposition, "No Y's are X's," gives

$$x = \frac{0}{0}(1 - y),$$

the interpretation of which is, "No X's are Y's." This is an example of what is termed simple conversion; though it is in reality of the same kind as the conversion exhibited in the previous example. All the examples of conversion which have been noticed by logicians are either of the above kind, or of that which consists in the mere transposition of the terms of a proposition, without altering their quality, as when we change

$vy = vx$, representing, Some Y's are X's,

into

$vx = vy$, representing, Some X's are Y's;

or they involve a combination of those processes with some auxiliary process of limitation, as when from the equation

$y = vx$, representing, All Y's are X's,

we deduce on multiplication by v,

$vy = vx$, representing, Some Y's are X's,

and hence

$vx = vy$, representing, Some X's are Y's.

In this example, the process of limitation precedes that of transposition.

From these instances it is seen that conversion is a particular application of a much more general process in Logic, of which many examples have been given in this work. That process has for its object the determination of any element in any proposition, however complex, as a logical function of the remaining elements. Instead of confining our attention to the subject and predicate, regarded as simple terms, we can take any element or any combination of elements entering into either of them; make that element, or that combination, the "subject" of a new proposition; and determine what its predicate shall be, in accordance with the data afforded to us. It may be remarked, that even the simple forms of propositions enumerated above afford some ground for the application of such a method, beyond what the received laws of conversion appear to recognise. Thus the equation

$y = vx$, representing, All *Y*'s are *X*'s,

gives us, in addition to the proposition before deduced, the three following:

1st. $y(1-x) = 0$. There are no *Y*'s that are not-*X*'s.

2nd. $1 - y = \frac{0}{0}x + (1-x)$. Things that are not-*Y*'s include all things that are not-*X*'s, and an indefinite remainder of things that are *X*'s.

3rd. $x = y + \frac{0}{0}(1-y)$. Things that are *X*'s include all things that are *Y*'s, and an indefinite remainder of things that are not-*Y*'s.

These conclusions, it is true, merely place the given proposition in other and equivalent forms,—but such and no more is the office of the received mode of "negative conversion."

Furthermore, these processes of conversion are not elementary, but they are combinations of processes more simple than they, more immediately dependent upon the ultimate laws and axioms which govern the use of the symbolical instrument of reasoning. This remark is equally applicable to the case of Syllogism, which we proceed next to consider.

5. The nature of syllogism is best seen in the particular instance. Suppose that we have the propositions,

All X's are Y's,

All Y's are Z's.

From these we may deduce the conclusion,

All X's are Z's.

This is a syllogistic inference. The terms X and Z are called the extremes, and Y is called the middle term. The function of the syllogism generally may now be defined. Given two propositions of the kind whose species are tabulated in (1), and involving one middle or common term Y, which is connected in one of the propositions with an extreme X, in the other with an extreme Z; required the relation connecting the extremes X and Z. The term Y may appear in its affirmative form, as, All Y's, Some Y's; or in its negative form, as, All not-Y's, Some not-Y's; in either proposition, without regard to the particular form which it assumes in the other.

Nothing is easier than in particular instances to resolve the Syllogism by the method of this treatise. Its resolution is, indeed, a particular application of the process for the reduction of systems of propositions. Taking the examples above given, we have,

$$x = vy,$$

$$y = vz;$$

whence by substitution,

$$x = vv'z,$$

which is interpreted into

All X's are Z's.

Or, proceeding rigorously in accordance with the method developed in (VIII. 7), we deduce

$$x(1-y) = 0, \qquad y(1-z) = 0.$$

Adding these equations, and eliminating y, we have

$$x(1-z) = 0;$$

whence $x = \frac{0}{0} z$, or, All X's are Z's.

And in the same way may any other case be treated.

6. Quitting, however, the consideration of special examples, let us examine the general forms to which all syllogism may be reduced.

PROPOSITION I.

To deduce the general rules of Syllogism.

By the general rules of Syllogism, I here mean the rules applicable to premises admitting of every variety both of quantity and of quality in their subjects and predicates, except the combination of two universal terms in the same proposition. The admissible forms of propositions are therefore those of which a tabular view is given in (1).

Let X and Y be the elements or things entering into the first premiss, Z and Y those involved in the second. Two cases, fundamentally different in character, will then present themselves. The terms involving Y will either be of *like* or of *unlike quality*, those terms being regarded as of like quality when they both speak of " Y's," or both of " Not-Y's," as of unlike quality when one of them speaks of " Y's," and the other of " Not-Y's." Any pair of premises, in which the former condition is satisfied, may be represented by the equations

$$vx = v'y, \quad (1)$$

$$wz = w'y; \quad (2)$$

for we can employ the symbol y to represent either " All Y's," or " All not-Y's," since the interpretation of the symbol is purely conventional. If we employ y in the sense of " All not-Y's," then $1 - y$ will represent " All Y's," and no other change will be introduced. An equal freedom is permitted with respect to the symbols x and z, so that the equations (1) and (2) may, by properly assigning the interpretations of x, y, and z, be made to represent all varieties in the combination of premises dependent upon the *quality* of the respective terms. Again, by assuming proper interpretations to the symbols v, v', w, w', in those equations, all varieties with reference to *quantity* may also be represented. Thus, if we take $v = 1$, and represent by v' a class indefinite, the equation (1) will represent a universal proposition according to the ordinary sense of that term, i. e., a proposition with universal subject and particular predicate. We may, in fact, give to subject and predicate in either premiss whatever *quantities* (using this term in the scholastic sense) we please, except that by hypothesis, they must not both be universal. The system (1), (2), represents, therefore, with perfect generality, the possible combinations of premises which have like middle terms.

7. That our analysis may be as general as the equations to which it is applied, let us, by the method of this work, eliminate y from (1) and (2), and seek the expressions for x, $1 - x$, and vx, in terms of z and of the symbols v, v', w, w'. The above will include all the possible forms of the subject of the conclusion. The form $v(1-x)$ is excluded, inasmuch as we cannot from the interpretation $vx =$ Some X's, given in the premises, interpret $v(1-x)$ as Some not-X's. The symbol v, when used in the sense of "some," applies to that term only with which it is connected in the premises.

The results of the analysis are as follows:

$$x = \left[vv'ww' + \frac{0}{0}\{vv'(1-w)(1-w') + ww'(1-v)(1-v') + (1-v)(1-w)\}\right]z$$
$$+ \frac{0}{0}\{vv'(1-w') + 1 - v\}(1-z), \qquad \text{(I.)}$$

$$1 - x = \left[v(1-v')\{ww' + (1-w)(1-w')\} + v(1-w)w'\right.$$
$$\left. + \frac{0}{0}\{vv'(1-w)(1-w') + ww'(1-v)(1-v') + (1-v)(1-w)\}\right]z$$
$$+ \left[v(1-w)w' + \frac{0}{0}\{vv'(1-w') + 1 - v\}\right](1-z), \qquad \text{(II.)}$$

$$vx = \left\{vv'ww' + \frac{0}{0}vv'(1-w)(1-w')\right\}z + \frac{0}{0}(1-w')(1-z). \qquad \text{(III.)}$$

Each of these expressions involves in its second member two terms, of one of which z is a factor, of the other $1 - z$. But syllogistic inference does not, as a matter of form, admit of contrary classes in its conclusion, as of Z's and not-Z's together.

We must, therefore, in order to determine the rules of that species of inference, ascertain under what conditions the second members of any of our equations are reducible to a single term.

The simplest form is (III.), and it is reducible to a single term if $w' = 1$. The equation then becomes

$$vx = vv'wz, \tag{3}$$

the first member is identical with the extreme in the first premiss; the second is of the same quantity and quality as the extreme in the second premiss. For since $w' = 1$, the second member of (2), involving the middle term y, is universal; therefore, by the hypothesis, the first member is particular, and therefore, the second member of (3), involving the same symbol w in its coefficient, is particular also. Hence we deduce the following law.

CONDITION OF INFERENCE.—One middle term, at least, universal.

RULE OF INFERENCE.—Equate the extremes.

From an analysis of the equations (I.) and (II.), it will further appear, that the above is the only condition of syllogistic inference when the middle terms are of like quality. Thus the second member of (I.) reduces to a single term, if $w' = 1$ and $v = 1$; and the second member of (II.) reduces to a single term, if $w' = 1$, $v = 1$, $w = 1$. In each of these cases, it is necessary that $w' = 1$, the solely sufficient condition before assigned.

Consider, secondly, the case in which the middle terms are of unlike quality. The premises may then be represented under the forms

$$vx = v'y, \tag{4}$$

$$wz = w'(1 - y); \tag{5}$$

and if, as before, we eliminate y, and determine the expressions of x, $1 - x$, and vx, we get

$$x = [vv'(1-w)w' + \frac{0}{0}\{ww'(1-v) + (1-v)(1-v')(1-w) + v'(1-w)(1-w')\}]z + [vv'w' + \frac{0}{0}\{(1-v)(1-v') + v'(1-w')\}](1-z). \tag{IV.}$$

$$1 - x = [ww'v + v(1 - v')(1 - w) + \frac{0}{0}\{ww'(1 - v) + (1 - v)(1 - v')(1 - w) + v'(1 - w)(1 - w')\}]z + [v(1 - v') + \frac{0}{0}\{v'(1 - w') + (1 - v)(1 - v')\}](1 - z). \quad \text{(V.)}$$

$$vx = \{vv'(1 - w)w' + \frac{0}{0}vv'(1 - w)(1 - w')\}z + \{vv'w' + \frac{0}{0}vv'(1 - w')\}(1 - z). \quad \text{(VI.)}$$

Now the second member of (VI.) reduces to a single term relatively to z, if $w = 1$, giving

$$vx = \{vv'w' + \frac{0}{0}vv'(1 - w')\}(1 - z);$$

the second member of which is opposite, both in quantity and quality, to the corresponding extreme, wz, in the second premiss. For since $w = 1$, wz is universal. But the factor vv' indicates that the term to which it is attached is particular, since by hypothesis v and v' are not both equal to 1. Hence we deduce the following law of inference in the case of like middle terms:

FIRST CONDITION OF INFERENCE.—*At least one universal extreme.*

RULE OF INFERENCE.—*Change the quantity and quality of that extreme, and equate the result to the other extreme.*

Moreover, the second member of (V.) reduces to a single term if $v' = 1$, $w' = 1$; it then gives

$$1 - x = \{vw + \frac{0}{0}(1 - v)w\}z.$$

Now since $v' = 1$, $w' = 1$, the middle terms of the premises are both universal, therefore the extremes vx, wz, are particular. But in the conclusion the extreme involving x is opposite, both in quantity and quality, to the extreme vx in the first premiss, while the extreme involving z agrees both in quantity and quality with the corresponding extreme wz in the second premiss. Hence the following general law:

SECOND CONDITION OF INFERENCE.—*Two universal middle terms.*

RULE OF INFERENCE.—*Change the quantity and quality of either extreme, and equate the result to the other extreme unchanged.*

There are in the case of unlike middle terms no other conditions or rules of syllogistic inference than the above. Thus the equation (IV.), though reducible to the form of a syllogistic conclusion, when $w = 1$ and $v = 1$, does not thereby establish a new condition of inference; since, by what has preceded, the single condition $v = 1$, or $w = 1$, would suffice.

8. The following examples will sufficiently illustrate the general rules of syllogism above given:

1. All *Y*'s are *X*'s.
All *Z*'s are *Y*'s.

This belongs to Case 1. All *Y*'s is the universal middle term. The extremes equated give as the conclusion

All *Z*'s are *X*'s;

the universal term, All *Z*'s, becoming the subject; the particular term (some) *X*'s, the predicate.

2. All *X*'s are *Y*'s.
No *Z*'s are *Y*'s.

The proper expression of these premises is

All *X*'s are *Y*'s.
All *Z*'s are not-*Y*'s.

They belong to Case 2, and satisfy the first condition of inference. The middle term, *Y*'s, in the first premiss, is particular-affirmative; that in the second premiss, not-*Y*'s, particular-negative. If we take All *Z*'s as the universal extreme, and change its quantity and quality according to the rule, we obtain the term Some not-*Z*'s, and this equated with the other extreme, All *X*'s, gives,

All *X*'s are not-*Z*'s, i. e., No *X*'s are *Z*'s.

If we commence with the other universal extreme, and proceed similarly, we obtain the equivalent result,

No *Z*'s are *X*'s.

3. All Y's are X's.
All not-Y's are Z's.

Here also the middle terms are unlike in quality. The premises therefore belong to Case 2, and there being two universal middle terms, the second condition of inference is satisfied. If by the rule we change the quantity and quality of the first extreme, (some) X's, we obtain All not-X's, which, equated with the other extreme, gives

All not-X's are Z's.

The reverse order of procedure would give the equivalent result,

All not-Z's are X's.

The conclusions of the two last examples would not be recognised as valid in the scholastic system of Logic, which virtually requires that the subject of a proposition should be affirmative. They are, however, perfectly legitimate in themselves, and the rules by which they are determined form undoubtedly the most general canons of syllogistic inference. The process of investigation by which they are deduced will probably appear to be of needless complexity; and it is certain that they might have been obtained with greater facility, and without the aid of any symbolical instrument whatever. It was, however, my object to conduct the investigation in the most general manner, and by an analysis thoroughly exhaustive. With this end in view, the brevity or prolixity of the method employed is a matter of indifference. Indeed the analysis is not properly that of the syllogism, but of a much more general combination of propositions; for we are permitted to assign to the symbols v, v', w, w', any class-interpretations that we please. To illustrate this remark, I will apply the solution (I.) to the following imaginary case:

Suppose that a number of pieces of cloth striped with different colours were submitted to inspection, and that the two following observations were made upon them:

1st. That every piece striped with white and green was also striped with black and yellow, and *vice versâ*.

2nd. That every piece striped with red and orange was also striped with blue and yellow, and *vice versâ*.

Suppose it then required to determine how the pieces marked with green stood affected with reference to the colours white, black, red, orange, and blue.

Here if we assume v = white, x = green, v' = black, y = yellow, w = red, z = orange, w' = blue, the expression of our premises will be

$$vx = v'y,$$
$$wz = w'y,$$

agreeing with the system (1) (2). The equation (I.) then leads to the following conclusion:

Pieces striped with green are either striped with orange, white, black, red, and blue, together, all pieces possessing which character are included in those striped with green; or they are striped with orange, white, and black, but not with red or blue; or they are striped with orange, red, and blue, but not with white or black; or they are striped with orange, but not with white or red; or they are striped with white and black, but not with blue or orange; or they are striped neither with white nor orange.

Considering the nature of this conclusion, neither the symbolical expression (I.) by which it is conveyed, nor the analysis by which that expression is deduced, can be considered as needlessly complex.

9. The form in which the doctrine of syllogism has been presented in this chapter affords ground for an important observation. We have seen that in each of its two great divisions the entire discussion is reducible, so far, at least, as concerns the determination of rules and methods, to the analysis of a pair of equations, viz., of the system (1), (2), when the premises have like middle terms, and of the system (4), (5), when the middle terms are unlike. Moreover, that analysis has been actually conducted by a method founded upon certain general laws deduced immediately from the constitution of language, Chap. II., confirmed by the study of the operations of the human mind, Chap. III., and proved to be applicable to the analysis of all systems of equations whatever, by which propositions, or combinations of propositions, can be represented, Chap. VIII. Here, then, we have the means of definitely resolving the question, whether syllogism is indeed the fundamental type of reasoning,—whether

the study of its laws is co-extensive with the study of deductive logic. For if it be so, some indication of the fact must be given in the systems of equations upon the analysis of which we have been engaged. It cannot be conceived that syllogism should be the one essential process of reasoning, and yet the manifestation of that process present nothing indicative of this high quality of pre-eminence. No sign, however, appears that the discussion of all systems of equations expressing propositions is involved in that of the particular system examined in this chapter. And yet writers on Logic have been all but unanimous in their assertion, not merely of the supremacy, but of the universal sufficiency of syllogistic inference in deductive reasoning. The language of Archbishop Whately, always clear and definite, and on the subject of Logic entitled to peculiar attention, is very express on this point. "For Logic," he says, "which is, as it were, the Grammar of Reasoning, does not bring forward the regular Syllogism as a *distinct mode of argumentation*, designed to be *substituted* for any other mode; but as the form to which *all* correct reasoning may be ultimately reduced."* And Mr. Mill, in a chapter of his System of Logic, entitled, "Of Ratiocination or Syllogism," having enumerated the ordinary forms of syllogism, observes, "All valid ratiocination, all reasoning by which from general propositions previously admitted, other propositions, equally or less general, are inferred, may be exhibited in some of the above forms." And again: "We are therefore at liberty, in conformity with the general opinion of logicians, to consider the two elementary forms of the first figure as the universal types of all correct ratiocination." In accordance with these views it has been contended that the science of Logic enjoys an immunity from those conditions of imperfection and of progress to which all other sciences are subject;† and its origin from the travail of one mighty mind of old has, by a somewhat daring metaphor, been compared to the mythological birth of Pallas.

As Syllogism is a species of elimination, the question before us manifestly resolves itself into the two following ones:—1st. Whether all elimination is reducible to Syllogism; 2ndly. Whe-

* Elements of Logic, p. 13, ninth edition.

† Introduction to Kant's "Logik."

*

ther deductive reasoning can with propriety be regarded as consisting only of elimination. I believe, upon careful examination, the true answer to the former question to be, that it is always theoretically possible so to resolve and combine propositions that elimination may subsequently be effected by the syllogistic canons, but that the process of reduction would in many instances be constrained and unnatural, and would involve operations which are not syllogistic. To the second question I reply, that reasoning cannot, except by an arbitrary restriction of its meaning, be confined to the process of elimination. No definition can suffice which makes it less than the aggregate of the methods which are founded upon the laws of thought, as exercised upon propositions; and among those methods, the process of elimination, eminently important as it is, occupies only a place.

Much of the error, as I cannot but regard it, which prevails respecting the nature of the Syllogism and the extent of its office, seems to be founded in a disposition to regard all those truths in Logic as *primary* which possess the character of simplicity and intuitive certainty, without inquiring into the relation which they sustain to other truths in the Science, or to general methods in the Art, of Reasoning. Aristotle's *dictum de omni et nullo* is a self-evident principle, but it is not found among those *ultimate* laws of the reasoning faculty to which all other laws, however plain and self-evident, admit of being traced, and from which they may in strictest order of scientific evolution be deduced. For though of every science the fundamental truths are usually the most simple of apprehension, yet is not that simplicity the *criterion* by which their title to be regarded as fundamental must be judged. This must be sought for in the nature and extent of the structure which they are capable of supporting. Taking this view, Leibnitz appears to me to have judged correctly when he assigned to the "principle of contradiction" a fundamental place in Logic;* for we have seen the consequences of that law of thought of which it is the axiomatic expression (III. 15). But enough has been said upon the nature of deductive inference and upon its constitutive elements. The subject of

* Nouveaux Essais sur l'entendement humain. Liv. IV. cap. 2. Theodicée Pt. I. sec. 44.

induction may probably receive some attention in another part of this work.

10. It has been remarked in this chapter that the ordinary treatment of hypothetical, is much more defective than that of categorical, propositions. What is commonly termed the hypothetical syllogism appears, indeed, to be no syllogism at all. Let the argument—

If A is B, C is D,
But A is B,
Therefore C is D,

be put in the form—

If the proposition X is true, Y is true,
But X is true,
Therefore Y is true;

wherein by X is meant the proposition A is B, and by Y, the proposition C is D. It is then seen that the premises contain only two terms or elements, while a syllogism essentially involves three. The following would be a genuine hypothetical syllogism:

If X is true, Y is true;
If Y is true, Z is true;
$\therefore$ If X is true, Z is true.

After the discussion of secondary propositions in a former part of this work, it is evident that the forms of hypothetical syllogism must present, in every respect, an exact counterpart to those of categorical syllogism. *Particular* Propositions, such as, "Sometimes if X is true, Y is true," may be introduced, and the conditions and rules of inference deduced in this chapter for categorical syllogisms may, without abatement, be interpreted to meet the corresponding cases in hypotheticals.

11. To what final conclusions are we then led respecting the nature and extent of the scholastic logic? I think to the following: that it is not a science, but a collection of scientific truths, too incomplete to form a system of themselves, and not sufficiently fundamental to serve as the foundation upon which a perfect system may rest. It does not, however, follow, that because the logic of the schools has been invested with attributes to which it

has no just claim, it is therefore undeserving of regard. A system which has been associated with the very growth of language, which has left its stamp upon the greatest questions and the most famous demonstrations of philosophy, cannot be altogether unworthy of attention. Memory, too, and usage, it must be admitted, have much to do with the intellectual processes; and there are certain of the canons of the ancient logic which have become almost inwoven in the very texture of thought in cultured minds. But whether the mnemonic forms, in which the particular rules of conversion and syllogism have been exhibited, possess any real utility,—whether the very skill which they are supposed to impart might not, with greater advantage to the mental powers, be acquired by the unassisted efforts of a mind left to its own resources,—are questions which it might still be not unprofitable to examine. As concerns the particular results deduced in this chapter, it is to be observed, that they are solely designed to aid the inquiry concerning the nature of the ordinary or scholastic logic, and its relation to a more perfect theory of deductive reasoning.

CHAPTER XVI.

ON THE THEORY OF PROBABILITIES.

1. BEFORE the expiration of another year just two centuries will have rolled away since Pascal solved the first known question in the theory of Probabilities, and laid, in its solution, the foundations of a science possessing no common share of the attraction which belongs to the more abstract of mathematical speculations. The problem which the Chevalier de Méré, a reputed gamester, proposed to the recluse of Port Royal (not yet withdrawn from the interests of science* by the more distracting contemplation of the "greatness and the misery of man"), was the first of a long series of problems, destined to call into existence new methods in mathematical analysis, and to render valuable service in the practical concerns of life. Nor does the interest of the subject centre merely in its mathematical connexion, or its associations of utility. The attention is repaid which is devoted to the theory of Probabilities as an independent object of speculation,—to the fundamental modes in which it has been conceived,—to the great secondary principles which, as in the contemporaneous science of Mechanics, have gradually been annexed to it,—and, lastly, to the estimate of the measure of perfection which has been actually attained. I speak here of that perfection which consists in unity of conception and harmony of processes. Some of these points it is designed very briefly to consider in the present chapter.

2. A distinguished writer† has thus stated the fundamental definitions of the science:

* See in particular a letter addressed by Pascal to Fermat, who had solicited his attention to a mathematical problem (Port Royal, par M. de Sainte Beuve); also various passages in the collection of Fragments published by M. Prosper Faugère.

† Poisson, Recherches sur la Probabilitè des Jugemens.

"The probability of an event is the reason we have to believe that it has taken place, or that it will take place."

"The measure of the probability of an event is the ratio of the number of cases favourable to that event, to the total number of cases favourable or contrary, and all equally possible" (equally likely to happen).

From these definitions it follows that the word *probability*, in its mathematical acceptation, has reference to the state of our knowledge of the circumstances under which an event may happen or fail. With the degree of information which we possess concerning the circumstances of an event, the reason we have to think that it will occur, or, to use a single term, our *expectation* of it, will vary. Probability is expectation founded upon partial knowledge. A perfect acquaintance with *all* the circumstances affecting the occurrence of an event would change expectation into certainty, and leave neither room nor demand for a theory of probabilities.

3. Though our expectation of an event grows stronger with the increase of the ratio of the number of the known cases favourable to its occurrence to the whole number of equally possible cases, favourable or unfavourable, it would be unphilosophical to affirm that the strength of that expectation, viewed as an emotion of the mind, is capable of being referred to any numerical standard. The man of sanguine temperament builds high hopes where the timid despair, and the irresolute are lost in doubt. As subjects of scientific inquiry, there is some analogy between *opinion* and *sensation*. The thermometer and the carefully prepared photographic plate indicate, not the intensity of the sensations of heat and light, but certain physical circumstances which accompany the production of those sensations. So also the theory of probabilities contemplates the numerical measure of the circumstances upon which expectation is founded; and this object embraces the whole range of its legitimate applications. The rules which we employ in life-assurance, and in the other statistical applications of the theory of probabilities, are altogether independent of the *mental* phænomena of expectation. They are founded upon the assumption that the future will bear a resem-

blance to the past; that under the same circumstances the same event will tend to recur with a definite numerical frequency; not upon any attempt to submit to calculation the strength of human hopes and fears.

Now experience actually testifies that events of a given species do, under given circumstances, tend to recur with definite frequency, whether their true causes be known to us or unknown. Of course this tendency is, in general, only manifested when the area of observation is sufficiently large. The judicial records of a great nation, its registries of births and deaths, in relation to age and sex, &c., present a remarkable uniformity from year to year. In a given language, or family of languages, the same sounds, and successions of sounds, and, if it be a written language, the same characters and successions of characters recur with determinate frequency. The key to the rude Ogham inscriptions, found in various parts of Ireland, and in which no distinction of words could at first be traced, was, by a strict application of this principle, recovered.* The same method, it is understood, has been applied† to the deciphering of the cuneiform records recently disentombed from the ruins of Nineveh by the enterprise of Mr. Layard.

4. Let us endeavour from the above statements and definitions to form a conception of the legitimate object of the theory of Probabilities.

Probability, it has been said, consists in the expectation founded upon a particular kind of knowledge, viz., the knowledge of the relative frequency of occurrence of events. Hence the probabilities of events, or of combinations of events, whether deduced from a knowledge of the particular constitution of things under which they happen, or derived from the long-continued observation of a past series of their occurrences and failures, constitute, in all cases, our data. The probability of some

---

* The discovery is due to the Rev. Charles Graves, Professor of Mathematics in the University of Dublin.—*Vide* Proceedings of the Royal Irish Academy, Feb. 14, 1848. Professor Graves informs me that he has verified the principle by constructing sequence tables for all the European languages.

† By the learned Orientalist, Dr. Edward Hincks.

connected event, or combination of events, constitutes the corresponding *quæsitum*, or object sought. Now in the most general, yet strict meaning of the term "event," every combination of events constitutes also an event. The simultaneous occurrence of two or more events, or the occurrence of an event under given conditions, or in any conceivable connexion with other events, is still an event. Using the term in this liberty of application, the object of the theory of probabilities might be thus defined. Given the probabilities of any events, of whatever kind, to find the probability of some other event connected with them.

5. Events may be distinguished as simple or compound, the latter term being applied to such events as consist in a combination of simple events (I. 13). In this manner we might define it as the practical end of the theory under consideration to determine the probability of some event, simple or compound, from the given probabilities of other events, simple or compound, with which, by the terms of its definition, it stands connected.

Thus if it is known from the constitution of a die that there is a probability, measured by the fraction $\frac{1}{6}$, that the result of any particular throw will be an ace, and if it is required to determine the probability that there shall occur one ace, and only one, in two successive throws, we may state the problem in the order of its *data* and its *quæsitum*, as follows:

FIRST DATUM.—Probability of the event that the first throw will give an ace $= \frac{1}{6}$.

SECOND DATUM.—Probability of the event that the second throw will give an ace $= \frac{1}{6}$.

QUÆSITUM.—Probability of the event that either the first throw will give an ace, and the second not an ace; or the first will not give an ace, and the second will give one.

Here the two data are the probabilities of simple events defined as the first throw giving an ace, and the second throw giving an ace. The quæsitum is the probability of a compound event,—a certain disjunctive combination of the simple events involved or implied in the data. Probably it will generally happen, when the numerical conditions of a problem are capable of being deduced, as above, from the constitution of things under which they exist, that the data will be the probabilities of *simple* events, and the quæsitum the probability of a *compound* event dependent upon the said simple events. Such is the case with a class of problems which has occupied perhaps an undue share of the attention of those who have studied the theory of probabilities, viz., games of chance and skill, in the former of which some physical circumstance, as the constitution of a die, determines the probability of each possible step of the game, its issue being some definite combination of those steps; while in the latter, the relative dexterity of the players, supposed to be known *à priori*, equally determines the same element. But where, as in statistical problems, the elements of our knowledge are drawn, not from the study of the constitution of things, but from the registered observations of Nature or of human society, there is no reason why the data which such observations afford should be the probabilities of simple events. On the contrary, the occurrence of events or conditions in marked combinations (indicative of some secret connexion of a causal character) suggests to us the propriety of making such concurrences, profitable for future instruction by a numerical record of their frequency. Now the data which observations of this kind afford are the probabilities of compound events. The solution, by some general method, of problems in which such data are involved, is thus not only essential to the perfect development of the theory of probabilities, but also a perhaps necessary condition of its application to a large and practically important class of inquiries.

6. Before we proceed to estimate to what extent known methods may be applied to the solution of problems such as the above, it will be advantageous to notice, that there is another form under which all questions in the theory of probabilities may be viewed; and this form consists in substituting for *events* the propositions which assert that those events have occurred, or will occur; and viewing the element of numerical probability as having reference to the *truth* of those *propositions*, not to the *oc-*

currence of the *events* concerning which they make assertion. Thus, instead of considering the numerical fraction p as expressing the probability of the occurrence of an event E, let it be viewed as representing the probability of the truth of the proposition X, which asserts that the event E will occur. Similarly, instead of any probability, q, being considered as referring to some compound event, such as the concurrence of the events E and F, let it represent the probability of the truth of the proposition which asserts that E and F will jointly occur; and in like manner, let the transformation be made from disjunctive and hypothetical combinations of events to disjunctive and conditional propositions. Though the new application thus assigned to probability is a necessary concomitant of the old one, its adoption will be attended with a practical advantage drawn from the circumstance that we have already discussed the theory of propositions, have defined their principal varieties, and established methods for determining, in every case, the amount and character of their mutual dependence. Upon this, or upon some equivalent basis, any general theory of probabilities must rest. I do not say that other considerations may not in certain cases of applied theory be requisite. The data may prove insufficient for *definite* solution, and this defect it may be thought necessary to supply by hypothesis. Or, where the statement of large numbers is involved, difficulties may arise *after the solution*, from this source, for which special methods of treatment are required. But in every instance, some form of the general problem as above stated (Art. 4) is involved, and in the discussion of that problem the proper and peculiar work of the theory consists. I desire it to be observed, that to this object the investigations of the following chapters are mainly devoted. It is not intended to enter, except incidentally, upon questions involving supplementary hypotheses, because it is of primary importance, even with reference to such questions (I. 17), that a general method, founded upon a solid and sufficient basis of theory, be first established.

7. The following is a summary, chiefly taken from Laplace, of the principles which have been applied to the solution of questions of probability. They are consequences of its fundamental defini-

tions already stated, and may be regarded as indicating the degree in which it has been found possible to render those definitions available.

PRINCIPLE 1st. If p be the probability of the occurrence of any event, $1 - p$ will be the probability of its non-occurrence.

2nd. The probability of the concurrence of two independent events is the product of the probabilities of those events.

3rd. The probability of the concurrence of two dependent events is equal to the product of the probability of one of them by the probability that if that event occur, the other will happen also.

4th. The probability that if an event, E, take place, an event, F, will also take place, is equal to the probability of the concurrence of the events E and F, divided by the probability of the occurrence of E.

5th. The probability of the occurrence of one or the other of two events which cannot concur is equal to the sum of their separate probabilities.

6th. If an observed event can only result from some one of n different causes which are *à priori* equally probable, the probability of any one of the causes is a fraction whose numerator is the probability of the event, on the hypothesis of the existence of that cause, and whose denominator is the sum of the similar probabilities relative to all the causes.

7th. The probability of a future event is the sum of the products formed by multiplying the probability of each cause by the probability that if that cause exist, the said future event will take place.

8. Respecting the extent and the relative sufficiency of these principles, the following observations may be made.

1st. It is always possible, by the due combination of these principles, to express the probability of a compound event, dependent in any manner upon independent simple events whose distinct probabilities are given. A very large proportion of the problems which have been actually solved are of this kind, and the difficulty attending their solution has not arisen from the insufficiency of the indications furnished by the theory of probabilities, but from the need of an analysis which should render those indications available when functions of large numbers, or series consisting of many and complicated terms, are thereby introduced. It may, therefore, be fully conceded, that all problems having for their data the probabilities of independent simple events fall within the scope of received methods.

2ndly. Certain of the principles above enumerated, and especially the sixth and seventh, do not presuppose that all the data are the probabilities of simple events. In their peculiar application to questions of causation, they do, however, assume, that the causes of which they take account are mutually exclusive, so that no combination of them in the production of an effect is possible. If, as before explained, we transfer the numerical probabilities from the events with which they are connected to the propositions by which those events are expressed, the most general problem to which the aforesaid principles are applicable may be stated in the following order of *data* and *quæsita*.

DATA.

1st. The probabilities of the n conditional propositions:

If the cause A_1 exist, the event E will follow;
" A_2 " E "
. . . .
" A_n " E "

2nd. The condition that the antecedents of those propositions are mutually conflicting.

REQUIREMENTS.

The probability of the truth of the proposition which declares the occurrence of the event E; also, when that proposition is known to be true, the probabilities of truth of the several propositions which affirm the respective occurrences of the causes $A_1, A_2 \ldots A_n$.

Here it is seen, that the data are the probabilities of a series of compound events, expressed by *conditional* propositions. But the system is obviously a very limited and particular one. For the antecedents of the propositions are subject to the condition of being mutually exclusive, and there is but one consequent, the event E, in the whole system. It does not follow, from our ability to discuss such a system as the above, that we are able to resolve problems whose data are the probabilities of *any* system of conditional propositions; far less that we can resolve problems whose data are the probabilities of *any system of propositions whatever*. And, viewing the subject in its material rather than its formal aspect, it is evident, that the hypothesis of *exclusive* causation is one which is not often realized in the actual world, the phænomena of which seem to be, usually, the products of complex causes, the amount and character of whose co-operation is unknown. Such is, without doubt, the case in nearly all departments of natural or social inquiry in which the doctrine of probabilities holds out any new promise of useful applications.

9. To the above principles we may add another, which has been stated in the following terms by the Savilian Professor of Astronomy in the University of Oxford.*

"Principle 8. If there be any number of mutually exclusive hypotheses, h_1, h_2, h_3, . . of which the probabilities relative to a particular state of information are p_1, p_2, p_3, . . and if new information be given which changes the probabilities of some of them, suppose of h_{m+1} and all that follow, without having otherwise *any reference to the rest*; then the probabilities of these latter have the same ratios to one another, after the new information, that they had before, that is,

$$p'_1 : p'_2 : p'_3 \ldots : p'_m = p_1 : p_2 : p_3 \ldots : p_m,$$

where the accented letters denote the values after the new information has been acquired."

This principle is apparently of a more fundamental character than the most of those before enumerated, and perhaps it might, as has been suggested by Professor Donkin, be regarded as axiomatic. It seems indeed to be founded in the very definition of the measure of probability, as "the ratio of the number of cases favourable to an event to the total number of cases favourable or contrary, and all equally possible." For, adopting this definition, it is evident that in whatever proportion the number of equally

* On certain Questions relating to the Theory of Probabilities; by W. F. Donkin, M. A., F. R. S., &c. Philosophical Magazine, May, 1851.

possible cases is diminished, while the number of favourable cases remains unaltered, in exactly the same proportion will the probabilities of any events to which these cases have reference be increased. And as the new hypothesis, viz., the diminution of the number of possible cases without affecting the number of them which are favourable to the events in question, increases the probabilities of those events in a constant ratio, the relative measures of those probabilities remain unaltered. If the principle we are considering be then, as it appears to be, inseparably involved in the very definition of probability, it can scarcely, *of itself*, conduct us further than the attentive study of the definition would alone do, in the solution of problems. From these considerations it appears to be doubtful whether, without some aid of a different kind from any that has yet offered itself to our notice, any considerable advance, either in the theory of probabilities as a branch of speculative knowledge, or in the practical solution of its problems can be hoped for. And the establishment, *solely* upon the basis of any such collection of principles as the above, of a method universally applicable to the solution of problems, without regard either to the number or to the nature of the propositions involved in the expression of their data, seems to be impossible. For the attainment of such an object other elements are needed, the consideration of which will occupy the next chapter.

CHAPTER XVII.

DEMONSTRATION OF A GENERAL METHOD FOR THE SOLUTION OF PROBLEMS IN THE THEORY OF PROBABILITIES.

1. IT has been defined (XVI. 2), that "the measure of the probability of an event is the ratio of the number of cases favourable to that event, to the total number of cases favourable or unfavourable, and all equally possible." In the following investigations the term probability will be used in the above sense of "measure of probability."

From the above definition we may deduce the following conclusions.

I. When it is certain that an event will occur, the probability of that event, in the above mathematical sense, is 1. For the cases which are favourable to the event, and the cases which are possible, are in this instance the same.

Hence, also, if p be the probability that an event x will happen, $1 - p$ will be the probability that the said event will not happen. To deduce this result directly from the definition, let m be the number of cases favourable to the event x, n the number of cases possible, then $n - m$ is the number of cases unfavourable to the event x. Hence, by definition,

$$\frac{m}{n} = \text{probability that } x \text{ will happen.}$$

$$\frac{n-m}{n} = \text{probability that } x \text{ will not happen.}$$

But
$$\frac{n-m}{n} = 1 - \frac{m}{n} = 1 - p.$$

II. The probability of the concurrence of any two events is the product of the probability of either of those events by the probability that if that event occur, the other will occur also.

Let m be the number of cases favourable to the happening of the first event, and n the number of equally possible cases unfavourable to it; then the probability of the first event is, by definition, $\frac{m}{m+n}$. Of the m cases favourable to the first event, let l cases be favourable to the conjunction of the first and second events, then, by definition, $\frac{l}{m}$ is the probability that if the first event happen, the second also will happen. Multiplying these fractions together, we have

$$\frac{m}{m+n} \times \frac{l}{m} = \frac{l}{m+n}.$$

But the resulting fraction $\frac{l}{m+n}$ has for its numerator the number of cases favourable to the conjunction of events, and for its denominator, the number $m+n$ of possible cases. Therefore, it represents the probability of the joint occurrence of the two events.

Hence, if p be the probability of any event x, and q the probability that if x occur y will occur, the probability of the conjunction xy will be pq.

III. The probability that if an event x occur, the event y will occur, is a fraction whose numerator is the probability of their joint occurrence, and denominator the probability of the occurrence of the event x.

This is an immediate consequence of Principle 2nd.

IV. The probability of the occurrence of some one of a series of exclusive events is equal to the sum of their separate probabilities.

For let n be the number of possible cases; m_1 the number of those cases favourable to the first event; m_2 the number of cases favourable to the second, &c. Then the separate probabilities of the events are $\frac{m_1}{n}$, $\frac{m_2}{n}$, &c. Again, as the events are exclusive, none of the cases favourable to one of them is favourable to another; and, therefore, the number of cases favourable to some one of the series will be $m_1 + m_2 \ldots$, and the probability of some one of the series happening will be $\frac{m_1 + m_2 \ldots}{n}$. But this is the sum of the previous fractions, $\frac{m_1}{n}$, $\frac{m_2}{n}$, &c. Whence the principle is manifest.

2. DEFINITION.—Two events are said to be independent when the probability of the happening of either of them is unaffected by our expectation of the occurrence or failure of the other.

From this definition, combined with Principle II., we have the following conclusion :

V. The probability of the concurrence of two independent events is equal to the product of the separate probabilities of those events.

For if p be the probability of an event x, q that of an event y regarded as quite independent of x, then is q also the probability that if x occur y will occur. Hence, by Principle II., pq is the probability of the concurrence of x and y.

Under the same circumstances, the probability that x will occur and y not occur will be $p(1-q)$. For p is the probability that x will occur, and $1-q$ the probability that y will not occur. In like manner $(1-p)(1-q)$ will be the probability that both the events fail of occurring.

3. There exists yet another principle, different in kind from the above, but necessary to the subsequent investigations of this chapter, before proceeding to the explicit statement of which I desire to make one or two preliminary observations.

I would, in the first place, remark that the distinction between simple and compound events is not one founded in the nature of events themselves, but upon the mode or connexion in which they are presented to the mind. How many separate particulars, for instance, are implied in the terms "To be in health," "To prosper," &c., each of which might still be regarded as expressing a "simple event"? The prescriptive usages of language, which have assigned to particular combinations of events single and definite appellations, and have left unnumbered other combinations to be expressed by corresponding combinations of distinct terms or phrases, is essentially arbitrary. When, then, we designate as simple events those which are expressed by a single verb, or by what grammarians term a simple sentence, we do not thereby imply any real simplicity in the events themselves, but use the term solely with reference to grammatical expression.

4. Now if this distinction of events, as simple or compound, is not founded in their real nature, but rests upon the accidents of language, it cannot affect the question of their mutual dependence or independence. If my knowledge of two simple events is confined to this particular fact, viz., that the probability of the occurrence of one of them is p, and that of the other q; then I regard the events as independent, and thereupon affirm that the probability of their joint occurrence is pq. But the ground of this affirmation is not that the events are simple ones, but that the data afford no information whatever concerning any connexion or dependence between them. When the probabilities of events are given, but all information respecting their dependence withheld, the mind regards them as independent. And this mode of thought is equally correct whether the events, judged according to actual expression, are simple or compound, i. e., whether each of them is expressed by a single verb or by a combination of verbs.

5. Let it, however, be supposed that, together with the probabilities of certain events, we possess some definite information respecting their possible combinations. For example, let it be known that certain combinations are excluded from happening, and therefore that the remaining combinations alone are possible. Then still is the same general principle applicable. The mode in which we avail ourselves of this information in the calculation of the probability of any conceivable issue of events depends not upon the nature of the events whose probabilities and whose limits of possible connexion are given. It matters not whether they are simple or compound. It is indifferent from what source, or by what methods, the knowledge of their probabilities and of their connecting relations has been derived. We must regard the events as independent of any connexion beside that of which we have information, *deeming it of no consequence whether such information has been explicitly conveyed to us in the data, or thence deduced by logical inference*. And this leads us to the statement of the general principle in question, viz.:

VI. The events whose probabilities are given are to be regarded as independent of any connexion but such as is either expressed, or necessarily implied, in the data; and the mode in which our knowledge of that connexion is to be employed is independent of the nature of the source from which such knowledge has been derived.

The practical importance of the above principle consists in the circumstance, that whatever may be the nature of the events whose probabilities are given,—whatever the nature of the event whose probability is sought, we are always able, by an application of the Calculus of Logic, to determine the expression of the latter event as a definite combination of the former events, and definitely to assign the whole of the implied relations connecting the former events with each other. In other words, we can determine what that combination of the given events is whose probability is required, and what combinations of them are alone possible. It follows then from the above principle, that we can reason upon those events as if they were simple events, whose conditions of possible combination had been directly given by experience, and of which the probability of some definite combination is sought. The possibility of a general method in probabilities depends upon this reduction.

6. As the investigations upon which we are about to enter are based upon the employment of the Calculus of Logic, it is necessary to explain certain terms and modes of expression which are derived from this application.

By the event x, I mean that event of which the proposition which affirms the occurrence is symbolically expressed by the equation

$$x = 1.$$

By the event $\phi(x, y, z, ..)$, I mean that event of which the occurrence is expressed by the equation

$$\phi(x, y, z, ..) = 1.$$

Such an event may be termed a compound event, in relation to the simple events x, y, z, which its conception involves. Thus, if x represent the event "It rains," y the event "It thunders," the separate occurrences of those events being expressed by the logical equations

$$x = 1, \qquad y = 1,$$

then will $x(1 - y) + y(1 - x)$ represent the event or state of things denoted by the Proposition, "It either rains or thunders, but not both;" the expression of that state of things being

$$x(1-y)+y(1-x)=1.$$

If for brevity we represent the function $\phi(x, y, z, \ldots)$, used in the above acceptation by V, it is evident (VI. 13) that the law of duality

$$V(1-V)=0,$$

will be identically satisfied.

The simple events x, y, z will be said to be "conditioned" when they are not free to occur in every possible combination; in other words, when some compound event depending upon them is precluded from occurring. Thus the events denoted by the propositions, "It rains," "It thunders," are "conditioned" if the event denoted by the proposition, "It thunders, but does not rain," is excluded from happening, so that the range of possible is less than the range of conceivable combination. Simple unconditioned events are by definition independent.

Any compound event is similarly said to be conditioned if it is assumed that it can only occur under a certain condition, that is, in combination with some other event constituting, by its presence, that condition.

7. We shall proceed in the natural order of thought, from simple and unconditioned, to compound and conditioned events.

PROPOSITION I.

1st. *If p, q, r are the respective probabilities of any unconditioned simple events x, y, z, the probability of any compound event V will be $[V]$, this function $[V]$ being formed by changing, in the function V, the symbols x, y, z into p, q, r, &c.*

2ndly. *Under the same circumstances, the probability that if the event V occur, any other event V' will also occur, will be $\dfrac{[VV']}{[V]}$, wherein $[VV']$ denotes the result obtained by multiplying together the logical functions V and V', and changing in the result x, y, z, &c. into p, q, r, &c.*

Let us confine our attention in the first place to the possible combinations of the two simple events, x and y, of which the respective probabilities are p and q. The primary combinations of those events (V. 11), and their corresponding probabilities, are as follows:

EVENTS.		PROBABILITIES.
xy,	Concurrence of x and y,	pq.
$x(1-y)$,	Occurrence of x without y,	$p(1-q)$.
$(1-x)y$,	Occurrence of y without x,	$(1-p)q$.
$(1-x)(1-y)$,	Conjoint failure of x and y,	$(1-p)(1-q)$.

We see that in these cases the probability of the compound event represented by a constituent is the same function of p and q as the logical expression of that event is of x and y; and it is obvious that this remark applies, whatever may be the number of the simple events whose probabilities are given, and whose *joint existence or failure* is involved in the compound event of which we seek the probability.

Consider, in the second place, any *disjunctive* combination of the above constituents. The compound event, expressed in ordinary language as the occurrence of "either the event x without the event y, or the event y without the event x," is symbolically expressed in the form $x(1-y)+y(1-x)$, and its probability, determined by Principles IV. and V., is $p(1-q)+q(1-p)$. The latter of these expressions is the same function of p and q as the former is of x and y. And it is obvious that this is also a particular illustration of a general rule. The events which are expressed by any two or more constituents are mutually exclusive. The only possible combination of them is a *disjunctive* one, expressed in ordinary language by the conjunction *or*, in the language of symbolical logic by the sign +. Now the probability of the occurrence of some one out of a set of mutually exclusive events is the sum of their separate probabilities, and is expressed by connecting the expressions for those separate probabilities by the sign +. Thus the law above exemplified is seen to be *general.* The probability of any unconditioned event V will be found by changing in V the symbols $x, y, z, \ldots$ into $p, q, r, \ldots$

8. Again, by Principle III., the probability that if the event V occur, the event V' will occur with it, is expressed by a frac-

tion whose numerator is the probability of the joint occurrence of V and V', and denominator the probability of the occurrence of V.

Now the expression of that event, or state of things, which is constituted by the joint occurrence of the events V and V', will be formed by multiplying together the expressions V and V' according to the rules of the Calculus of Logic; since whatever constituents are found in both V and V' will appear in the product, and no others. Again, by what has just been shown, the probability of the event represented by that product will be determined by changing therein x, y, z into $p, q, r, ..$ Hence the numerator sought will be what $[VV']$ by definition represents. And the denominator will be $[V]$, wherefore

$$\text{Probability that if } V \text{ occur, } V' \text{ will occur with it} = \frac{[VV']}{[V]}.$$

9. For example, if the probabilities of the simple events x, y, z are p, q, r respectively, and it is required to find the probability that if either x or y occur, then either y or z will occur, we have for the logical expressions of the antecedent and consequent—

1st. Either x or y occurs, $x(1-y)+y(1-x)$.

2nd. Either y or z occurs, $y(1-z)+z(1-y)$.

If now we multiply these two expressions together according to the rules of the Calculus of Logic, we shall have for the expression of the concurrence of antecedent and consequent,

$$xz(1-y)+y(1-x)(1-z).$$

Changing in this result x, y, z into p, q, r, and similarly transforming the expression of the antecedent, we find for the probability sought the value

$$\frac{pr(1-q)+q(1-p)(1-r)}{p(1-q)+q(1-p)}.$$

The special function of the calculus, in a case like the above, is to supply the office of the reason in determining what are the conjunctures involved at once in the consequent and the antecedent. But the advantage of this application is almost entirely prospective, and will be made manifest in a subsequent proposition.

PROPOSITION II.

10. *It is known that the probabilities of certain simple events* $x, y, z, \ldots$ *are* $p, q, r, \ldots$ *respectively when a certain condition* V *is satisfied;* V *being in expression a function of* $x, y, z, \ldots$ *Required the absolute probabilities of the events* $x, y, z, \ldots$, *that is, the probabilities of their respective occurrence independently of the condition* V.

Let p', q', r', &c., be the probabilities required, i. e. the probabilities of the events $x, y, z, \ldots$, regarded not only as simple, but as independent events. Then by Prop. I. the probabilities that these events will occur when the condition V, represented by the logical equation $V = 1$, is satisfied, are

$$\frac{[xV]}{[V]}, \quad \frac{[yV]}{[V]}, \quad \frac{[zV]}{[V]}, \text{ \&c.,}$$

in which $[xV]$ denotes the result obtained by multiplying V by x, according to the rules of the Calculus of Logic, and changing in the result x, y, z, into p', q', r', &c. But the above conditioned probabilities are by hypothesis equal to $p, q, r, \ldots$ respectively. Hence we have,

$$\frac{[xV]}{[V]} = p, \quad \frac{[yV]}{[V]} = q, \quad \frac{[zV]}{[V]} = r, \text{ \&c.,}$$

from which system of equations equal in number to the quantities $p', q', r', \ldots$, the values of those quantities may be determined.

Now xV consists simply of those constituents in V of which x is a factor. Let this sum be represented by V_x, and in like manner let yV be represented by V_y, &c. Our equations then assume the form

$$\frac{[V_x]}{[V]} = p, \quad \frac{[V_y]}{[V]} = q, \text{ \&c.;} \qquad (1)$$

where $[V_x]$ denotes the results obtained by changing in V_x the symbols x, y, z, &c., into p, q, r, &c.

To render the meaning of the general problem and the principle of its solution more evident, let us take the following example. Suppose that in the drawing of balls from an urn attention had only been paid to those cases in which the balls drawn were either of a particular colour, "white," or of a particular composition, "marble," or were marked by both these characters, no record having been kept of those cases in which a ball that was neither white nor of marble had been drawn. Let it then have been found, that whenever the supposed condition was satisfied, there was a probability p that a white ball would be drawn, and a probability q that a marble ball would be drawn: and from these data alone let it be required to find the probability that in the next drawing, without reference at all to the condition above mentioned, a white ball will be drawn; also the probability that a marble ball will be drawn.

Here if x represent the drawing of a white ball, y that of a marble ball, the condition V will be represented by the logical function

$$xy + x(1-y) + (1-x)y.$$

Hence we have

$$V_x = xy + x(1-y) = x, \qquad V_y = xy + (1-x)y = y;$$

whence

$$[V_x] = p', \qquad [V_y] = q';$$

and the final equations of the problem are

$$\frac{p'}{p'q' + p'(1-q') + q'(1-p')} = p, \qquad \frac{q'}{p'q' + p'(1-q') + q'(1-p')} = q;$$

from which we find

$$p' = \frac{p+q-1}{q}, \qquad q' = \frac{p+q-1}{p}.$$

It is seen that p' and q' are respectively proportional to p and q, as by Professor Donkin's principle they ought to be. The solution of this class of problems might indeed, by a direct application of that principle, be obtained.

To meet a possible objection, I here remark, that the above reasoning does not require that the drawings of a white and a marble ball should be independent, in virtue of the physical constitution of the balls. The assumption of their independence is indeed involved in the solution, but it does not rest upon any prior assumption as to the nature of the balls, and their relations, or freedom from relations, of form, colour, structure, &c. It is founded upon our total ignorance of all these things. Probability always has reference to the state of our *actual* knowledge, and its numerical value varies with varying information.

Proposition III.

11. *To determine in any question of probabilities the logical connexion of the quæsitum with the data; that is, to assign the event whose probability is sought, as a logical function of the event whose probabilities are given.*

Let S, T, &c., represent any compound events whose probabilities are given, S and T being in expression known functions of the symbols x, y, z, &c., representing simple events. Similarly let W represent any event whose probability is sought, W being also a known function of x, y, z, &c. As S, T, .. W must satisfy the fundamental law of duality, we are permitted to replace them by single logical symbols, s, t, .. w. Assume then

$$s = S, \quad t = T, \quad w = W.$$

These, by the definition of S, T, .. W, will be a series of logical equations connecting the symbols s, t, .. w, with the symbols x, y, z ..

By the methods of the Calculus of Logic we can eliminate from the above system any of the symbols x, y, z, .., representing events whose probabilities are not given, and determine w as a developed function of s, t, &c., and of such of the symbols x, y, z, &c., if any such there be, as correspond to events whose probabilities are given. The result will be of the form

$$w = A + 0\,B + \frac{0}{0}\,C + \frac{1}{0}\,D,$$

where A, B, C, and D comprise among them all the possible *constituents* which can be formed from the symbols s, t, &c., i. e. from all the symbols representing events whose probabilities are given.

The above will evidently be the complete expression of the relation sought. For it fully determines the event W, represented by the single symbol w, as a function or combination of the events similarly denoted by the symbols s, t, &c., and it assigns by the laws of the Calculus of Logic the condition

$$D = 0,$$

as connecting the events s, t, &c., among themselves. We may, therefore, by Principle VI., regard s, t, &c., as *simple* events, of which the combination w, and the condition with which it is associated D, are definitely determined.

Uniformity in the logical processes of reduction being desirable, I shall here state the order which will generally be pursued.

12. By (VIII. 8), the primitive equations are reducible to the forms

$$\begin{aligned} s(1-S)+S(1-s)&=0;\\ t(1-T)+T(1-t)&=0; \qquad (1)\\ \cdots\cdots\cdots&\\ w(1-W)+W(1-w)&=0; \end{aligned}$$

under which they can be added together without impairing their significance. We can then eliminate the symbols x, y, z, either separately or together. If the latter course is chosen, it is necessary, after adding together the equations of the system, to develop the result with reference to all the symbols to be eliminated, and equate to 0 the product of all the coefficients of the constituents (VII. 9).

As w is the symbol whose expression is sought, we may also, by Prop. III. Chap. IX., express the result of elimination in the form

$$Ew + E'(1-w) = 0.$$

E and E' being successively determined by making in the general system (1), $w = 1$ and $w = 0$, and eliminating the symbols x, y, z, .. Thus the single equations from which E and E' are to be respectively determined become

$$s(1-S)+S(1-s)+t(1-T)+T(1-t)\,..+1-W=0;$$
$$s(1-S)+S(1-s)+t(1-T)+T(1-t)+W=0.$$

From these it only remains to eliminate x, y, z, &c., and to determine w by subsequent development.

In the process of elimination we may, if needful, avail ourselves of the simplifications of Props. I. and II. Chap. IX.

13. Should the data, beside informing us of the probabilities of events, further assign among them any explicit connexion, such connexion must be logically expressed, and the equation or equations thus formed be introduced into the general system.

Proposition IV.

14. *Given the probabilities of any system of events; to determine by a general method the consequent or derived probability of any other event.*

As in the last Proposition, let S, T, &c., be the events whose probabilities are given, W the event whose probability is sought, these being known functions of x, y, z, &c. Let us represent the data as follows:

$$\left.\begin{aligned}\text{Probability of } S &= p;\\ \text{Probability of } T &= q;\end{aligned}\right\} \qquad (1)$$

and so on, p, q, &c., being known numerical values. If then we represent the compound event S by s, T by t, and W by w, we find by the last proposition,

$$w = A + 0\,B + \frac{0}{0}\,C + \frac{1}{0}\,D; \qquad (2)$$

A, B, C, and D being functions of s, t, &c. Moreover the data (1) are transformed into

$$\text{Prob. } s = p, \quad \text{Prob. } t = q, \quad \text{\&c.} \qquad (3)$$

Now the equation (2) is resolvable into the system

$$\left.\begin{aligned} w &= A + qC \\ D &= 0, \end{aligned}\right\} \qquad (4)$$

q being an indefinite class symbol (VI. 12). But since by the properties of constituents (V. Prop. III.), we have

$$A + B + C + D = 1,$$

the second equation of the above system may be expressed in the form

$$A + B + C = 1.$$

If we represent the function $A + B + C$ by V, the system (4) becomes

$$w = A + qC; \tag{5}$$

$$V = 1. \tag{6}$$

Let us for a moment consider this result. Since V is the sum of a series of constituents of s, t, &c., it represents the compound event in which the simple events involved are those denoted by s, t, &c. Hence (6) shows that the events denoted by s, t, &c., and whose probabilities are p, q, &c., have such probabilities not as *independent events*, but as events subject to a certain condition V. Equation (5) expresses w as a similarly conditioned combination of the same events.

Now by Principle VI. the mode in which this knowledge of the connexion of events has been obtained does not influence the mode in which, when obtained, it is to be employed. We must reason upon it as if experience had presented to us the events s, t, &c., as simple events, free to enter into every combination, but possessing, when actually subject to the condition V, the probabilities p, q, &c., respectively.

Let then p', q', . . , be the corresponding probabilities of such events, when the restriction V is removed. Then by Prop. II. of the present chapter, these quantities will be determined by the system of equations,

$$\frac{[V_s]}{[V]} = p, \quad \frac{[V_t]}{[V]} = q, \text{ \&c.}; \tag{7}$$

and by Prop. I. the probability of the event w under the same condition V will be

$$\text{Prob. } w = \frac{[A + cC]}{[V]}; \tag{8}$$

wherein V_s denotes the sum of those constituents in V of which s is a factor, and $[V_s]$ what that sum becomes when s, t, . . , are changed into p', q', . . , respectively. The constant c represents the probability of the indefinite event q; it is, therefore, arbitrary, and admits of any value from 0 to 1.

Now it will be observed, that the values of p', q', &c., are determined from (7) only in order that they may be substituted in (8), so as to render Prob. w a function of known quantities, p, q,

&c. It is obvious, therefore, that instead of the letters p', q', &c., we might employ any others as s, t, &c., in the same *quantitative* acceptations. This particular step would simply involve a change of meaning of the symbols s, t, &c.—their ceasing to be *logical*, and becoming *quantitative*. The systems (7) and (8) would then become

$$\frac{V_s}{V} = p, \qquad \frac{V_t}{V} = q, \text{ \&c.}; \tag{9}$$

$$\text{Prob. } w = \frac{A + cC}{V}. \tag{10}$$

In employing these, it is only necessary to determine from (9) s, t, &c., regarded as quantitative symbols, in terms of p, q, &c., and substitute the resulting values in (10). It is evident, that s, t, &c., inasmuch as they represent *probabilities*, will be positive proper fractions.

The system (9) may be more symmetrically expressed in the form

$$\frac{V_s}{p} = \frac{V_t}{q} \ldots = V. \tag{11}$$

Or we may express both (9) and (10) together in the symmetrical system

$$\frac{V_s}{p} = \frac{V_t}{q} \ldots = \frac{A + cC}{u} = V; \tag{12}$$

wherein u represents Prob. w.

15. It remains to interpret the constant c assumed to represent the probability of the indefinite event q. Now the *logical* equation

$$w = A + qC,$$

interpreted in the reverse order, implies that if either the event A take place, or the event C in connexion with the event q, the event w will take place, and not otherwise. Hence q represents that condition under which, if the event C take place, the event w will take place. But the probability of q is c. Hence, therefore, c = probability that if the event C take place the event w will take place.

Wherefore by Principle II.,

$$c = \frac{\text{Probability of concurrence of } C \text{ and } w}{\text{Probability of } C}.$$

We may hence determine the nature of that new experience from which the actual value of c may be obtained. For if we substitute in C for s, t, &c., their original expressions as functions of the simple events x, y, z, &c., we shall form the expression of that event whose probability constitutes the denominator of the above value of c; and if we multiply that expression by the original expression of w, we shall form the expression of that event whose probability constitutes the numerator of c, and *the ratio of the frequency of this event to that of the former one, determined by new observations*, will give the value of c. Let it be remarked here, that the constant c does not *necessarily* make its appearance in the solution of a problem. It is only when the data are insufficient to render determinate the probability sought, that this arbitrary element presents itself, and in this case it is seen that the final logical equation (2) or (5) informs us how it is to be determined.

If that new experience by which c may be determined cannot be obtained, we can still, by assigning to c its limiting values 0 and 1, determine the limits of the probability of w. These are

$$\text{Minor limit of Prob. } w = \frac{A}{V}.$$

$$\text{Superior limit} \qquad = \frac{A+C}{V}.$$

Between these limits, it is certain that the probability sought must lie independently of all new experience which does not absolutely contradict the past.

If the expression of the event C consists of many constituents, the logical value of w being of the form

$$w = A + \frac{0}{0} C_1 + \frac{0}{0} C_2 + \&c.,$$

we can, instead of employing their aggregate as above, present the final solution in the form

$$\text{Prob. } w = \frac{A + c_1 C_1 + c_2 C_2 + \&c.}{V}.$$

Here c_1 = probability that if the event C_1 occur, the event w will occur, and so on for the others. Convenience must decide which form is to be preferred.

16. The above is the complete theoretical solution of the problem proposed. It may be added, that it is applicable equally to the case in which any of the events mentioned in its original statement are conditioned. Thus, if one of the data is the probability p, that if the event x occur the event y will occur; the probability of the occurrence of x not being given, we must assume Prob. $x = c$ (an arbitrary constant), then Prob. $xy = cp$, and these two conditions must be introduced into the data, and employed according to the previous method. Again, if it is sought to determine the probability that if an event x occur an event y will occur, the solution will assume the form

$$\text{Prob. sought} = \frac{\text{Prob. } xy}{\text{Prob. } x},$$

the numerator and denominator of which must be separately determined by the previous general method.

17. We are enabled by the results of these investigations to establish a general rule for the solution of questions in probabilities.

General Rule.

Case I.—*When all the events are unconditioned.*

Form the symbolical expressions of the events whose probabilities are given or sought.

Equate such of those expressions as relate to compound events to a new series of symbols, s, t, &c., which symbols regard as representing the events, no longer as compound but simple, to whose expressions they have been equated.

Eliminate from the equations thus formed all the logical symbols, except those which express events, s, t, &c., whose respective probabilities p, q, &c. are given, or the event w whose probability is sought, and determine w as a developed function of s, t, &c. in the form

$$w = A + 0B + \frac{0}{0}C + \frac{1}{0}D.$$

Let $A + B + C = V$, and let V_s represent the aggregate of those constituents in V which contain s as a factor, V_t of those which contain t as a factor, and thus for all the symbols whose probabilities are given.

Then, passing from Logic to Algebra, form the equations

$$\frac{V_s}{p} = \frac{V_t}{q} = V, \tag{1}$$

$$\text{Prob. } w = \frac{A + cC}{V}, \tag{2}$$

from (1) determine s, t, &c. as functions of p, q, &c., and substitute their values in (2). The result will express the solution required.

Or form the symmetrical system of equations

$$\frac{V_s}{p} = \frac{V_t}{q} \,.\,. = \frac{A + cC}{u} = \frac{V}{1}, \tag{3}$$

where u represents the probability sought.

If c appear in the solution, its interpretation will be

$$c = \frac{\text{Prob. } Cw}{\text{Prob. } c},$$

and this interpretation indicates the nature of the experience which is necessary for its discovery.

CASE II.—*When some of the events are conditioned.*

If there be given the probability p that if the event X occur, the event Y will occur, and if the probability of the antecedent X be not given, resolve the proposition into the two following, viz.:

$$\text{Probability of } X = c,$$
$$\text{Probability of } Xy = cp.$$

If the quæsitum be the probability that if the event W occur, the event Z will occur, determine separately, by the previous case, the terms of the fraction

$$\frac{\text{Prob. } WZ}{\text{Prob. } W},$$

and the fraction itself will express the probability sought.

It is understood in this case that X, Y, W, Z may be any compound events whatsoever. The expressions XY and WZ represent the products of the symbolical expressions of X and Y and of W and Z, formed according to the rules of the Calculus of Logic.

The determination of the single constant c may in certain cases be resolved into, or replaced by, the determination of a series of arbitrary constants c_1, c_2. . according to convenience, as previously explained.

18. It has been stated (I. 12) that there exist two distinct definitions, or modes of conception, upon which the theory of probabilities may be made to depend, one of them being connected more immediately with Number, the other more directly with Logic. We have now considered the consequences which flow from the numerical definition, and have shown how it conducts us to a point in which the necessity of a connexion with Logic obviously suggests itself. We have seen to some extent what is the nature of that connexion; and further, in what manner the peculiar processes of Logic, and the more familiar ones of quantitative Algebra, are involved in the same general method of solution, each of these so accomplishing its own object that the two processes may be regarded as supplementary to each other. It remains to institute the reverse order of investigation, and, setting out from a definition of probability in which the logical relation is more immediately involved, to show how the numerical definition would thence arise, and how the same general method, equally dependent upon both elements, would finally, but by a different order of procedure, be established.

That between the symbolical expressions of the logical calculus and those of Algebra there exists a close analogy, is a fact to which attention has frequently been directed in the course of the present treatise. It might even be said that they possess a community of forms, and, to a very considerable degree, a community of laws. With a single exception in the latter respect, their difference is only one of interpretation. Thus the same expression admits of a logical or of a quantitative interpretation, according to the particular meaning which we attach to the symbols it involves. The expression xy represents, under the former condition, a concurrence of the events denoted by x and y; under the latter, the product of the numbers or quantities denoted by x and y. And thus every expression denoting an event, simple or compound, admits, under another system of interpretation, of a meaning purely quantitative. Here then arises the question, whether there exists any principle of transition, in accordance with which the logical and the numerical interpretations of the same symbolical expression shall have an intelligible connexion. And to this question the following considerations afford an answer.

19. Let it be granted that there exists such a feeling as expectation, a feeling of which the object is the occurrence of events, and which admits of differing degrees of intensity. Let it also be granted that this feeling of expectation accompanies our knowledge of the circumstances under which events are produced, and that it varies with the degree and kind of that knowledge. Then, without assuming, or tacitly implying, that the intensity of the feeling of expectation, viewed as a mental emotion, admits of precise numerical measurement, it is perfectly legitimate to inquire into the possibility of a mode of numerical estimation which shall, at least, satisfy these following conditions, viz., that the numerical value which it assigns shall increase when the known circumstances of an event are felt to justify a stronger expectation, shall diminish when they demand a weaker expectation, and shall remain constant when they obviously require an equal degree of expectation.

Now these conditions at least will be satisfied, if we assume the fundamental principle of expectation to be this, viz., that the laws for the expression of expectation, viewed as a numerical element, shall be the same as the laws for the expression of the expected event viewed as a logical element. Thus if $\phi(x, y, z)$ represent any unconditional event compounded in any manner of the events x, y, z, let the same expression $\phi(x, y, z)$, according to the above principle, denote the expectation of that event; x, y, z representing no longer the simple events involved, but the expectations of those events.

For, in the first place, it is evident that, under this hypothesis, the probability of the occurrence of some one of a set of mutually exclusive events will be equal to the sum of the separate probabilities of those events. Thus if the alternation in question consist of n mutually exclusive events whose expressions are

$$\phi_1(x, y, z), \quad \phi_2(x, y, z), \ldots \phi_n(x, y, z),$$

the expression of that alternation will be

$$\phi_1(x, y, z) + \phi_2(x, y, z) \ldots + \phi_n(x, y, z) = 1\,;$$

the literal symbols x, y, z being logical, and relating to the simple events of which the three alternatives are compounded: and, by hypothesis, the expression of the probability that some one of those alternatives will occur is

$$\phi_1(x, y, z) + \phi_2(x, y, z) \ldots + \phi_n(x, y, z),$$

x, y, z here denoting the probabilities of the above simple events. Now this expression increases, *cæteris paribus*, with the increase of the number of the alternatives which are involved, and diminishes with the diminution of their number; which is agreeable to the condition stated.

Furthermore, if we set out from the above hypothetical definition of the measure of probability, we shall be conducted, either by necessary inference or by successive steps of suggestion, which might perhaps be termed *necessary*, to the received numerical definition. We are at once led to recognise unity (1) as the proper numerical measure of certainty. For it is certain that any event x or its contrary $1 - x$ will occur. The expression of this proposition is

$$x + (1 - x) = 1,$$

whence, by hypothesis, $x + (1 - x)$, the measure of the probability of the above proposition, becomes the measure of certainty. But the value of that expression is 1, whatever the particular value of x may be. Unity, or 1, is therefore, on the hypothesis in question, the measure of certainty.

Let there, in the next place, be n mutually exclusive, but equally possible events, which we will represent by $t_1, t_2, \ldots t_n$.

The proposition which affirms that some one of these must occur will be expressed by the equation

$$t_1 + t_2 \ldots + t_n = 1\,;$$

and, as when we pass in accordance with the reasoning of the last section to numerical probabilities, the same equation remains true in form, and as the probabilities t_1, $t_2 \ldots t_n$ are equal, we have

$$nt_1 = 1,$$

whence $t_1 = \frac{1}{n}$, and similarly $t_2 = \frac{1}{n}$, $t_n = \frac{1}{n}$. Suppose it then required to determine the probability that some one event of the partial series t_2, $t_2 \ldots t_m$ will occur, we have for the expression required

$$t_1 + t_2 \ldots + t_m = \frac{1}{n} + \frac{1}{n} \ldots \text{to } m \text{ terms}$$

$$= \frac{m}{n}.$$

Hence, therefore, if there are m cases favourable to the occurrence of a particular alternation of events out of n possible and equally probable cases, the probability of the occurrence of that alternation will be expressed by the fraction $\frac{m}{n}$.

Now the occurrence of any event which may happen in different equally possible ways is really equivalent to the occurrence of an alternation, i. e., of some one out of a set of alternatives. Hence the probability of the occurrence of any event may be expressed by a fraction whose numerator represents the number of cases favourable to its occurrence, and denominator the total number of equally possible cases. But this is the rigorous numerical definition of the measure of probability. That definition is therefore involved in the more peculiarly *logical* definition, the consequences of which we have endeavoured to trace.

20. From the above investigations it clearly appears, 1st, that whether we set out from the ordinary numerical definition of the measure of probability, or from the definition which assigns to the numerical measure of probability such a law of value as shall establish a formal identity between the logical expressions

of events and the algebraic expressions of their values, we shall be led to the same system of practical results. 2ndly, that either of these definitions pursued to its consequences, and considered in connexion with the relations which it inseparably involves, conducts us, by inference or suggestion, to the other definition. To a scientific view of the theory of probabilities it is essential that both principles should be viewed together, in their mutual bearing and dependence.

CHAPTER XVIII.

ELEMENTARY ILLUSTRATIONS OF THE GENERAL METHOD IN PROBABILITIES.

1. IT is designed here to illustrate, by elementary examples, the general method demonstrated in the last chapter. The examples chosen will be chiefly such as, from their simplicity, permit a ready verification of the solutions obtained. But some intimations will appear of a higher class of problems, hereafter to be more fully considered, the analysis of which would be incomplete without the aid of a distinct method determining the necessary conditions among their data, in order that they may represent a possible experience, and assigning the corresponding limits of the final solutions. The fuller consideration of that method, and of its applications, is reserved for the next chapter.

2. Ex. 1.—The probability that it thunders upon a given day is p, the probability that it both thunders and hails is q, but of the connexion of the two phenomena of thunder and hail, nothing further is supposed to be known. Required the probability that it hails on the proposed day.

Let x represent the event—It thunders.
Let y represent the event—It hails.

Then xy will represent the event—It thunders and hails; and the data of the problem are

$$\text{Prob. } x = p, \quad \text{Prob. } xy = q.$$

There being here but one compound event xy involved, assume, according to the rule,

$$xy = u. \tag{1}$$

Our data then become

$$\text{Prob. } x = p, \quad \text{Prob. } u = q; \tag{2}$$

and it is required to find Prob. y. Now (1) gives

$$y = \frac{u}{x} = ux + \frac{1}{0} u(1-x) + 0\,(1-u)\,x + \frac{0}{0}(1-u)\,(1-x).$$

Hence (XVII. 17) we find

$$V = ux + (1-u)\,x + (1-u)\,(1-x),$$

$$V_x = ux + (1-u)\,x = x, \qquad V_u = ux\,;$$

and the equations of the General Rule, viz.,

$$\frac{V_x}{p} = \frac{V_u}{q} = V.$$

$$\text{Prob. } y = \frac{A + cC}{V}$$

become, on substitution, and observing that $A = ux$, $C = (1-u)(1-x)$, and that V reduces to $x + (1-u)\,(1-x)$,

$$\frac{x}{p} = \frac{ux}{q} = x + (1-u)\,(1-x), \tag{3}$$

$$\text{Prob. } y = \frac{ux + c(1-u)\,(1-x)}{x + (1-u)\,(1-x)}, \tag{4}$$

from which we readily deduce, by elimination of x and u,

$$\text{Prob. } y = q + c(1-p). \tag{5}$$

In this result c represents the unknown probability that if the event $(1-u)\,(1-x)$ happen, the event y will happen. Now $(1-u)\,(1-x) = (1-xy)\,(1-x) = 1-x$, on actual multiplication. Hence c is the unknown probability that if it do not thunder, it will hail.

The general solution (5) may therefore be interpreted as follows:—The probability that it hails is equal to the probability that it thunders and hails, q, together with the probability that it does not thunder, $1-p$, multiplied by the probability c, that if it does not thunder it will hail. And common reasoning verifies this result.

If c cannot be numerically determined, we find, on assigning to it the limiting values 0 and 1, the following limits of Prob. y, viz.:

$$\text{Inferior limit} = q.$$

$$\text{Superior limit} = q + 1 - p.$$

3. Ex. 2.—The probability that one or both of two events happen is p, that one or both of them fail is q. What is the probability that only one of these happens?

Let x and y represent the respective events, then the data are—

$$\text{Prob. } xy + x(1-y) + (1-x)y = p,$$
$$\text{Prob. } x(1-y) + (1-x)y + (1-x)(1-y) = q;$$

and we are to find

$$\text{Prob. } x(1-y) + y(1-x).$$

Here all the events concerned being compound, assume

$$xy + x(1-y) + (1-x)y = s,$$
$$x(1-y) + (1-x)y + (1-x)(1-y) = t,$$
$$x(1-y) + (1-x)y = w.$$

Then eliminating x and y, and determining w as a developed function of s and t, we find

$$w = st + 0\,s(1-t) + 0\,(1-s)t + \frac{1}{0}(1-s)(1-t).$$

Hence $A = st$, $C = 0$, $V = st + s(1-t) + (1-s)t = s + (1-s)t$, $V_s = s$, $V_t = t$; and the equations of the General Rule (XVII. 17) become

$$\frac{s}{p} = \frac{t}{q} = s + (1-s)t, \qquad (1)$$

$$\text{Prob. } w = \frac{st}{s + (1-s)t};$$

whence we find, on eliminating s and t,

$$\text{Prob. } w = p + q - 1.$$

Hence $p + q - 1$ is the measure of the probability sought. This result may be verified as follows:—Since p is the probability that one or both of the given events occur, $1 - p$ will be the probability that they both fail; and since q is the probability that one or both fail, $1 - q$ is the probability that they both happen. Hence $1 - p + 1 - q$, or $2 - p - q$, is the probability that they either both happen or both fail. But the only remaining alternative which is possible is that one alone of the events happens. Hence the probability of this occurrence is $1 - (2 - p - q)$, or $p + q - 1$, as above.

4. Ex. 3.—The probability that a witness A speaks the truth is p, the probability that another witness B speaks the truth is q, and the probability that they disagree in a statement is r. What is the probability that if they agree, their statement is true?

Let x represent the hypothesis that A speaks truth; y that B speaks truth; then the hypothesis that A and B disagree in their statement will be represented by $x(1-y)+y(1-x)$; the hypothesis that they agree in statement by $xy+(1-x)(1-y)$, and the hypothesis that they agree in the truth by xy. Hence we have the following data:

$$\text{Prob. } x = p, \quad \text{Prob. } y = q, \quad \text{Prob. } x(1-y)+y(1-x) = r,$$

from which we are to determine

$$\frac{\text{Prob. } xy}{\text{Prob. } xy+(1-x)(1-y)}.$$

But as Prob. $x(1-y)+y(1-x)=r$, it is evident that Prob. $xy+(1-x)(1-y)$ will be $1-r$; we have therefore to seek

$$\frac{\text{Prob. } xy}{1-r}.$$

Now the compound events concerned being in expression, $x(1-y)+y(1-x)$ and xy, let us assume

$$\left.\begin{aligned} x(1-y)+y(1-x) &= s \\ xy &= w \end{aligned}\right\} \qquad (1)$$

Our data then are Prob. $x=p$, Prob. $y=q$, Prob. $s=r$, and we are to find Prob. w.

The system (1) gives, on reduction,

$$\{x(1-y)+y(1-x)\}(1-s)+s\{xy+(1-x)(1-y)\} \\ +xy(1-w)+w(1-xy)=0;$$

whence

$$w = \frac{x(1-y)(1-s)+y(1-x)(1-s)+sxy+s(1-x)(1-y)+xy}{2xy-1}$$

$$\begin{aligned} &= \frac{1}{0}xys+xy(1-s)+0x(1-y)s+\frac{1}{0}x(1-y)(1-s) \\ &\quad +0(1-x)ys+\frac{1}{0}(1-x)(1-y)s+\frac{1}{0}(1-x)y(1-s) \qquad (2) \\ &\quad +0(1-x)(1-y)(1-s). \end{aligned}$$

In the expression of this development, the coefficient $\frac{1}{0}$ has been made to replace every equivalent form (X. 6). Here we have

$V = xy(1-s) + x(1-y)s + (1-x)ys + (1-x)(1-y)(1-s)$;

whence, passing from Logic to Algebra,

$$\frac{xy(1-s) + x(1-y)s}{p} = \frac{xy(1-s) + (1-x)ys}{q}$$

$$= \frac{x(1-y)s + (1-x)ys}{r}$$

$$= xy(1-s) + x(1-y)s + (1-x)ys + (1-x)(1-y)(1-s).$$

$$\text{Prob. } w = \frac{xy(1-s)}{xy(1-s) + x(1-y)s + (1-x)ys + (1-x)(1-y)(1-s)},$$

from which we readily deduce

$$\text{Prob. } w = \frac{p+q-r}{2};$$

whence we have

$$\frac{\text{Prob. } xy}{1-r} = \frac{p+q-r}{2(1-r)} \qquad (3)$$

for the value sought.

If in the same way we seek the probability that if A and B agree in their statement, that statement will be false, we must replace the second equation of the system (1) by the following, viz.:

$$(1-x)(1-y) = w;$$

the final logical equation will then be

$$w = \frac{1}{0}xys + 0xy(1-s) + 0x(1-y)s + \frac{1}{0}x(1-y)(1-s)$$

$$+ 0(1-x)ys + \frac{1}{0}(1-x)y(1-s) + \frac{1}{0}(1-x)(1-y)s$$

$$+ (1-x)(1-y)(1-s); \qquad (4)$$

whence, proceeding as before, we finally deduce

$$\text{Prob. } w = \frac{2-p-q-r}{2}. \qquad (5)$$

Wherefore we have

$$\frac{\text{Prob.}\,(1-x)(1-y)}{1-r} = \frac{2-p-q-r}{2(1-r)} \qquad (6)$$

for the value here sought.

These results are mutually consistent. For since it is certain that the joint statement of A and B must be either true or false, the second members of (3) and (5) ought by addition to make 1. Now we have identically,

$$\frac{p+q-r}{2(1-r)} + \frac{2-p-q-r}{2(1-r)} = 1.$$

It is probable, from the simplicity of the results (5) and (6), that they might easily be deduced by the application of known principles; but it is to be remarked that they do not fall directly within the scope of known *methods*. The number of the data exceeds that of the simple events which they involve. M. Cournot, in his very able work, "Exposition de la Theorie des Chances," has proposed, in such cases as the above, to select from the original premises different sets of data, each set equal in number to the simple events which they involve, to assume that those simple events are independent, determine separately from the respective sets of the data their probabilities, and comparing the different values thus found for the same elements, judge how far the assumption of independence is justified. This method can only approach to correctness when the said simple events prove, according to the above criterion, to be nearly or quite independent; and in the questions of testimony and of judgment, in which such an hypothesis is adopted, it seems doubtful whether it is justified by actual experience of the ways of men.

5. Ex. 4.—From observations made during a period of general sickness, there was a probability p that any house taken at random in a particular district was visited by fever, a probability q that it was visited by cholera, and a probability r that it escaped both diseases, and was not in a defective sanitary condition as regarded cleanliness and ventilation. What is the probability that any house taken at random was in a defective sanitary condition?

With reference to any house, let us appropriate the symbols x, y, z, as follows, viz.:

The symbol x to the visitation of fever.
y ,, cholera.
z defective sanitary condition.

The events whose probabilities are given are then denoted by x, y, and $(1-x)(1-y)(1-z)$, the event whose probability is sought is z. Assume then,

$$(1-x)(1-y)(1-z)=w;$$

then our data are,

$$\text{Prob. } x=p, \quad \text{Prob. } y=q, \quad \text{Prob. } w=r,$$

and we are to find Prob. z. Now (1) gives

$$z=\frac{(1-x)(1-y)-w}{(1-x)(1-y)}$$

$$=\frac{1}{0}xyw+\frac{0}{0}xy(1-w)+\frac{1}{0}x(1-y)w+\frac{0}{0}x(1-y)(1-w)$$

$$+\frac{1}{0}(1-x)yw+\frac{0}{0}(1-x)y(1-w)+0(1-x)(1-y)w$$

$$+(1-x)(1-y)(1-w). \quad (1)$$

The value of V deduced from the above is

$$V=xy(1-w)+x(1-y)(1-w)+(1-x)y(1-w)$$
$$+(1-x)(1-y)w+(1-x)(1-y)(1-w)=1-w+w(1-x)(1-y);$$

and similarly reducing V_x, V_y, V_w, we get

$$V_x=x(1-w), \quad V_y=y(1-w), \quad V_w=w(1-x)(1-y);$$

furnishing the algebraic equations

$$\frac{x(1-w)}{p}=\frac{y(1-w)}{q}=\frac{w(1-x)(1-y)}{r}=1-w+w(1-x)(1-y). \quad (2)$$

As respects those terms of the development characterized by the coefficients $\frac{0}{0}$, I shall, instead of collecting them into a single term, present them, for the sake of variety (XVII. 18), in the form

$$\frac{0}{0}x(1-w)+\frac{0}{0}(1-x)y(1-w); \quad (3)$$

the value of Prob. z will then be

$$\text{Prob.}\, z = \frac{(1-x)(1-y)(1-w) + cx(1-w) + c'(1-x)y(1-w)}{1-w+w(1-x)(1-y)} \quad (4)$$

From (2) and (4) we deduce

$$\text{Prob.}\, z = \frac{(1-p-r)(1-q-r)}{1-r} + cp + c'\frac{q(1-p-r)}{1-r},$$

as the expression of the probability required. If in this result we make $c = 0$, and $c' = 0$, we find for an inferior limit of its value $\frac{(1-p-r)(1-q-r)}{1-r}$; and if we make $c = 1$, $c' = 1$, we obtain for its superior limit $1 - r$.

6. It appears from inspection of this solution, that the premises chosen were exceedingly defective. The constants c and c' indicate this, and the corresponding terms (3) of the final logical equation show how the deficiency is to be supplied. Thus, since

$$x(1-w) = x\{1-(1-x)(1-y)(1-z)\} = x,$$

$$(1-x)y(1-w) = (1-x)y\{1-(1-x)(1-y)(1-z)\} = (1-x)y,$$

we learn that c is the probability that if any house was visited by fever its sanitary condition is defective, and that c' is the probability that if any house was visited by cholera without fever, its sanitary condition was defective.

If the terms of the logical development affected by the coefficient $\frac{0}{0}$ had been collected together as in the direct statement of the general rule, the final solution would have assumed the following form:

$$\text{Prob.}\, z = \frac{(1-p-r)(1-q-r)}{1-r} + c\left(p+q-\frac{pq}{1-r}\right)$$

c here representing the probability that if a house was visited by either or both of the diseases mentioned, its sanitary condition was defective. This result is perfectly consistent with the former one, and indeed the *necessary* equivalence of the different forms of solution presented in such cases may be formally established.

The above solution may be verified in particular cases. Thus, taking the second form, if $c = 1$ we find Prob. $z = 1 - r$, a correct result. For if the presence of either fever or cholera *certainly* indicated a defective sanitary condition, the probability that any house would be in a defective sanitary state would be simply equal to the probability that it was *not* found in that category denoted by z, the probability of which would, by the data, be $1-r$. Perhaps the general verification of the above solution would be difficult.

The constants p, q, and r in the above solution are subject to the conditions

$$p + r \leqq 1, \qquad q + r \leqq 1.$$

7. Ex. 5.—Given the probabilities of the premises of a hypothetical syllogism to find the probability of the conclusion.

Let the syllogism in its naked form be as follows:

Major premiss: If the proposition Y is true X is true.
Minor premiss: If the proposition Z is true Y is true.
Conclusion: If the proposition Z is true X is true.

Suppose the probability of the major premiss to be p, that of the minor premiss q.

The data then are as follows, representing the proposition X by x, &c., and assuming c and c' as arbitrary constants:

$$\text{Prob. } y = c, \qquad \text{Prob. } xy = cp;$$
$$\text{Prob. } z = c', \qquad \text{Prob. } yz = c'q;$$

from which we are to determine,

$$\frac{\text{Prob. } xz}{\text{Prob. } z} \text{ or } \frac{\text{Prob. } xz}{c'}.$$

Let us assume,

$$xy = u, \qquad yz = v, \qquad xz = w;$$

then, proceeding according to the usual method to determine w as a developed function of y, z, u, and v, the symbols corresponding to propositions whose probabilities are given, we find

$$w = uzvy + 0u(1-z)(1-v)y + 0(1-u)zvy$$
$$+ \frac{0}{0}(1-u)z(1-v)(1-y) + 0(1-u)(1-z)(1-v)y$$
$$+ 0(1-u)(1-z)(1-v)(1-y) + \text{terms whose coefficients are } \frac{1}{0};$$

and passing from Logic to Algebra,

$$\frac{uzvy+u(1-z)(1-v)y}{cp}=\frac{uzvy+(1-u)zvy+(1-u)z(1-v)(1-y)}{c'}$$

$$=\frac{uzvy+(1-u)zvy}{c'q}$$

$$=\frac{uzvy+u(1-z)(1-v)y+(1-u)zvy+(1-u)(1-z)(1-v)y}{c}=V.$$

$$\text{Prob. } w=\frac{uzvy+a(1-u)z(1-v)(1-y)}{V},$$

wherein

$$V=uzvy+u(1-z)(1-v)y+(1-u)zvy+(1-u)z(1-v)(1-y)$$
$$+(1-u)(1-z)(1-v)y+(1-u)(1-z)(1-v)(1-y),$$

the solution of this system of equations gives

$$\text{Prob. } w=c'pq+ac'(1-q),$$

whence

$$\frac{\text{Prob. } xy}{c'}=pq+a(1-q),$$

the value required. In this expression the arbitrary constant a is the probability that if the proposition Z is true and Y false, X is true. In other words, it is the probability, that if the minor premiss is false, the conclusion is true.

This investigation might have been greatly simplified by assuming the proposition Z to be true, and then seeking the probability of X. The data would have been simply

$$\text{Prob. } y=q, \qquad \text{Prob. } xy=pq;$$

whence we should have found Prob. $x=pq+a(1-q)$. It is evident that under the circumstances this mode of procedure would have been allowable, but I have preferred to deduce the solution by the direct and unconditioned application of the method. The result is one which ordinary reasoning verifies, and which it does not indeed require a calculus to obtain. General methods are apt to appear most cumbrous when applied to cases in which their aid is the least required.

Let it be observed, that the above method is equally applicable to the categorical syllogism, and not to the syllogism only, but to every form of deductive ratiocination. Given the probabilities separately attaching to the premises of *any* train of argument; it is always possible by the above method to determine the consequent probability of the truth of a conclusion legitimately drawn from such premises. It is not needful to remind the reader, that the truth and the correctness of a conclusion are different things.

8. One remarkable circumstance which presents itself in such applications deserves to be specially noticed. It is, that propositions which, when true, are equivalent, are not necessarily equivalent when regarded only as probable. This principle will be illustrated in the following example.

Ex. 6.—Given the probability p of the disjunctive proposition "Either the proposition Y is true, or both the propositions X and Y are false," required the probability of the conditional proposition, "If the proposition X is true, Y is true."

Let x and y be appropriated to the propositions X and Y respectively. Then we have

$$\text{Prob.}\, y + (1 - x)(1 - y) = p,$$

from which it is required to find the value of $\dfrac{\text{Prob.}\, xy}{\text{Prob.}\, x}$.

Assume $$y + (1 - x)(1 - y) = t. \qquad (1)$$

Eliminating y we get

$$(1 - x)(1 - t) = 0.$$

whence $$x = \frac{0}{0}t + 1 - t;$$

and proceeding in the usual way,

$$\text{Prob.}\, x = 1 - p + cp. \qquad (2)$$

Where c is the probability that if either Y is true, or X and Y false, X is true.

Next to find Prob. xy. Assume

$$xy = w. \qquad (3)$$

Eliminating y from (1) and (3) we get

$$w(1 - t) = 0;$$

whence, proceeding as above,

$$\text{Prob. } z = cp,$$

c having the same interpretation as before. Hence

$$\frac{\text{Prob. } xy}{\text{Prob. } x} = \frac{cp}{1 - p + cp},$$

for the probability of the truth of the conditional proposition given.

Now in the science of pure Logic, which, as such, is conversant only with truth and with falsehood, the above disjunctive and conditional propositions are equivalent. They are true and they are false together. It is seen, however, from the above investigation, that when the disjunctive proposition has a probability p, the conditional proposition has a different and partly indefinite probability $\frac{cp}{1 - p + cp}$. Nevertheless these expressions are such, that when *either of them becomes* 1 *or* 0, *the other assumes the same value.* The results are, therefore, perfectly consistent, and the logical transformation serves to verify the formula deduced from the theory of probabilities.

The reader will easily prove by a similar analysis, that if the probability of the conditional proposition were given as p, that of the disjunctive proposition would be $1 - c + cp$, where c is the arbitrary probability of the truth of the proposition X.

9. Ex. 7.—Required to determine the probability of an event x, having given either the first, or the first and second, or the first, second, and third of the following data, viz.:

1st. The probability that the event x occurs, or that it alone of the three events x, y, z, fails, is p.

2nd. The probability that the event y occurs, or that it alone of the three events x, y, z, fails, is q.

3rd. The probability that the event z occurs, or that it alone of the three events x, y, z, fails, is r.

SOLUTION OF THE FIRST CASE.

Here we suppose that only the first of the above data is given.

We have then,

$$\text{Prob. } \{x + (1 - x)\, yz\} = p,$$

to find Prob. x.

Let $$x + (1 - x)\, yz = s,$$

then eliminating yz as a single symbol, we get,

$$x\,(1 - s) = 0.$$

Hence

$$x = \frac{0}{1 - s} = \frac{0}{0}\, s + 0\,(1 - s),$$

whence, proceeding according to the rule, we have

$$\text{Prob. } x = cp, \qquad (1)$$

where c is the probability that if x occurs, or alone fails, the former of the two alternatives is the one that will happen. The limits of the solution are evidently 0 and p.

This solution appears to give us no information beyond what unassisted good sense would have conveyed. It is, however, all that the single datum here assumed really warrants us in inferring. We shall in the next solution see how an addition to our data restricts within narrower limits the final solution.

SOLUTION OF THE SECOND CASE.

Here we assume as our data the equations

$$\text{Prob. } \{x + (1 - x)\, yz\} = p,$$
$$\text{Prob. } \{y + (1 - y)\, xz\} = q.$$

Let us write

$$x + (1 - x)\, yz = s,$$
$$y + (1 - y)\, xz = t;$$

from the first of which we have, by (VIII. 7),

$$\{x + (1 - x)\, yz\}\ (1 - s) + s\,\{1 - x - (1 - x)\, yz\} = 0,$$

or $$(x + \bar{x}yz)\,\bar{s} + s\bar{x}\,(1 - yz) = 0;$$

provided that for simplicity we write $\bar{x}$ for $1 - x$, $\bar{y}$ for $1 - y$, and so on. Now, writing for $1 - yz$ its value in constituents, we have

$$(x + \bar{x}yz)\,\bar{s} + s\bar{x}\,(y\bar{z} + \bar{y}z + \bar{y}\,\bar{z}) = 0,$$

an equation consisting solely of positive terms.

In like manner we have from the second equation,

$$(y + \bar{y}xz)\bar{t} + t\bar{y}(x\bar{z} + \bar{x}z + \bar{x}\bar{z}) = 0;$$

and from the sum of these two equations we are to eliminate y and z.

If in that sum we make $y = 1$, $z = 1$, we get the result $\bar{s} + \bar{t}$.

If in the same sum we make $y = 1$, $z = 0$, we get the result

$$x\bar{s} + s\bar{x} + \bar{t}.$$

If in the same sum we make $y = 0$, $z = 1$, we get

$$x\bar{s} + s\bar{x} + x\bar{t} + t\bar{x}.$$

And if, lastly, in the same sum we make $y = 0$, $z = 0$, we find

$$x\bar{s} + s\bar{x} + tx + t\bar{x}, \text{ or } x\bar{s} + s\bar{x} + t.$$

These four expressions are to be multiplied together. Now the first and third may be multiplied in the following manner:

$$(\bar{s} + \bar{t})(x\bar{s} + s\bar{x} + x\bar{t} + t\bar{x})$$

$$= x\bar{s} + x\bar{t} + (\bar{s} + \bar{t})(s\bar{x} + t\bar{x}) \text{ by (IX. Prop. II.)}$$

$$= x\bar{s} + x\bar{t} + \bar{s}\bar{x}t + s\bar{x}\bar{t}. \qquad (2)$$

Again, the second and fourth give by (IX. Prop. I.)

$$(x\bar{s} + s\bar{x} + \bar{t})(x\bar{s} + s\bar{x} + t)$$

$$= x\bar{s} + s\bar{x}. \qquad (3)$$

Lastly, (2) and (3) multiplied together give

$$(x\bar{s} + s\bar{x})(x\bar{s} + s\bar{x}\bar{t} + x\bar{t} + t\bar{x}\bar{s})$$

$$= x\bar{s} + s\bar{x}(s\bar{x}\bar{t} + x\bar{t} + t\bar{x}\bar{s})$$

$$= x\bar{s} + s\bar{x}\bar{t}.$$

Whence the final equation is

$$(1 - s)x + s(1 - t)(1 - x) = 0,$$

which, solved with reference to x, gives

$$x = \frac{s(1 - t)}{s(1 - t) - (1 - s)}$$

$$= \frac{0}{0}st + s(1 - t) + 0(1 - s)t + 0(1 - s)(1 - t),$$

and, proceeding with this according to the rule, we have, finally,

$$\text{Prob. } x = p(1-q) + cpq. \tag{4}$$

where c is the probability that if the event st happen, x will happen. Now if we form the developed expression of st by multiplying the expressions for s and t together, we find—

$c =$ Prob. that if x and y happen together, or x and z happen together, and y fail, or y and z happen together, and x fail, the event x will happen.

The limits of Prob. x are evidently $p(1-q)$ and p.

This solution is more definite than the former one, inasmuch as it contains a term unaffected by an arbitrary constant.

SOLUTION OF THE THIRD CASE.

Here the data are—

$$\text{Prob. } \{x + (1-x)yz\} = p,$$
$$\text{Prob. } \{y + (1-y)xz\} = q,$$
$$\text{Prob. } \{z + (1-z)xy\} = r.$$

Let us, as before, write $\bar{x}$ for $1-x$, &c., and assume

$$x + \bar{x}yz = s,$$
$$y + \bar{y}xz = t,$$
$$z + \bar{z}xy = u.$$

On reduction by (VIII. 8) we obtain the equation

$$(x + \bar{x}yz)\bar{s} + s\bar{x}(y\bar{z} + \bar{y}z + \bar{y}\bar{z})$$
$$+ (y + \bar{y}xz)\bar{t} + t\bar{y}(z\bar{x} + x\bar{z} + \bar{x}\bar{z})$$
$$+ (z + \bar{z}xy)\bar{u} + u\bar{z}(x\bar{y} + \bar{x}y + \bar{x}\bar{y}) = 0. \tag{5}$$

Now instead of directly eliminating y and z from the above equation, let us, in accordance with (IX. Prop. III.), assume the result of that elimination to be

$$Ex + E'(1-x) = 0,$$

then E will be found by making in the given equation $x = 1$, and eliminating y and z from the resulting equation, and E' will be found by making in the given equation $x = 0$, and eliminating y and z from the result. First, then, making $x = 1$, we have

$$\bar{s} + (y + \bar{y}\bar{z})\,\bar{t} + t\bar{y}\bar{z} + (z + y\bar{z})\,\bar{u} + u\bar{y}\bar{z} = 0,$$

and making in the first member of this equation successively $y = 1$, $z = 1$, $y = 1$, $z = 0$, &c., and multiplying together the results, we have the expression

$$(\bar{s} + \bar{t} + \bar{u}\ )\ (\bar{s} + \bar{t} + \bar{u})\,(\bar{s} + \bar{t} + \bar{u})\ (\bar{s} + t + u),$$

which is equivalent to

$$(\bar{s} + \bar{t} + \bar{u})\ (\bar{s} + t + u).$$

This is the expression for E. We shall retain it in its present form. It has already been shown by example (VIII. 3), that the actual reduction of such expressions by multiplication, though convenient, is not necessary.

Again in (5), making $x = 0$, we have

$$yz\bar{s} + s\,(y\bar{z} + \bar{y}z + \bar{y}\,\bar{z}) + y\bar{t} + t\bar{y} + z\bar{u} + u\bar{z} = 0\,;$$

from which, by the same process of elimination, we find for E' the expression

$$(\bar{s} + \bar{t} + \bar{u})\ (s + \bar{t} + u)\ (s + t + \bar{u})\ (s + t + u).$$

The final result of the elimination of y and z from (5) is therefore

$$(\bar{s}+\bar{t}+\bar{u})(\bar{s}+t+u)x+(\bar{s}+\bar{t}+\bar{u})(s+\bar{t}+u)(s+t+\bar{u})(s+t+u)(1-x)=0.$$

Whence we have

$$x = \frac{(\bar{s}+\bar{t}+\bar{u})\ (s+\bar{t}+u)\,(s+t+\bar{u})\,(s+t+u)}{(\bar{s}+\bar{t}+\bar{u})\,(s+\bar{t}+u)\,(s+t+\bar{u})\ (s+t+u)-(\bar{s}+\bar{t}+\bar{u})(\bar{s}+t+u)}\,;$$

or, developing the second member,

$$x = \frac{0}{0}\,stu + \frac{1}{0}\,s\bar{t}u + \frac{1}{0}\,st\bar{u} + s\bar{t}\bar{u} \qquad (6)$$

$$+ \frac{1}{0}\,\bar{s}tu + 0\bar{s}\bar{t}u + 0\bar{s}t\bar{u} + 0\,\bar{s}\bar{t}\bar{u}.$$

Hence, passing from Logic to Algebra,

$$\frac{stu + s\bar{t}\bar{u}}{p} = \frac{stu + \bar{s}t\bar{u}}{q} = \frac{stu + \bar{s}\bar{t}u}{r} \qquad (7)$$

$$= stu + s\bar{t}\bar{u} + \bar{s}\bar{t}u + \bar{s}t\bar{u} + \bar{s}\bar{t}\bar{u}.$$

$$\text{Prob. } x = \frac{s\bar{t}\bar{u} + cstu}{stu + s\bar{t}\bar{u} + \bar{s}\bar{t}u + \bar{s}t\bar{u} + \bar{s}\bar{t}\bar{u}}. \tag{8}$$

To simplify this system of equations, change $\frac{s}{\bar{s}}$ into s, $\frac{t}{\bar{t}}$ into t, &c., and after the change let λ stand for $stu + s + t + 1$. We then have

$$\text{Prob. } x = \frac{s + cstu}{\lambda}, \tag{9}$$

with the relations

$$\frac{stu + s}{p} = \frac{stu + t}{q} = \frac{stu + u}{r} = stu + s + t + u + 1 = \lambda. \tag{10}$$

From these equations we get

$$stu + s = \lambda p, \tag{11}$$

$$stu + s = \lambda - t - u - 1,$$

$$\therefore \lambda p = \lambda - u - t - 1,$$

$$u + t = \lambda (1 - p) - 1.$$

Similarly,

$$u + s = \lambda (1 - q) - 1,$$

and

$$s + t = \lambda (1 - r) - 1.$$

From which equations we find

$$s = \frac{\lambda (1 + p - q - r) - 1}{2}, \qquad t = \frac{\lambda (1 + q - r - p) - 1}{2},$$

$$u = \frac{\lambda (1 + r - p - q) - 1}{2}. \tag{12}$$

Now, by (10),

$$stu = \lambda p - s.$$

Substitute in this equation the values of s, t, and u above determined, and we have

$$\{(1 + p - q - r) \lambda - 1\} \{(1 + q - p - r) \lambda - 1\} \{(1 + r - p - q) \lambda - 1\}$$
$$= 4 \{(p + q + r - 1) \lambda + 1\}, \tag{13}$$

an equation which determines λ. The values of s, t, and u, are then given by (12), and their substitution in (9) completes the solution of the problem.

10. Now a difficulty, the bringing of which prominently before the reader has been one object of this investigation, here arises. How shall it be determined, which root of the above equation ought to taken for the value of λ. To this difficulty some reference was made in the opening of the present chapter, and it was intimated that its fuller consideration was reserved for the next one; from which the following results are taken.

In order that the data of the problem may be derived from a possible experience, the quantities p, q, and r must be subject to the following conditions:

$$\begin{aligned} 1+p-q-r &\overset{=}{>} 0, \\ 1+q-p-r &\overset{=}{>} 0, \\ 1+r-p-q &\overset{=}{>} 0. \end{aligned} \qquad (14)$$

Moreover, the value of λ to be employed in the general solution must satisfy the following conditions:

$$\lambda \overset{=}{>} \frac{1}{1+p-q-r}, \quad \lambda \overset{=}{>} \frac{1}{1+q-p-r}, \quad \lambda \overset{=}{>} \frac{1}{1+r-p-q}. \qquad (15)$$

Now these two sets of conditions suffice for the limitation of the general solution. It may be shown, that the central equation (13) furnishes but one value of λ, which does satisfy these conditions, and that value of λ is the one required.

Let $1+p-q-r$ be the least of the three coefficients of λ given above, then $\frac{1}{1+p-q-r}$ will be the greatest of those values, above which we are to show that there exists but one value of λ. Let us write (13) in the form

$$\{(1+p-q-r)\lambda-1\}\{(1+q-p-r)\lambda-1\}\{(1+r-p-q)\lambda-1\} \\ -4\{(p+q+r-1)\lambda+1\}=0; \qquad (16)$$

and represent the first member by V.

Assume $\lambda = \frac{1}{1+p-q-r}$, then V becomes

$$-4\left(\frac{p+q+r-1}{1+p-q-r}+1\right) = -4\left(\frac{2p}{1+p-q-r}\right),$$

which is *negative*.

Let $\lambda = \infty$, then V is *positive* and infinite.

Again,

$$\frac{d^2 V}{d\lambda^2} = (1 + p - q - r)(1 + q - p - r)\{(1 + r - p - q)\lambda - 1\}$$

+ similar positive terms,

which expression is positive between the limits $\lambda = \frac{1}{1 + p - q - r}$ and $\lambda = \infty$.

If then we construct a curve whose abscissa shall be measured by λ, and whose ordinates by V, that curve will, between the limits specified, pass from below to above the abscissa λ, its convexity always being downwards. Hence it will but once intersect the abscissa λ within those limits; and the equation (16) will, therefore, have but one root thereto corresponding.

The solution is, therefore, expressed by (9), λ being that root of (13) which satisfies the conditions (15), and s, t, and u being given by (12). The interpretation of c may be deduced in the usual way.

It appears from the above, that the problem is, in all cases, more or less indeterminate.

CHAPTER XIX.

OF STATISTICAL CONDITIONS.

1. BY the term statistical conditions, I mean those conditions which must connect the numerical data of a problem in order that those data may be consistent with each other, and therefore such as statistical observations might actually have furnished. The determination of such conditions constitutes an important problem, the solution of which, to an extent sufficient at least for the requirements of this work, I purpose to undertake in the present chapter, regarding it partly as an independent object of speculation, but partly also as a necessary supplement to the theory of probabilities already in some degree exemplified. The nature of the connexion between the two subjects may be stated as follows:

2. There are innumerable instances, and one of the kind presented itself in the last chapter, Ex. 7, in which the solution of a question in the theory of probabilities is finally dependent upon the solution of an algebraic equation of an elevated degree. In such cases the selection of the proper root must be determined by certain conditions, partly relating to the numerical values assigned in the data, partly to the due limitation of the element required. The discovery of such conditions may sometimes be effected by unaided reasoning. For instance, if there is a probability p of the occurrence of an event A, and a probability q of the concurrence of the said event A, and another event B, it is evident that we must have

$$p \overset{=}{>} q.$$

But for the *general* determination of such relations, a distinct method is required, and this we proceed to establish.

As derived from actual experience, the probability of any event is the result of a process of approximation. It is the limit of the ratio of the number of cases in which the event is observed to occur, to the whole number of equally possible cases which observation records,—a limit to which we approach the more nearly as the number of observations is increased. Now let the symbol n, prefixed to the expression of any class, represent the number of individuals contained in that class. Thus, x representing men, and y white beings, let us assume

nx = number of men.

nxy = number of white men.

$nx(1-y)$ = number of men who are not white; and so on.

In accordance with this notation $n(1)$ will represent the number of individuals contained in the universe of discourse, and $\frac{n(x)}{n(1)}$ will represent the probability that any individual being, selected out of that universe of being denoted by $n(1)$, is a man. If observation has not made us acquainted with the *total* values of $n(x)$ and $n(1)$, then the probability in question is the limit to which $\frac{n(x)}{n(1)}$ approaches as the number of individual observations is increased.

In like manner if, as will generally be supposed in this chapter, x represent an event of a particular kind observed, $n(x)$ will represent the number of occurrences of that event, $n(1)$ the number of observed events (equally probable) of all kinds, and $\frac{n(x)}{n(1)}$, or its limit, the probability of the occurrence of the event x.

Hence it is clear that any conclusions which may be deduced respecting the ratios of the quantities $n(x)$, $n(y)$, $n(1)$, &c. may be converted into conclusions respecting the probabilities of the events represented by x, y, &c. Thus, if we should find such a relation as the following, viz.,

$$n(x) + n(y) < n(1),$$

expressing that the number of times in which the event x occurs and the number of times in which the event y occurs, are together less than the number of possible occurrences $n(1)$, we might thence deduce the relation,

$$\frac{n(x)}{n(1)} + \frac{n(y)}{n(1)} < 1,$$

or

$$\text{Prob. } x + \text{Prob. } y < 1.$$

And generally any such statistical relations as the above will be converted into relations connecting the probabilities of the events concerned, by changing $n(1)$ into 1, and any other symbol $n(x)$ into Prob. x.

3. First, then, we shall investigate a method of determining the numerical relations of classes or events, and more particularly the major and minor limits of numerical value. Secondly, we shall apply the method to the limitation of the solutions of questions in the theory of probabilities.

It is evident that the symbol n is distributive in its operation. Thus we have

$$n\{xy + (1-x)(1-y)\} = nxy + n(1-x)(1-y)$$

$$nx(1-y) = nx - nxy,$$

and so on. The number of things contained in any class resolvable into distinct groups or portions is equal to the sum of the numbers of things found in those separate portions. It is evident, further, that any expression formed of the logical symbols x, y, &c. may be developed or expanded in any way consistent with the laws of the symbols, and the symbol n applied to each term of the result, provided that any constant multiplier which may appear, be placed outside the symbol n; without affecting the value of the result. The expression $n(1)$, should it appear, will of course represent the number of individuals contained in the universe. Thus,

$$n(1-x)(1-y) = n(1-x-y+xy)$$

$$n(1) - n(x) - n(y) + n(xy).$$

Again, $$n\{xy + (1-x)(1-y)\} = n(1-x-y+2xy)$$

$$= n(1) - nx - ny + 2nxy).$$

In the last member the term $2nxy$ indicates *twice* the number of individuals contained in the class xy.

4. We proceed now to investigate the numerical limits of classes whose logical expression is given. In this inquiry the following principles are of fundamental importance:

1st. If all the members of a given class possess a certain property x, the total number of individuals contained in the class x

will be a superior limit of the number of individuals contained in the given class.

2nd. A minor limit of the number of individuals in any class y will be found by subtracting a major numerical limit of the contrary class, $1 - y$, from the number of individuals contained in the universe.

To exemplify these principles, let us apply them to the following problem:

PROBLEM.—Given, $n(1)$, $n(x)$, and $n(y)$, required the superior and inferior limits of nxy.

Here our data are the number of individuals contained in the universe of discourse, the number contained in the class x, and the number in the class y, and it is required to determine the limits of the number contained in the class composed of the individuals that are found at once in the class x and in the class y.

By Principle I. this number cannot exceed the number contained in the class x, nor can it exceed the number contained in the class y. Its major limit will then be the least of the two values $n(x)$ and (y).

By Principle II. a minor limit of the class xy will be given by the expression

$$n(1) - \text{major limit of } \{x(1-y) + y(1-x) + (1-x)(1-y)\}, \quad (1)$$

since $x(1-y) + y(1-x) + (1-x)(1-y)$ is the complement of the class xy, i. e. what it wants to make up the universe.

Now $x(1-y) + (1-x)(1-y) = 1 - y$. We have therefore for (1),

$$n(1) - \text{major limit of } \{1 - y + y(1-x)\}$$
$$= n(1) - n(1-y) - \text{major limit of } y(1-x). \quad (2)$$

The major limit of $y(1-x)$ is the least of the two values $n(y)$ and $n(1-x)$. Let $n(y)$ be the least, then (2) becomes

$$n(1) - n(1-y) - n(y)$$
$$= n(1) - n(1) + n(y) - n(y) = 0.$$

Secondly, let $n(1-x)$ be less than $n(y)$, then

$$\text{major limit of } ny(1-x) = n(1-x);$$

therefore (2) becomes

$$n(1) - n(1-y) - n(1-x)$$
$$= n(1) - n(1) + n(y) - n(1) + n(x)$$
$$= nx + ny - n(1).$$

The minor limit of nxy is therefore either 0 or $n(x) + n(y) - n(1)$, according as $n(y)$ is less or greater than $n(1-x)$, or, which is an equivalent condition, according as $n(x)$ is greater or less than $n(1-y)$.

Now as 0 is necessarily a minor limit of the numerical value of *any* class, it is sufficient to take account of the second of the above expressions for the minor limit of $n(xy)$. We have, therefore,

Major limit of $n(xy)$ = least of values $n(x)$ and $n(y)$.
Minor limit of $n(xy) = n(x) + n(y) - n(1)$.*

Proposition I.

5. *To express the major and minor limits of a class represented by any constituent of the symbols x, y, z, &c., having given the values of $n(x)$, $n(y)$, $n(z)$, &c., and $n(1)$.*

Consider first the constituent xyz.

It is evident that the major numerical limit will be the least of the values $n(x)$, $n(y)$, $n(z)$.

The minor numerical limit may be deduced as in the previous problem, but it may also be deduced from the solution of that problem. Thus:

$$\text{Minor limit of } n(xyz) = n(xy) + n(z) - n(1). \qquad (1)$$

Now this means that $n(xyz)$ is at least as great as the expression $n(xy) + n(z) - n(1)$. But $n(xy)$ is at least as great as $n(x) + n(y) - n(1)$. Therefore $n(xyz)$ is at least as great as

$$n(x) + n(y) - n(1) + n(z) - n(1),$$

or $$n(x) + n(y) + n(z) - 2n(1).$$

* The above expression for the minor limit of nxy is applied by Professor De Morgan, by whom it appears to have been first given, to the syllogistic form:

Most men in a certain company have coats.
Most men in the same company have waistcoats.
Therefore some in the company have coats and waistcoats.

Hence we have

$$\text{Minor limit of } n(xyz) = n(x) + n(y) + n(z) - 2n(1).$$

By extending this mode of reasoning we shall arrive at the following conclusions:

1st. The major numerical limit of the class represented by any constituent will be found by prefixing n separately to each factor of the constituent, and taking the least of the resulting values.

2nd. The minor limit will be found by adding all the values above mentioned together, and subtracting from the result as many, less one, times the value of $n(1)$.

Thus we should have

Major limit of $nxy(1-z)$ = least of the values nx, ny, and $n(1-z)$.

$$\begin{aligned}\text{Minor limit of } nxy(1-z) &= n(x) + n(y) + n(1-z) - 2n(1)\\ &= nx + n(y) - n(z) - n(1).\end{aligned}$$

In the use of general symbols it is perhaps better to regard all the values $n(x)$, $n(y)$, $n(1-z)$, as major limits of $n\{xy(1-z)\}$, since, in fact, it cannot exceed any of them. I shall in the following investigations adopt this mode of expression.

Proposition II.

6. *To determine the major numerical limit of a class expressed by a series of constituents of the symbols x, y, z, &c., the values of $n(x)$, $n(y)$, $n(z)$, &c., and $n(1)$, being given.*

Evidently one mode of determining such a limit would be to form the least possible sum of the major limits of the several constituents. Thus a major limit of the expression

$$n\{xy + (1-x)(1-y)\}$$

would be found by adding the least of the two values nx, ny, furnished by the first constituent, to the least of the two values $n(1-x)$, $n(1-y)$, furnished by the second constituent. If we do not know which is in each case the least value, we must form the four possible sums, and reject any of these which are equal to or exceed $n(1)$. Thus in the above example we should have

$$nx \quad + n(1-x) = n(1).$$
$$n(x) + n(1-y) = n(1) + n(x) - n(y).$$
$$n(y) + n(1-y) = n(1) + n(y) - n(x).$$
$$n(y) + n(1-y) = n(1).$$

Rejecting the first and last of the above values, we have

$$n(1) + n(x) - n(y), \text{ and } n(1) + n(y) - n(x),$$

for the expressions required, one of which will (unless $nx = ny$) be less than $n(1)$, and the other greater. The least must of course be taken.

When two or more of the constituents possess a common factor, as x, that factor can only, as is obvious from Principle I., furnish a single term $n(x)$ in the final expression of the major limit. Thus if $n(x)$ appear as a major limit in two or more constituents, we must, in adding those limits together, replace $nx + nx$ by nx, and so on. Take, for example, the expression $n\{xy + x(1-y)z\}$. The major limits of this expression, immediately furnished by addition, would be—

1. nx.
2. $nx + n(1-y)$.
3. $nx + n(z)$.
4. $ny + nx$.
5. $ny + n(1-y)$.
6. $ny + nz$.

Of these the first and sixth only need be retained; the second, third, and fourth being greater than the first; and the fifth being equal to $n(1)$. The limits are therefore

$$n(x) \text{ and } n(y) + n(z),$$

and of these two values the last, supposing it to be less than $n(1)$, must be taken.

These considerations lead us to the following Rule:

RULE.—*Take one factor from each constituent, and prefix to it the symbol n, add the several terms or results thus formed together, rejecting all repetitions of the same term; the sum thus obtained will be a major limit of the expression, and the least of all such sums will be the major limit to be employed.*

Thus the major limits of the expression

$$xyz + x(1-y)(1-z) + (1-x)(1-y)(1-z)$$

would be

$$n(x) + n(1-y), \quad \text{and } n(x) + n(1-z),$$

or $$n(x) + n(1) - n(y), \text{ and } n(x) + n(1) - n(z).$$

If we began with $n(y)$, selected from the first term, and took $n(x)$ from the second, we should have to take $n(1-y)$ from the third term, and this would give

$$n(y) + n(x) + n(1-y), \text{ or } n(1) + n(x).$$

But as this result exceeds $n(1)$, which is an obvious major limit to *every* class, it need not be taken into account.

PROPOSITION III.

7. *To find the minor numerical limit of any class expressed by constituents of the symbols x, y, z, having given $n(x)$, $n(y)$, $n(z)$.. $n(1)$.*

This object may be effected by the application of the preceding Proposition, combined with Principle II., but it is better effected by the following method:

Let any two constituents, which differ from one another only by a single factor, be added, so as to form a single class term as $x(1-y) + xy$ form x, and this species of aggregation having been carried on as far as possible, i. e., there having been selected out of the given series of constituents as many sums of this kind as can be formed, each such sum comprising as many constituents as can be collected into a single term, without regarding whether any of the said constituents enter into the composition of other terms, let these ultimate aggregates, together with those constituents which do not admit of being thus added together, be written down as distinct terms. Then the several minor limits of those terms, deduced by Prop. I., will be the minor limits of the expression given, and one only of those minor limits will at the same time be positive.

Thus from the expression $xy + (1-x)y + (1-x)(1-y)$ we can form the aggregates y and $1-x$, by respectively adding the first and second terms together, and the second and third. Hence $n(y)$ and $n(1-x)$ will be the minor limits of the expression given. Again, if the expression given were

$$xyz + x(1-y)z + (1-x)yz + (1-x)(1-y)z + xy(1-z) + (1-x)(1-y)(1-z),$$

we should obtain by addition of the first four terms the single term z, by addition of the first and fifth term the single term xy, and by addition of the fourth and sixth terms the single term $(1-x)(1-y)$; and there is no other way in which constituents can be collected into single terms, nor are there are any constituents left which have not been thus taken account of. The three resulting terms give, as the minor limits of the given expression, the values

$$n(z), \ n(x) + n(y) - n(1),$$

and $\quad n(1-x) + n(1-y) - n(1)$, or $n(1) - n(x) - n(y)$.

8. The proof of the above rule consists in the proper application of the following principles:—1st. The minor limit of any collection of constituents which admit of being added into a single term, will obviously be the minor limit of that single term. This explains the first part of the rule. 2nd. The minor limit of the sum of any two terms which either are distinct constituents, or consist of distinct constituents, but do not admit of being added together, will be the sum of their respective minor limits, if those minor limits are both positive; but if one be positive, and the other negative, it will be equal to the positive minor limit alone. For if the negative one were added, the value of the limit would be diminished, i. e. it would be less for the sum of two terms than for a single term. Now whenever two constituents differ in more than one factor, so as not to admit of being added together, the minor limits of the two cannot be both positive. Thus let the terms be xyz and $(1-x)(1-y)z$, which differ in two factors, the minor limit of the first is $n(x+y+z-2)$, that of the second $n(1-x+1-y+z-2)$, or,

1st. $n\{x+y-1-(1-z)\}$. 2nd. $n\{1-x-y-(1-z)\}$.

If $n(x+y-1)$ is positive, $n(1-x-y)$ is negative, and the second must be negative. If $n(x+y-1)$ is negative, the first is negative; and similarly for cases in which a larger number of factors are involved. It may in this manner be shown that, according to the mode in which the aggregate terms are formed in the application of the rule, no two minor limits of distinct terms can be added together, for either those terms will involve some common constituent, in which case it is clear that we cannot add their minor limits together,—or the minor limits of the two will not be both positive, in which case the addition would be useless.

PROPOSITION IV.

9. *Given the respective numbers of individuals comprised in any classes, s, t, &c. logically defined, to deduce a system of numerical limits of any other class w, also logically defined.*

As this is the most general problem which it is meant to discuss in the present chapter, the previous inquiries being merely introductory to it, and the succeeding ones occupied with its application, it is desirable to state clearly its nature and design.

When the classes $s, t \ldots w$ are said to be logically defined, it is meant that they are classes so defined as to enable us to write down their symbolical expressions, whether the classes in question be simple or compound. By the general method of this treatise, the symbol w can then be determined directly as a developed function of the symbols s, t, &c. in the form

$$w = A + 0B + \frac{0}{0}C + \frac{1}{0}D, \qquad (1)$$

wherein A, B, C, and D are formed of the constituents of s, t, &c. How from such an expression the numerical limits of w may in the most general manner be determined, will be considered hereafter. At present we merely purpose to show how far this object can be accomplished on the principles developed in the previous propositions; such an inquiry being sufficient for the purposes of this work. For simplicity, I shall found my argument upon the particular development,

$$w = st + 0s(1-t) + \frac{1}{0}(1-s)t + \frac{0}{0}(1-s)(1-t), \qquad (2)$$

in which all the varieties of coefficients present themselves.

Of the constituent $(1-s)(1-t)$, which has for its coefficient $\frac{0}{0}$, it is implied that some, none, or all of the class denoted by that constituent are found in w. It is evident that $n(w)$ will have its highest numerical value when all the members of the class denoted by $(1-s)(1-t)$ are found in w. Moreover, as none of the individuals contained in the classes denoted by $s(1-t)$ and $(1-s)t$ are found in w, the superior numerical limits of w will be identical with those of the class $st+(1-s)(1-t)$. They are, therefore,

$$ns+n(1-t) \text{ and } nt+n(1-s).$$

In like manner a system of superior numerical limits of the development $A+0B+\frac{0}{0}C+\frac{1}{0}D$, *may be found from those of* $A+C$ *by Prop.* 2.

Again, any minor numerical limit of w will, by Principle II., be given by the expression

$$n(1) - \text{major limit of } n(1-w),$$

but the development of w being given by (1), that of $1-w$ will obviously be

$$1-w=0A+B+\frac{0}{0}C+\frac{1}{0}D.$$

This may be directly proved by the method of Prop. 2, Chap. X. Hence

$$\text{Minor limit of } n(w) = n(1) - \text{major limit } (B+C)$$
$$= \text{minor limit of } (A+D),$$

by Principle II., since the classes $A+D$ and $B+C$ are supplementary. Thus the minor limit of the second member of (2) would be $n(t)$, and, generalizing this mode of reasoning, we have the following result:

A system of minor limits of the development

$$A+0B+\frac{0}{0}C+\frac{1}{0}D$$

will be given by the minor limits of $A+D$.

This result may also be directly inferred. For of minor numerical limits we are bound to seek the greatest. Now we obtain in general a higher minor limit by connecting the class D

with A in the expression of w, a combination which, as shown in various examples of the Logic we are permitted to make, than we otherwise should obtain.

Finally, as the concluding term of the development of w indicates the equation $D = 0$, it is evident that $n(D) = 0$. Hence we have

$$\text{Minor limit of } n(D) \leqq 0,$$

and this equation, treated by Prop. 3, gives the requisite conditions among the numerical elements $n(s)$, $n(t)$, &c., in order that the problem may be real, and may embody in its data the results of a possible experience.

Thus from the term $\frac{1}{0}(1-s)t$ in the second member of (2) we should deduce

$$n(1-s) + n(t) - n(1) \leqq 0,$$
$$\therefore n(t) \leqq n(s).$$

These conclusions may be embodied in the following rule:

10. RULE.—*Determine the expression of the class w as a developed logical function of the symbols s, t, &c. in the form*

$$w = A + 0B + \frac{0}{0}C + \frac{1}{0}D.$$

Then will

$$\text{Maj. lim. } w = \text{Maj. lim. } A + C.$$
$$\text{Min. lim. } w = \text{Min. lim. } A + D.$$

The necessary numerical conditions among the data being given by the inequality

$$\text{Min. lim. } D \leqq n(1).$$

To apply the above method to the limitation of the solutions of questions in probabilities, it is only necessary to replace in each of the formulæ $n(x)$ by Prob. x, $n(y)$ by Prob. y, &c., and, finally, $n(1)$ by 1. The application being, however, of great importance, it may be desirable to exhibit in the form of a rule the chief results of transformation.

11. Given the probabilities of any events s, t, &c., whereof another event w is a developed logical function, in the form

$$w = A + 0B + \frac{0}{0}C + \frac{1}{0}D,$$

required the systems of superior and inferior limits of Prob. w, and the conditions among the data.

SOLUTION.—The superior limits of Prob. $(A + C)$, and the inferior limits of Prob. $(A + D)$ will form two such systems as are sought. The conditions among the constants in the data will be given by the inequality,

$$\text{Inf. lim. Prob. } D \overline{\gtrless} 0.$$

In the application of these principles we have always

$$\text{Inf. lim. Prob. } x_1 x_2 \ldots x_n = \text{Prob.} x_1 + \text{Prob. } x_2 \ldots + \text{Prob.} x_n - (n-1).$$

Moreover, the inferior limits can only be determined from *single* terms, either given or formed by aggregation. Superior limits are included in the form Σ Prob. x, Prob. x applying only to symbols which are different, and are taken from different terms in the expression whose superior limit is sought. Thus the superior limits of Prob. $xyz + x(1-y)(1-z)$ are

$$\text{Prob. } x, \quad \text{Prob. } y + \text{Prob. } (1-z), \text{ and Prob. } z + \text{Prob. } (1-y).$$

Let it be observed, that if in the last case we had taken Prob. z from the first term, and Prob. $(1-z)$ from the second,—a connexion not forbidden,—we should have had as their sum 1, which as a result would be useless because *à priori* necessary. It is obvious that we may reject any limits which do not fall between 0 and 1.

Let us apply this method to Ex. 7, Case III. of the last chapter.

The final logical solution is

$$x = \frac{0}{0} stu + \frac{1}{0} s\bar{t}u + \frac{1}{0} st\bar{u} + s\bar{t}\bar{u}$$

$$+ \frac{1}{0} \bar{s}tu + 0\bar{s}\bar{t}u + 0\bar{s}t\bar{u} + 0\bar{s}\bar{t}\bar{u},$$

the data being

$$\text{Prob. } s = p, \quad \text{Prob. } t = q, \quad \text{Prob. } u = r.$$

We shall seek both the numerical limits of x, and the conditions connecting p, q, and r.

The superior limits of x are, according to the rule, given by those of $stu + s\bar{t}\bar{u}$. They are, therefore,

$$p, q + 1 - r, \quad r + 1 - q.$$

The inferior limit of x are given by those of

$$s\bar{t}u + st\bar{u} + s\bar{t}\bar{u} + \bar{s}tu.$$

We may collect the first and third of these constituents in the single term $s\bar{t}$, and the second and third in the single term $s\bar{u}$. The inferior limits of x must then be deduced separately from the terms $s(1 - t)$, $s(1 - u)$, $(1 - s)\,tu$, which give

$$p + 1 - q - 1, \quad p + 1 - r - 1, \quad 1 - p + q + r - 2,$$

or $$p - q, \quad p - r, \quad \text{and } q + r - p - 1.$$

Finally, the conditions among the constants p, q, and r, are given by the terms

$$s\bar{t}u, \quad st\bar{u}, \quad \bar{s}tu,$$

from which, by the rule, we deduce

$$p + 1 - q + r - 2 \stackrel{=}{<} 0, \quad p + q + 1 - r - 2 \stackrel{=}{<} 0, \quad 1 - p + q + r - 2 \stackrel{=}{<} 0.$$

or $$1 + q - p - r \stackrel{=}{>} 0, \qquad 1 + r - p - q \stackrel{=}{>} 0, \quad 1 + p - q - r \stackrel{=}{>} 0.$$

These are the limiting conditions employed in the analysis of the final solution. The conditions by which in that solution λ is limited, were determined, however, simply from the conditions that the quantities s, t, and u should be positive. Narrower limits of that quantity might, in all probability, have been deduced from the above investigation.

12. The following application is taken from an important problem, the solution of which will be given in the next chapter. There are given,

$$\text{Prob. } x = c_1, \quad \text{Prob. } y = c_2, \quad \text{Prob. } s = c_1 p_1, \quad \text{Prob. } t = c_2 p_2,$$

together with the logical equation

$$z = stxy + s\bar{t}x\bar{y} + \bar{s}t\bar{x}y + 0\bar{s}\bar{t}$$

$$+ \frac{1}{0}\left\{\begin{array}{l} stx\bar{y} + st\bar{x}y + st\bar{x}\bar{y} + s\bar{t}xy + s\bar{t}\bar{x}y \\ + s\bar{t}\bar{x}\bar{y} + \bar{s}txy + \bar{s}tx\bar{y} + \bar{s}t\bar{x}\bar{y}; \end{array}\right.$$

and it is required to determine the conditions among the constants c_1, c_2, p_1, p_2, and the major and minor limits of z.

First let us seek the conditions among the constants. Confining our attention to the terms whose coefficients are $\frac{1}{0}$, we readily form, by the aggregation of constituents, the following terms, viz.:

$$s(1-x), \quad t(1-y), \quad sq(1-t), \quad tx(1-s);$$

nor can we form any other terms which are not included under these. Hence the conditions among the constants are,

$$n(s) + n(1-x) - n(1) \leqq 0,$$
$$n(t) + n(1-y) - n(1) \leqq 0,$$
$$n(s) + n(y) + n(1-t) - 2n(1) \leqq 0,$$
$$n(t) + n(x) + n(1-s) - 2n(1) \leqq 0.$$

Now replace $n(x)$ by c_1, $n(y)$ by c_2, $n(s)$ by $c_1 p_1$, $n(t)$ by $c_2 p_2$, and $n(1)$ by 1, and we have, after slight reductions,

$$c_1 p_1 \leqq c_1, \qquad c_2 p_2 \leqq c_2,$$
$$c_1 p_1 \leqq 1 - c_2(1-p_2), \qquad c_2 p_2 \leqq 1 - c_1(1-p_1).$$

Such are, then, the requisite conditions among the constants.

Again, the major limits of z are identical with those of the expression

$$stxy + s(1-t)x(1-y) + (1-s)t(1-x)y;$$

which, if we bear in mind the conditions

$$n(s) \leqq n(x), \qquad n(t) \leqq n(y),$$

above determined, will be found to be

$$n(s) + n(t), \qquad \text{or, } c_1 p_1 + c_2 p_2,$$
$$n(s) + n(1-x), \text{ or, } 1 - c_1(1-p_1)$$
$$n(t) + n(1-y), \text{ or, } 1 - c_2(1-p_2).$$

Lastly, to ascertain the minor limits of z, we readily form from the constituents, whose coefficients are 1 or $\frac{1}{0}$, the single terms s and t, nor can any other terms not included under these be formed by selection or aggregation. Hence, for the minor limits of z we have the values $c_1 p_1$ and $c_2 p_2$.

13. It is to be observed, that the method developed above does not always assign the narrowest limits which it is possible to determine. But it in all cases, I believe, sufficiently limits the solutions of questions in the theory of probabilities.

The problem of the determination of the *narrowest* limits of numerical extension of a class is, however, always reducible to a purely algebraical form.* Thus, resuming the equations

$$w = A + 0B + \frac{0}{0} C + \frac{1}{0} D,$$

let the highest inferior numerical limit of w be represented by the formula $an(s) + bn(t) \ldots + dn(1)$, wherein $a, b, c, \ldots d$ are numerical constants to be determined, and s, t, &c., the logical symbols of which A, B, C, D are constituents. Then

$$an(s) + bn(t) \ldots + dn(1) = \text{minor limit of } A \text{ subject to the condition } D = 0.$$

Hence if we develop the function

$$as + bt \ldots + d,$$

reject from the result all constituents which are found in D, the coefficients of those constituents which remain, and are found also in A, ought not individually to exceed unity in value, and the coefficients of those constituents which remain, and which are not found in A, should individually not exceed 0 in value. Hence we shall have a series of inequalities of the form $f \overline{\overline{<}} 1$, and another series of the form $g \overline{\overline{<}} 0$, f and g being linear functions of a, b, c, &c. Then those values of $a, b \ldots d$, which, while satisfying the above conditions, give to the function

$$an(s) + bn(t) \ldots + dn(1),$$

its highest value must be determined, and the highest value in

* The author regrets the loss of a manuscript, written about four years ago, in which this method, he believes, was developed at considerable length. His recollection of the contents is almost entirely confined to the *impression* that the principle of the method was the same as above described, and that its sufficiency was proved. The prior methods of this chapter are, it is almost needless to say, easier, though certainly less general.

question will be the highest minor limit of w. To the above we may add the relations similarly formed for the determination of the relations among the given constants $n(s)$, $n(t)$ . . $n(1)$.

14. The following somewhat complicated example will show how the limitation of a solution is effected, when the problem involves an arbitrary element, constituting it the representative of a system of problems agreeing in their data, but unlimited in their quæsita.

PROBLEM.—Of n events $x_1\, x_2 \ldots x_n$, the following particulars are known:

1st. The probability that either the event x_1 will occur, or all the events fail, is p_1.

2nd. The probability that either the event x_2 will occur, or all the events fail, is p_2. And so on for the others.

It is required to find the probability of any single event, or combination of events, represented by the general functional form $\phi(x_1 \ldots x_n)$, or ϕ.

Adopting a previous notation, the data of the problem are

$$\text{Prob.}\,(x_1 + \bar{x}_1 \ldots \bar{x}_n) = p_1 \ \ldots \ \text{Prob.}\,(x_n + \bar{x}_1 \ldots \bar{x}_n) = p_n.$$

And Prob. $\phi\,(x_1 \ldots x_n)$ is required.

Assume generally

$$x_r + \bar{x}_1 \ldots \bar{x}_n = s_r, \tag{1}$$

$$\phi = w. \tag{2}$$

We hence obtain the collective logical equation of the problem

$$\Sigma\{(x_r + \bar{x}_1 \ldots \bar{x}_n)\,\bar{s}_r + s_r\,(\bar{x}_r - \bar{x}_1 \ldots \bar{x}_n)\} + \phi\bar{w} + w\bar{\phi} = 0. \tag{3}$$

From this equation we must eliminate the symbols $x_1, \ldots x_n$, and determine w as a developed logical function of $s_1 \ldots s_n$.

Let us represent the result of the aforesaid elimination in the form

$$Ew + E'(1 - w) = 0;$$

then will E be the result of the elimination of the same symbols from the equation

$$\Sigma\,\{(x_r + \bar{x}_1 \ldots \bar{x}_n)\,\bar{s}_r + s_r\,(\bar{x}_r - \bar{x}_1 \ldots \bar{x}_n)\} + 1 - \phi = 0. \tag{4}$$

Now E will be the product of the coefficients of all the constituents (considered with reference to the symbols x_1, $x_2 \ldots x_n$)

which are found in the development of the first member of the above equation. Moreover, ϕ, and therefore $1-\phi$, will consist of a series of such constituents, having unity for their respective coefficients. In determining the forms of the coefficients in the development of the first member of (4), it will be convenient to arrange them in the following manner:

1st. The coefficients of constituents found in $1-\phi$.

2nd. The coefficient of $\bar{x}_1, \bar{x}_2 \ldots \bar{x}_n$, if found in ϕ.

3rd. The coefficients of constituents found in ϕ, excluding the constituent $\bar{x}_1, \bar{x}_2 \ldots \bar{x}_n$.

The above is manifestly an exhaustive classification.

First then; the coefficient of any constituent found in $1-\phi$, will, in the development of the first member of (4), be of the form

$$1 + \text{positive terms derived from } \Sigma.$$

Hence, every such coefficient may be replaced by unity, Prop. I. Chap. IX.

Secondly; the coefficient of $\bar{x}_1 \ldots \bar{x}_n$, if found in ϕ, in the development of the first member of (4) will be

$$\Sigma \bar{s}_r, \text{ or } \bar{s}_1 + \bar{s}_2 \ldots + \bar{s}_n$$

Thirdly; the coefficient of any other constituent, $x_1 \ldots x_i$, $\bar{x}_{i+1} \ldots \bar{x}_n$, found in ϕ, in the development of the first member of (4) will be $\bar{s}_1 \ldots + \bar{s}_i + s_{i+1} \ldots + s_n$.

Now it is seen, that E is the product of all the coefficients above determined; but as the coefficients of those constituents which are not found in ϕ reduce to unity, E may be regarded as the product of the coefficients of those constituents which are found in ϕ. From the mode in which those coefficients are formed, we derive the following rule for the determination of E, viz., in each constituent found in ϕ, except the constituent $\bar{x}_1 \bar{x}_2 \ldots \bar{x}_n$, for x_1 write $\bar{s}_1$, for $\bar{x}_1$ write s_1, and so on, and add the results; but for the constituent $\bar{x}_1, \bar{x}_2 \ldots \bar{x}_n$, if it occur in ϕ, write $\bar{s}_1 + \bar{s}_2 \ldots + \bar{s}_n$; the product of all these sums is E.

To find E' we must in (3) make $w=0$, and eliminate $x_1, x_2 \ldots x_n$ from the reduced equation. That equation will be

$$\Sigma \{ (x_r + \bar{x}_1 \ldots + \bar{x}_n) \bar{s}_r + s_r (\bar{x}_r - \bar{x}_1 \ldots \bar{x}_n) \} + \phi = 0. \qquad (5)$$

Hence E' will be formed from the constituents in $1 - \phi$, i.e. from the constituents *not* found in ϕ in the same way as E is formed from the constituents found in ϕ.

Consider next the equation

$$Ew + E'(1 - w) = 0.$$

This gives

$$w = \frac{E'}{E' - E}. \qquad (6)$$

Now E and E' are functions of the symbols $s_1, s_2 \ldots s_n$. The expansion of the value of w will, therefore, consist of all the constituents which can be formed out of those symbols, with their proper coefficients annexed to them, as determined by the rule of development.

Moreover, E and E' are each formed by the multiplication of factors, and neither of them can vanish unless some one of the factors of which it is composed vanishes. Again, any factor, as $\bar{s}_1 \ldots + \bar{s}_n$ can only vanish when all the terms by the addition of which it is formed vanish together, since in development we attribute to these terms the values 0 and 1, only. It is further evident, that no two factors differing from each other can vanish together. Thus the factors $\bar{s}_1 + \bar{s}_2 \ldots + \bar{s}_n$, and $s_1 + \bar{s}_2 \ldots + \bar{s}_n$, cannot simultaneously vanish, for the former cannot vanish unless $s_1 = 0$, or $s_1 = 1$; but the latter cannot vanish unless $s_1 = 0$.

First, let us determine the coefficient of the constituent $\bar{s}_1 \bar{s}_2 \ldots \bar{s}_n$ in the development of the value of w.

The simultaneous assumption $\bar{s}_1 = 1$, $\bar{s}_2 = 1 \ldots \bar{s}_n = 1$, would cause the factor $s_1 + s_2 \ldots + s_n$ to vanish if this should occur in E or E'; and no other factor under the same assumption would vanish; but $s_1 + s_2 \ldots + s_n$ does not occur as a factor of either E or E'; neither of these quantities, therefore, can vanish; and, therefore, the expression $\frac{E'}{E' - E}$, is neither 1, 0, nor $\frac{0}{0}$.

Wherefore the coefficient of $\bar{s}_1 \bar{s}_2 \ldots \bar{s}_n$ *in the expanded value of* w, may be represented by $\frac{1}{0}$.

Secondly, let us determine the coefficient of the constituent $s_1 s_2 \ldots s_n$.

The assumptions $s_1 = 1$, $s_2 = 1$, .. $s_n = 1$, would cause the factor $\bar{s}_1 + \bar{s}_2 \ldots + \bar{s}_n$ to vanish. Now this factor is found in E and not in E' whenever ϕ contains both the constituents $x_1\,x_2 \ldots x_n$ and $\bar{x}_1\,\bar{x}_2 \ldots \bar{x}_n$. Here then $\frac{E'}{E'-E}$ becomes $\frac{E'}{E'}$ or 1. The factor $\bar{s}_1 + \bar{s}_2 \ldots + \bar{s}_n$ is found in E' and not in E, if ϕ contains neither of the constituents $x_1\,x_2 \ldots x_n$ and $\bar{x}_1\,\bar{x}_2 \ldots \bar{x}_n$. Here then $\frac{E'}{E'-E}$ becomes $\frac{0}{-E}$ or 0. Lastly, the factor $\bar{s}_1 + \bar{s}_2 \ldots + \bar{s}_n$ is contained in both E and E', if one of the constituents $x_1\,x_2 \ldots x_n$ and $\bar{x}_1\bar{x}_2 \ldots \bar{x}_n$ is found in ϕ, and one is not. Here then $\frac{E'}{E'-E}$ becomes $\frac{0}{0}$.

The coefficient of the constituent $s_1\,s_2 \ldots s_n$, *will therefore be* 1, 0, *or* $\frac{0}{0}$, *according as* ϕ *contains both the constituents* $x_1\,x_2 \ldots x_n$ *and* $\bar{x}_1\,\bar{x}_2 \ldots \bar{x}_n$, *or neither of them, or one of them and not the other.*

Lastly, to determine the coefficient of any other constituent as $s_1 \ldots s_i\,\bar{s}_{i+1} \ldots \bar{s}_n$.

The assumptions $s_1 = 1$, .. $s_i = 1$, $s_{i+1} = 0$, $s_n = 0$, would cause the factor $\bar{s}_1 \ldots + \bar{s}_i + s_{i+1} \ldots + s_n$ to vanish. Now this factor is found in E, if the constituent $x_1 \ldots x_i\,\bar{x}_{i+1} \ldots \bar{x}_n$ is found in ϕ and in E', if the said constituent is not found in ϕ. In the former case we have $\frac{E'}{E'-E} = \frac{E'}{E'} = 1$; in the latter case we have $\frac{E'}{E'-E} = \frac{0}{0-E} = 0$.

Hence the coefficient of any other constituent $s_1 \ldots s_i$, $\bar{s}_{i+1} \ldots \bar{s}_n$, *is* 1 *or* 0 *according as the similar constituent* $x_1 \ldots x_i\ \bar{x}_{i+1} \ldots \bar{x}_n$ *is or is not found in* ϕ.

We may, therefore, practically determine the value of w in the following manner. Rejecting from the given expression of ϕ the constituents $x_1\,x_2 \ldots x_n$ and and $\bar{x}_1\,\bar{x}_2 \ldots \bar{x}_n$, should both or either of them be contained in it, let the symbols x_1, x_2, .. x_n, in the result be changed into s_1, s_2, .. s_n respectively. Let the coefficients of the constituents $s_1\,s_2 \ldots s_n$ and $\bar{s}_1\,\bar{s}_2 \ldots \bar{s}_n$ be determined according to the special rules for those cases given above, and let every other constituent have for its coefficient 0. The result will be the value of w as a function of $s_1, s_2, \ldots s_n$.

As a particular case, let $\phi = x_1$. It is required from the given data to determine the probability of the event x_1.

The symbol x_1, expanded in terms of the entire series of symbols $x_1, x_2, \ldots x_n$, will generate all the constituents of those symbols which have x_1 as a factor. Among those constituents will be found the constituent $x_1 x_2 \ldots x_n$, but not the constituent $\bar{x}_1 \bar{x}_2 \ldots \bar{x}_n$.

Hence in the expanded value of x_1 as a function of the symbols $s_1, s_2, \ldots s_n$, the constituent $s_1 s_2 \ldots s_n$ will have the coefficient $\frac{0}{0}$, and the constituent $\bar{s}_1 \bar{s}_2 \ldots \bar{s}_n$ the coefficient $\frac{1}{0}$.

If from x_1 we reject the constituent $x_1 x_2 \ldots x_n$, the result will be $x_1 - x_1 x_2 \ldots x_n$, and changing therein x_1 into s_1, &c., we have $s_1 - s_1 s_2 \ldots s_n$ for the corresponding portion of the expression of x_1 as a function of $s_1, s_2, \ldots s_n$.

Hence the final expression for x_1 is

$$x_1 = s_1 - s_1 s_2 \ldots s_n + \frac{0}{0} s_1 s_2 \ldots s_n + \frac{1}{0} \bar{s}_1 \bar{s}_2 \ldots \bar{s}_n \qquad (7)$$
$$+ \text{ constituents whose coefficients are } 0.$$

The sum of all the constituents in the above expansion whose coefficients are either 1, 0, or $\frac{0}{0}$, will be $1 - \bar{s}_1 \bar{s}_2 \ldots \bar{s}_n$.

We shall, therefore, have the following *algebraic* system for the determination of Prob. x_1, viz.:

$$\text{Prob. } x_1 = \frac{s_1 - s_1 s_2 \ldots s_n + c s_1 s_2 \ldots s_n}{1 - \bar{s}_1 \bar{s}_2 \ldots \bar{s}_n}, \qquad (8)$$

with the relations

$$\frac{s_1}{p_1} = \frac{s_2}{p_2} \ldots = \frac{s_n}{p_n} \qquad (9)$$
$$= 1 - \bar{s}_1 s_2 \ldots s_n = \lambda.$$

It will be seen, that the relations for the determination of $s_1 s_2 \ldots s_n$ are quite independent of the form of the function ϕ, and the values of these quantities, determined once, will serve for all possible problems in which the data are the same, however the *quæsita* of those problems may vary. The nature of that event, or combination of events, whose probability is sought, will affect only the form of the function in which the determined values of $s_1\, s_2 \,..\, s_n$ are to be substituted.

We have from (9)

$$s_1 = p_1\lambda, \qquad s_2 = p_2\lambda, \;..\; s_n = p_n\lambda.$$

Whence

$$1 - (1 - p_1\lambda)\,(1 - p_2\lambda)\,..\,(1 - p_n\lambda) = \lambda.$$

Or,

$$1 - \lambda = (1 - p_1\lambda)\,(1 - p_2\lambda)\,..\,(1 - p_n\lambda)\,; \qquad (10)$$

from which equation the value of λ is to be determined.

Supposing this value determined, the value of Prob. x_1 will be

$$\frac{p_1\lambda - (1 - c)\,p_1\,p_2\,..\,p_n\,\lambda^n}{1 - (1 - p_1\lambda)\,(1 - p_2\lambda)\,..\,(1 - p_n\lambda)},$$

or, on reduction by (10),

$$\text{Prob. } x_1 = p_1 - (1 - c)\,p_1 p_2\,..\,p_n\lambda^{n-1}. \qquad (11)$$

Let us next seek the conditions which must be fulfilled among the constants $p_1, p_2, .. p_n$, and the limits of the value of Prob. x_1.

As there is but one term with the coefficient $\frac{1}{0}$, there is but one condition among the constants, viz.,

$$\text{Minor limit, } (1 - s_1)\,(1 - s_2)\,..\,(1 - s_n) \overline{\overline{<}}\, 0.$$

Or, $n\,(1 - s_1) + n\,(1 - s_2)\,..+ n\,(1 - s_n) - (n - 1)\,n\,(1) \overline{\overline{<}}\, 0.$

Or, $$n\,(1) - n\,(s_1) - n\,(s_2)\,..- n\,(s_n) \overline{\overline{<}}\, 0.$$

Whence $$p_1 + p_2\,..+ p_n \overline{\overline{>}}\, 1,$$

the condition required.

The major limit of Prob. x_1 is the major limit of the sum of those constituents whose coefficients are 1 or $\frac{0}{0}$. But that sum is s_1. Hence,

$$\text{Major limit, Prob. } x_1 = \text{major limit } s_1 = p_1.$$

The minor limit of Prob. x_1 will be identical with the minor limit of the expression

$$s_1 - s_1 s_2 \ldots s_n + (1 - s_1)(1 - s_2) \ldots (1 - s_n).$$

A little attention will show that the different aggregates, terms which can be formed out of the above, each including the greatest possible number of constituents, will be the following, viz.:

$$s_1(1 - s_2), \quad s_1(1 - s_3), \ldots s_1(1 - s_n), \quad (1 - s_2)(1 - s_3) \ldots (1 - s_n).$$

From these we deduce the following expressions for the minor limit, viz.:

$$p_1 - p_2, \quad p_1 - p_3 \ldots p_1 - p_n, \quad 1 - p_2 - p_3 \ldots - p_n.$$

The value of Prob. x_1 will, therefore, not fall short of any of these values, nor exceed the value of p_1.

Instead, however, of employing these conditions, we may directly avail ourselves of the principle stated in the demonstration of the general method in probabilities. The condition that $s_1, s_2, \ldots s_n$ must each be less than unity, requires that λ should be less than each of the quantities $\frac{1}{p_1}, \frac{1}{p_2}, \ldots \frac{1}{p_n}$. And the condition that $s_1, s_2, \ldots s_n$, must each be greater than 0, requires that λ should also be greater than 0. Now $p_1\, p_2 \ldots p_n$ being proper fractions satisfying the condition

$$p_1 + p_2 \ldots + p_n > 1,$$

it may be shown that but one positive value of λ can be deduced from the central equation (10) which shall be less than each of the quantities $\frac{1}{p_1}, \frac{1}{p_2}, \ldots \frac{1}{p_n}$. That value of λ is, therefore, the one required.

To prove this, let us consider the equation

$$(1 - p_1\lambda)(1 - p_2\lambda) \ldots (1 - p_n\lambda) - 1 + \lambda = 0.$$

When $\lambda = 0$ the first member vanishes, and the equation is satisfied. Let us examine the variations of the first member between the limits $\lambda = 0$ and $\lambda = \frac{1}{p_1}$, supposing p_1 the greatest of the values $p_1\, p_2 \ldots p_n$.

Representing the first member of the equation by V, we have

$$\frac{dV}{d\lambda} = -p_1(1-p_2\lambda)\ldots(1-p_n\lambda)\ldots -p_n(1-p_1\lambda)\ldots(1-p_{n-1}\lambda)+1,$$

which, when $\lambda = 0$, assumes the form $-p_1 - p_2 \ldots - p_n + 1$, and is negative in value.

Again, we have

$$\frac{d^2V}{d\lambda^2} = p_1p_2(1-p_3\lambda)(1-p_n\lambda) + \&c.,$$

consisting of a series of terms which, under the given restrictions with reference to the value of λ, are *positive*.

Lastly, when $\lambda = \frac{1}{p_1}$, we have

$$V = -1 + \frac{1}{p_1},$$

which is positive.

From all this it appears, that if we construct a curve, the ordinates of which shall represent the value of V corresponding to the abscissa λ, that curve will pass through the origin, and will for small values of λ lie beneath the abscissa. Its convexity will, between the limits $\lambda = 0$ and $\lambda = \frac{1}{p_1}$ be downwards, and at the extreme limit $\frac{1}{p_1}$ the curve will be above the abscissa, its ordinate being positive. It follows from this description, that it will intersect the abscissa once, and only once, within the limits specified, viz., between the values $\lambda = 0$, and $\lambda = \frac{1}{p_1}$.

The solution of the problem is, therefore, expressed by (11), the value of λ being that root of the equation (10), which lies within the limits 0 and $\frac{1}{p_1}, \frac{1}{p_2}, \ldots \frac{1}{p_n}$.

The constant c is obviously the probability, that if the events $x_1, x_2, \ldots x_n$, all happen, or all fail, they will all happen.

This determination of the value of λ suffices for all problems in which the data are the same as in the one just considered. It is, as from previous discussions we are prepared to expect, a determination independent of the form of the function ϕ.

Let us, as another example, suppose

$$\phi = \text{or } w = x_1(1-x_2)\,..\,(1-x_n)\,..+x_n(1-x_1)\,..\,(1-x_{n-1}).$$

This is equivalent to requiring the probability, that of the events $x_1, x_2, \ldots x_n$ one, and only one, will happen. The value of w will obviously be

$$w = s_1(1-s_2)\,..\,(1-s_n)\,..+s_n(1-s_1)\,..\,(1-s_{n-1})+\frac{1}{0}(1-s_1)\,..\,(1-s_n),$$

from which we should have

$$\text{Prob. }\{x_1(1-x_2)\,..\,(1-x_n)\,..+x_n(1-x_1)\,..\,(1-x_{n-1})\}$$

$$=\frac{s_1(1-s_2)\,..\,(1-s_n)\,..+s_n(1-s_1)\,..\,(1-s_{n-1})}{1-(1-s_1)\,..\,(1-s_n)}$$

$$=\frac{p_1\lambda(1-p_2\lambda)\,..\,(1-p_n\lambda)\,..+p_n\lambda(1-p_1\lambda)\,..\,(1-p_{n-1}\lambda)}{\lambda}$$

$$=\frac{p_1(1-\lambda)}{1-p_1\lambda}+\frac{p_2(1-\lambda)}{1-p_2\lambda}\,..+\frac{p_n(1-\lambda)}{1-p_n\lambda}.$$

This solution serves well to illustrate the remarks made in the introductory chapter (I. 16) The essential difficulties of the problem are founded in the nature of its data and not in that of its quæsita. The central equation by which λ is determined, and the peculiar discussions connected therewith, are equally pertinent to every form which that problem can be made to assume, by varying the interpretation of the arbitrary elements in its original statement.

CHAPTER XX.

PROBLEMS RELATING TO THE CONNEXION OF CAUSES AND EFFECTS.

1. SO to apprehend in all particular instances the relation of cause and effect, as to connect the two extremes in thought according to the order in which they are connected in nature (for the *modus operandi* is, and must ever be, unknown to us), is the final object of science. This treatise has shown, that there is special reference to such an object in the constitution of the intellectual faculties. There is a sphere of thought which comprehends things only as coexistent parts of a universe; but there is also a sphere of thought (Chap. XI.) in which they are apprehended as links of an unbroken, and, to human appearance, an endless chain—as having their place in an order connecting them both with that which has gone before, and with that which shall follow after. In the contemplation of such a series, it is impossible not to feel the pre-eminence which is due, above all other relations, to the relation of cause and effect.

Here I propose to consider, in their abstract form, some problems in which the above relation is involved. There exists among such problems, as might be anticipated from the nature of the relation with which they are concerned, a wide diversity. From the probabilities of causes assigned *à priori*, or given by experience, and their respective probabilities of association with an effect contemplated, it may be required to determine the probability of that effect; and this either, 1st, absolutely, or 2ndly, under given conditions. To such an object some of the earlier of the following problems relate. On the other hand, it may be required to determine the probability of a particular cause, or of some particular connexion among a system of causes, from observed effects, and the known tendencies of the said causes, singly or in connexion, to the production of such effects. This class of questions will be considered in a subsequent portion of the chapter, and other forms of the general inquiry will also be noticed. I would remark, that although these examples are designed chiefly as illustrations of a *method*, no regard has been paid to the question of ease or convenience in the application of that method. On the contrary, they have been devised, with whatever success, as types of the class of problems which might be expected to arise from the study of the relation of cause and effect in the more complex of its actual and visible manifestations.

2. PROBLEM I.—The probabilities of two causes A_1 and A_2 are c_1 and c_2 respectively. The probability that if the cause A_1 present itself, an event E will accompany it (whether as a consequence of the cause A_1 or not) is p_1, and the probability that if the cause A_2 present itself, that event E will accompany it, whether as a consequence of it or not, is p_2. Moreover, the event E cannot appear in the absence of both the causes A_1 and A_2.* Required the probability of the event E.

The solution of what this problem becomes in the case in which the causes A_1, A_2 are mutually exclusive, is well known to be

$$\text{Prob.}\, E = c_1 p_1 + c_2 p_2;$$

and it expresses a particular case of a fundamental and very important principle in the received theory of probabilities. Here it is proposed to solve the problem free from the restriction above stated.

---

* The mode in which such data as the above might be furnished by experience is easily conceivable. Opposite the window of the room in which I write is a field, liable to be overflowed from two causes, distinct, but capable of being combined, viz., floods from the upper sources of the River Lee, and tides from the ocean. Suppose that observations made on N separate occasions have yielded the following results: On A occasions the river was swollen by freshets, and on P of those occasions it was inundated, whether from this cause or not. On B occasions the river was swollen by the tide, and on Q of those occasions it was inundated, whether from this cause or not. Supposing, then, that the field cannot be inundated in the absence of *both* the causes above mentioned, let it be required to determine the total probability of its inundation.

Here the elements a, b, p, q of the general problem represent the ratios

$$\frac{A}{N}, \frac{P}{A}, \frac{B}{N}, \frac{Q}{B},$$

or rather the values to which those ratios approach, as the value of N is indefinitely increased.

Let us represent

The cause A_1 by x.
The cause A_2 by y.
The effect E by z.

Then we have the following numerical data:

$$\text{Prob. } x = c_1, \qquad \text{Prob. } y = c_2,$$
$$\text{Prob. } xz = c_1 p_1, \qquad \text{Prob. } yz = c_2 p_2. \qquad (1)$$

Again, it is provided that if the causes A_1, A_2 are both absent, the effect E does not occur; whence we have the logical equation

$$(1 - x)(1 - y) = v(1 - z).$$

Or, eliminating v,

$$z(1 - x)(1 - y) = 0. \qquad (2)$$

Now assume,

$$xz = s, \qquad yz = t. \qquad (3)$$

Then, reducing these equations (VIII. 7), and connecting the result with (2),

$$xz(1 - s) + s(1 - xz) + yz(1 - t) + t(1 - yz) + z(1 - x)(1 - y) = 0. \qquad (4)$$

From this equation, z must be determined as a developed logical function of x, y, s, and t, and its probability thence deduced by means of the data,

$$\text{Prob. } x = c_1, \quad \text{Prob. } y = c_2, \quad \text{Prob. } s = c_1 p_1, \quad \text{Prob. } t = c_2 p_2. \qquad (5)$$

Now developing (4) with respect to z, and putting $\bar{x}$ for $1 - x$, $\bar{y}$ for $1 = y$, and so on, we have

$$(x\bar{s} + s\bar{x} + y\bar{t} + t\bar{y} + \bar{x}\bar{y})z + (s + t)\bar{z} = 0,$$

$$\therefore\ z + \frac{s+t}{s + t - x\bar{s} - s\bar{x} - y\bar{t} - t\bar{y} - \bar{x}\bar{y}}$$

$$= stxy + \frac{1}{0}stx\bar{y} + \frac{1}{0}st\bar{x}y + \frac{1}{0}st\bar{x}\bar{y}$$

$$+ \frac{1}{0}s\bar{t}xy + s\bar{t}x\bar{y} + \frac{1}{0}s\bar{t}\bar{x}y + \frac{1}{0}s\bar{t}\bar{x}\bar{y}$$

$$+ \frac{1}{0}\bar{s}txy + \frac{1}{0}\bar{s}tx\bar{y} + \bar{s}t\bar{x}y + \frac{1}{0}\bar{s}t\bar{x}\bar{y}$$

$$+ 0\bar{s}\bar{t}xy + 0\bar{s}\bar{t}x\bar{y} + 0\bar{s}\bar{t}\bar{x}y + 0\bar{s}\bar{t}\bar{x}\bar{y}. \qquad (6)$$

From this result we find (XVII. 17),

$$V = stxy + s\bar{t}x\bar{y} + \bar{s}t\bar{x}y + \bar{s}\bar{t}xy + \bar{s}\bar{t}x\bar{y}$$
$$+ \bar{s}\bar{t}\bar{x}y + \bar{s}\bar{t}\bar{x}\bar{y}$$
$$= stxy + s\bar{t}x\bar{y} + \bar{s}t\bar{x}y + \bar{s}\bar{t}.$$

Whence, passing from Logic to Algebra, we have the following system of equations, u standing for the probability sought:

$$\frac{stxy + s\bar{t}x\bar{y} + \bar{s}\bar{t}x}{c_1} = \frac{stxy + \bar{s}t\bar{x}y + \bar{s}\bar{t}y}{c_2}$$

$$= \frac{stxy + s\bar{t}x\bar{y}}{c_1 p_1} = \frac{stxy + \bar{s}t\bar{x}y}{c_2 p_2} \qquad (7)$$

$$= \frac{stxy + s\bar{t}x\bar{y} + \bar{s}t\bar{x}y}{u} = \frac{stxy + s\bar{t}x\bar{y} + \bar{s}t\bar{x}y + \bar{s}\bar{t}}{1} = V,$$

from which we must eliminate s, t, x, y, and V.

Now if we have any series of equal fractions, as

$$\frac{a}{a'} = \frac{b}{b'} = \frac{c}{c'} \ldots = \lambda,$$

we know that

$$\frac{la + mb + nc}{la' + mb' + nc'} = \lambda.$$

And thus from the above system of equations we may deduce

$$\frac{\bar{s}t\bar{x}y}{u - c_1 p_1} = \frac{s\bar{t}x\bar{y}}{u - c_2 p_2} = \frac{\bar{s}\bar{t}}{1 - u} = V;$$

whence we have, on equating the product of the three first members to the cube of the last,

$$\frac{s\bar{s}^2 t\bar{t}^2 x\bar{x}y\bar{y}}{(u - c_1 p_1)(u - c_2 p_2)(1 - u)} = V^3. \qquad (8)$$

Again, from the system (7) we have

$$\frac{\bar{s}\bar{t}\bar{x}}{1 - u - c_1 + c_1 p_1} = \frac{\bar{s}\bar{t}\bar{y}}{1 - u - c_2 + c_2 p_2} = \frac{stxy}{c_1 p_1 + c_2 p_2 - u} = V,$$

whence proceeding as before

$$\frac{s\bar{s}^2 t\bar{t}^2 x\bar{x}y\bar{y}}{(1 - c_1 + c_1 p_1 - u)(1 - c_2 + c_2 p_2 - u)(c_1 p_1 + c_2 p_2 - u)} = V^3. \qquad (9)$$

Equating the values of V^3 in (8) and (9), we have

$$(u - c_1 p_1)(u - c_2 p_2)(1 - u)$$
$$= \{1 - c_1(1 - p_1) - u\}\{1 - c_2(1 - p_2) - u\}(c_1 p_1 + c_2 p_2 - u),$$

which may be more conveniently written in the form

$$\frac{(u - c_1 p_1)(u - c_2 p_2)}{c_1 p_1 + c_2 p_2 - u} = \frac{\{1 - c_1(1 - p_1) - u\}\{1 - c_2(1 - p_2) - u\}}{1 - u}. \quad (10)$$

From this equation the value of u may be found. It remains only to determine which of the roots must be taken for this purpose.

3. It has been shown (XIX. 12) that the quantity u, in order that it may represent the probability required in the above case, must exceed each of the quantities $c_1 p_1$, $c_2 p_2$, and fall short of each of the quantities $1 - c_1(1 - p_1)$, $1 - c_2(1 - p_2)$, and $c_1 p_1 + c_2 p_2$; the condition among the constants, moreover, being that the three last quantities must individually exceed each of the two former ones. Now I shall show that these conditions being satisfied, the final equation (10) has but one root which falls within the limits assigned. That root will therefore be the required value of u.

Let us represent the lower limits $c_1 p_1$, $c_2 p_2$, by a, b respectively, and the upper limits $1 - c_1(1 - p_1)$, $1 - c_2(1 - p_2)$, and $c_1 p_1 + c_2 p_2$, by a', b', c' respectively. Then the general equation may be expressed in the form

$$(u - a)(u - b)(1 - u) - (a' - u)(b' - u)(c' - u) = 0, \quad (11)$$

or $$(1 - a' - b')u^2 - \{ab - a'b' + (1 - a' - b')c'\}u + ab - a'b'c' = 0.$$

Representing the first member of the above equation by V, we have

$$\frac{d^2 V}{du^2} = 2(1 - a' - b'). \quad (12)$$

Now let us suppose a the highest of the lower limits of u, a' the lowest of its higher limits, and trace the progress of the values of V between the limits $u = a$ and $u = a'$.

When $u = a$, we see from the form of the first member of (11) that V is negative, and when $u = a'$ we see that V is positive.

Between those limits V varies continuously without becoming infinite, and $\frac{d^2V}{du^2}$ is always of the same sign.

Hence if u represent the abscissa V the ordinate of a plane curve, it is evident that the curve will pass from a point below the axis of u corresponding to $u = a$, to a point above the axis of u corresponding to $u = a'$, the curve remaining continuous, and having its concavity or convexity always turned in the same direction. A little attention will show that, under these circumstances, it must cut the axis of u once, and only once.

Hence between the limits $u = a$, $u = a'$, there exists one value of u, and only one, which satisfies the equation (11). It will further appear, if in thought the curve be traced, that the other value of u will be less than a when the quantity $1 - a' - b'$ is positive and greater than any one of the quantities a', b', c' when $1 - a' - b'$ is negative. It hence follows that in the solution of (11) the positive sign of the radical must be taken. We thus find

$$u = \frac{ab - a'b' + (1 - a' - b')c' + \sqrt{Q}}{2(1 - a' - b')}, \qquad (13)$$

where $Q = \{ab - a'b' + (1 - a' - b')c'\}^2 - 4(1 - a' - b')(ab - a'b'c')$.

4. The results of this investigation may to some extent be verified. Thus, it is evident that the probability of the event E must in general exceed the probability of the concurrence of the event E and the cause A_1 or A_2. Hence we must have, as the solution indicates,

$$u > c_1 p_1, \quad u > c_2 p_2.$$

Again, it is clear that the probability of the effect E must in general be less than it would be if the causes A_1, A_2 were mutually exclusive. Hence

$$u \overline{\overline{<}} c_1 p_1 + c_2 p_2.$$

Lastly, since the probability of the failure of the effect E concurring with the presence of the cause A_1 must, in general, be less than the absolute probability of the failure of E, we have

$$c_1(1 - p_1) \overline{\overline{<}} 1 - u,$$

$$\therefore u \overline{\overline{<}} 1 - c_1(1 - p_1).$$

Similarly,

$$u \overline{\overline{<}} 1 - c_2 (1 - p_2).$$

And thus the conditions by which the general solution was limited are confirmed.

Again, let $p_1 = 1$, $p_2 = 1$. This is to suppose that when either of the causes A_1, A_2 is present, the event E will occur. We have then $a = c_1$, $b = c_2$, $a' = 1$, $b' = 1$, $c' = c_1 + c_2$, and substituting in (13) we get

$$u = \frac{c_1 c_2 - c_1 - c_2 - 1 + \sqrt{\{(c_1 c_2 - c_1 - c_2 - 1)^2 + 4 (c_1 c_2 - c_1 - c_2)\}}}{-2}$$

$$= c_1 + c_2 - c_1 c_2 \text{ on reduction}$$

$$= 1 - (1 - c_1)(1 - c_2).$$

Now this is the known expression for the probability that one cause at least will be present, which, under the circumstances, is evidently the probability of the event E.

Finally, let it be supposed that c_1 and c_2 are very small, so that their product may be neglected; then the expression for u reduces to $c_1 p_1 + c_2 p_2$. Now the smaller the probability of each cause, the smaller, in a much higher degree, is the probability of a conjunction of causes. Ultimately, therefore, such reduction continuing, the probability of the event E becomes the same as if the causes were mutually exclusive.

I have dwelt at greater length upon this solution, because it serves in some respect as a model for those which follow, some of which, being of a more complex character, might, without such preparation, appear difficult.

5. PROBLEM II.—In place of the supposition adopted in the previous problem, that the event E cannot happen when both the causes A_1, A_2 are absent, let it be assumed that the causes A_1, A_2 cannot both be absent, and let the other circumstances remain as before. Required, then, the probability of the event E.

Here, in place of the equation (2) of the previous solution, we have the equation

$$(1 - x)(1 - y) = 0.$$

The developed logical expression of z is found to be

$$z = stxy + \frac{1}{0}stx\bar{y} + \frac{1}{0}st\bar{x}y + \frac{1}{0}st\bar{x}\bar{y}$$

$$+ \frac{1}{0}s\bar{t}xy + s\bar{t}x\bar{y} + \frac{1}{0}s\bar{t}\bar{x}y + \frac{1}{0}s\bar{t}\bar{x}\bar{y}$$

$$+ \frac{1}{0}\bar{s}txy + \frac{1}{0}\bar{s}tx\bar{y} + \bar{s}t\bar{x}y + \frac{1}{0}\bar{s}t\bar{x}\bar{y}$$

$$+ 0\bar{s}\bar{t}xy + 0\bar{s}\bar{t}x\bar{y} + 0\bar{s}\bar{t}\bar{x}y + \frac{1}{0}\bar{s}\bar{t}\bar{x}\bar{y};$$

and the final solution is

$$\text{Prob. } E = u;$$

the quantity u being determined by the solution of the equation

$$\frac{(u-a)(u-b)}{a+b-u} = \frac{(a'-u)(b'-u)}{u-a'-b'+1}, \qquad (1)$$

wherein $a = c_1 p_1$, $b = c_2 p_2$, $a' = 1 - c_1(1 - p_1)$, $b' = 1 - c_2(1 - p_2)$.

The conditions of limitation are the following :—That value of u must be chosen which exceeds each of the three quantities

$$a, \; b, \text{ and } a' + b' - 1,$$

and which at the same time falls short of each of the three quantities

$$a', \; b', \text{ and } a + b.$$

Exactly as in the solution of the previous problem, it may be shown that the quadratic equation (1) will have one root, and only one root, satisfying these conditions. The conditions themselves were deduced by the same rule as before, excepting that the minor limit $a' + b' - 1$ was found by seeking the major limit of $1 - z$.

It may be added that the constants in the data, beside satisfying the conditions implied above, viz., that the quantities a', b', and $a + b$, must individually exceed a, b, and $a' + b' - 1$, must also satisfy the condition $c_1 + c_2 \overline{\overline{>}} 1$. This also appears from the application of the rule.

6. PROBLEM III.—The probabilities of two events A and B are a and b respectively, the probability that if the event A take place an event E will accompany it is p, and the probability that

if the event B take place, the same event E will accompany it is q. Required the probability that if the event A take place the event B will take place, or *vice versâ*, the probability that if B take place, A will take place.

Let us represent the event A by x, the event B by y, and the event E by z. Then the data are—

$$\text{Prob. } x = a, \qquad \text{Prob. } y = b.$$
$$\text{Prob. } xz = ap, \qquad \text{Prob. } yz = bq.$$

Whence it is required to find

$$\frac{\text{Prob. } xy}{\text{Prob. } x} \text{ or } \frac{\text{Prob. } xy}{\text{Prob. } y}.$$

Let $$xy = s, \quad yz = t, \quad xy = w.$$

Eliminating z, we have, on reduction,

$$s\bar{x} + t\bar{y} + sy\bar{t} + xt\bar{s} + xy\bar{w} + (1 - xy)\, w = 0,$$

$$\therefore w = \frac{s\bar{x} + t\bar{y} + sy\bar{t} + xt\bar{s} + xy}{2xy - 1}$$

$$= xyst + \frac{1}{0}\, x\bar{y}st + \frac{1}{0}\, \bar{x}yst + \frac{1}{0}\, \bar{x}\bar{y}st$$

$$+ \frac{1}{0}\, xys\bar{t} + 0x\bar{y}s\bar{t} + \frac{1}{0}\, \bar{x}ys\bar{t} + \frac{1}{0}\, \bar{x}\bar{y}s\bar{t}$$

$$+ \frac{1}{0}\, xy\bar{s}t + \frac{1}{0}\, x\bar{y}\bar{s}t + 0\bar{x}y\bar{s}t + \frac{1}{0}\, \bar{x}\bar{y}\bar{s}t$$

$$+ xy\bar{s}\bar{t} + 0x\bar{y}\bar{s}\bar{t} + 0\bar{x}y\bar{s}\bar{t} + 0\bar{x}\bar{y}\bar{s}\bar{t}. \qquad (1)$$

Hence, passing from Logic to Algebra,

$$\text{Prob. } xy = \frac{xyst + xy\bar{s}\bar{t}}{V},$$

x, y, s, and t being determined by the system of equations

$$\frac{xyst + x\bar{y}s\bar{t} + xy\bar{s}\bar{t} + x\bar{y}\bar{s}\bar{t}}{a} = \frac{xyst + \bar{x}y\bar{s}t + xy\bar{s}\bar{t} + \bar{x}y\bar{s}\bar{t}}{b}$$

$$= \frac{xyst + x\bar{y}s\bar{t}}{ap} = \frac{xyst + \bar{x}y\bar{s}t}{bq}$$

$$= xyst + x\bar{y}s\bar{t} + \bar{x}y\bar{s}t + xy\bar{s}\bar{t} + x\bar{y}\bar{s}\bar{t} + \bar{x}y\bar{s}\bar{t} + \bar{x}\bar{y}\bar{s}\bar{t} = V.$$

To reduce the above system to a more convenient form, let every member be divided by $\bar{x}\bar{y}\,\bar{s}\,\bar{t}$, and in the result let

$$\frac{xs}{\bar{x}\,\bar{s}} = m, \quad \frac{yt}{\bar{y}\,\bar{t}} = m', \quad \frac{x}{\bar{x}} = n, \quad \frac{y}{\bar{y}} = n'.$$

We then find

$$\frac{mm' + m + nn' + n}{a} = \frac{mm' + m' + nn' + n'}{b}$$

$$= \frac{mm' + m}{ap} = \frac{mm' + m'}{bq}$$

$$= mm' + m + m' + nn' + n + n' + 1.$$

Also,

$$\text{Prob. } xy = \frac{mm' + nn'}{mm' + m + m' + nn' + n + n' + 1}.$$

These equations may be reduced to the form

$$\frac{mm' + m}{ap} = \frac{mm' + m'}{bq} = \frac{nn' + n}{a(1-p)} = \frac{nn' + n'}{b(1-q)}$$

$$= (m+1)(m'+1) + (n+1)(n'+1) - 1.$$

$$\text{Prob. } xy = \frac{mm' + nn'}{(m+1)(m'+1) + (n+1)(n'+1) - 1}.$$

Now assume

$$(m+1)(m'+1) = \frac{\mu}{\nu + \mu - 1}, \quad (n+1)(n'+1) = \frac{\nu}{\nu + \mu - 1}. \quad (2)$$

Then since $mm' + m = \dfrac{m(m'+1)(m+1)}{m+1} = \dfrac{m\mu}{(m+1)(\nu + \mu - 1)}$, and so on for the other numerators of the system, we find, on substituting and multiplying each member of the system by $\nu + \mu - 1$, the following results:

$$\frac{m\mu}{(m+1)ap} = \frac{m'\mu}{(m'+1)bq} = \frac{n\nu}{(n+1)a(1-p)} = \frac{n'\nu}{(n'+1)b(1-q)} = 1.$$

$$\text{Prob. } xy = (mm' + nn')(\nu + \mu - 1). \quad (3)$$

From the above system we have

$$\frac{m}{m+1} = \frac{ap}{\mu}, \text{ whence } m = \frac{ap}{\mu - ap}.$$

Similarly

$$m' = \frac{bq}{\mu - bq}, \quad n = \frac{a(1-p)}{\nu - a(1-p)}, \quad n' = \frac{b(1-q)}{\nu - b(1-q)}.$$

Hence
$$m + 1 = \frac{\mu}{\mu - ap}, \quad n + 1 = \frac{\nu}{\nu - a(1-p)}, \text{ \&c.}$$

Substitute these values in (2) reduced to the form

$$\nu + \mu - 1 = \frac{\mu}{(m+1)(m'+1)} = \frac{\nu}{(n+1)(\nu'+1)},$$

and we have

$$\nu + \mu - 1 = \frac{(\mu - ap)(\mu - bq)}{\mu} = \frac{\{\nu - a(1-p)\}\{\nu - b(1-q)\}}{\nu}. \quad (4)$$

Substitute also for m, m', &c. their values in (3), and we have

$$\text{Prob. } xy$$

$$= \left[\frac{abpq}{(\mu - ap)(\mu - bq)} + \frac{ab(1-p)(1-q)}{\{\nu - a(1-p)\}\{\nu - b(1-q)\}}\right](\nu + \mu - 1)$$

$$= \frac{abpq}{\mu} + \frac{ab(1-p)(1-q)}{\nu} \text{ by (4).}$$

Now the first equation of the system (4) gives

$$\nu + \mu - 1 = \mu - ap - bq + \frac{apbq}{\mu}, \quad (5)$$

$$\therefore \frac{abpq}{\mu} = \nu - 1 + ap + bq.$$

Similarly,

$$\frac{ab(1-p)(1-q)}{\nu} = \mu - 1 + a(1-p) + b(1-q).$$

Adding these equations together, and observing that the first member of the result becomes identical with the expression just found for Prob. xy, we have

$$\text{Prob. } xy = \nu + \mu + a + b - 2.$$

Let us represent Prob. xy by u, and let $a + b - 2 = m$, then

$$\mu + \nu = u - m. \quad (6)$$

Again, from (5) we have

$$\mu\nu = abpq - (ap + bq - 1)\mu. \quad (7)$$

Similarly from the first and third members of (4) equated we have

$$\mu\nu = ab(1-p)(1-q) - \{a(1-p) + b(1-q) - 1\}\nu.$$

Let us represent $ap + bq - 1$ by h, and $a(1-p) + b(1-q) - 1$ by h'. We find on equating the above values of $\mu\nu$,

$$\begin{aligned} h\mu - h'\nu &= ab\{pq + (1-p)(1-q)\} \\ &= ab(p+q-1). \end{aligned}$$

Let $ab(p+q-1) = l$, then

$$h\mu - h'\nu = l. \qquad (8)$$

Now from (6) and (8) we get

$$\mu = \frac{h'(u-m)+l}{m}, \quad \nu = \frac{h(u-m)-l}{m}.$$

Substitute these values in (7) reduced to the form

$$\mu(\nu + h) = abpq,$$

and we have

$$(hu - l)\{h'(u-m) + l\} = abpqm^2, \qquad (9)$$

a quadratic equation, the solution of which determines u, the value of Prob. xy sought.

The solution may readily be put in the form

$$h = \frac{lh' + h(h'm - l) \pm \sqrt{[\{lh' - h(h'm-l)\}^2 + 4hh'abpqm^2]}}{2hh'}.$$

But if we further observe that

$$lh' - h(h'm - l) = l(h + h') - hh'm = (l - hh')m,$$

since $\qquad h = ap + bq - 1, \; h' = a(1-p) + b(1-q) - 1,$

whence $\qquad h + h' = a + b - 2 = m,$

we find

$$\text{Prob. } xy = \frac{lh' + h(p'm - l) \pm m\sqrt{\{(l-hh')^2 + 4hh'abpq\}}}{2hh'}. \qquad (10)$$

It remains to determine which sign must be given to the radical. We might ascertain this by the general method exemplified in the last problem, but it is far easier, and it fully suffices in the present instance, to determine the sign by a comparison of the above formula with the result proper to some known case. For instance, if it were certain that the event A is *always*, and the event B *never*, associated with the event E, then it is certain that the events A and B are never conjoined. Hence if $p = 1$, $q = 0$, we ought to have $u = 0$. Now the assumptions $p = 1$, $q = 0$, give

$$h = a - 1, \quad h' = b - 1, \quad l = 0, \quad m = a + b - 2.$$

Substituting in (10) we have

$$\text{Prob. } xy = \frac{(a-1)\, b-1)\,(a+b-2) \pm (a+b-2)\,(a-1)\,(b-1)}{2\,(a-1)\,(b-1)},$$

and this expression vanishes when the lower sign is taken. Hence the final solution of the general problem will be expressed in the form

$$\frac{\text{Prob. } xy}{\text{Prob. } x} = \frac{lh' + h\,(h'm - l) - m\sqrt{\{(l - hh')^2 + 4hh'abpq\}}}{2ahh'},$$

wherein $h = ap + bq - 1$, $h' = a\,(1-p) + b\,(1-q) - 1$,

$$l = ab\,(p + q - 1), \quad m = a + b - 2.$$

As the terms in the final logical solution affected by the coefficient $\frac{1}{0}$ are the same as in the first problem of this chapter, the conditions among the constants will be the same, viz.,

$$ap \leqq 1 - b\,(1-q), \quad bq \leqq 1 - a\,(1-p).$$

7. It is a confirmation of the correctness of the above solution that the expression obtained is symmetrical with respect to the two sets of quantities p, q, and $1-p$, $1-q$, i. e. that on changing p into $1-p$, and q into $1-q$, the expression is unaltered This is apparent from the equation

$$\text{Prob. } xy = ab\left\{\frac{pq}{\mu} + \frac{(1-p)\,(1-q)}{\nu}\right\}$$

employed in deducing the final result. Now if there exist probabilities p, q of the event E, as consequent upon a knowledge of the occurrences of A and B, there exist probabilities $1-p$, $1-q$ of the contrary event, that is, of the non-occurrence of E under the same circumstances. As then the data are unchanged in form, whether we take account in them of the occurrence or of the non-occurrence of E, it is evident that the solution ought to be, as it is, a symmetrical function of p, q and $1-p$, $1-q$.

Let us examine the particular case in which $p=1$, $q=1$. We find

$$h = a + b - 1, \quad h' = -1, \quad l = ab, \quad m = a + b - 2,$$

and substituting

$$\frac{\text{Prob. } xy}{\text{Prob. } x} = \frac{-ab + (a+b-1)(2-a-b-ab) - (a+b-2)(ab-a-b+1)}{-2a(a+b-1)}$$

$$= \frac{-2ab(a+b-1)}{-2a(a+b-1)} = b.$$

It would appear, then, that in this case the events A and B are virtually independent of each other. The supposition of their invariable association with some other event E, of the frequency of whose occurrence, except as it may be inferred from this particular connexion, absolutely nothing is known, does not establish any dependence between the events A and B themselves. I apprehend that this conclusion is agreeable to reason, though particular examples may appear at first sight to indicate a different result. For instance, if the probabilities of the casting up, 1st, of a particular species of weed, 2ndly, of a certain description of zoophytes upon the sea-shore, had been separately determined, and if it had also been ascertained that neither of these events could happen except during the agitation of the waves caused by a tempest, it would, I think, justly be concluded that the events in question were not independent. The picking up of a piece of seaweed of the kind supposed would, it is presumed, render more probable the discovery of the zoophytes than it would otherwise have been. But I apprehend that this fact is due to our knowledge of another circumstance not implied in the actual conditions of the problem, viz., that the occurrence of a tempest is but an *occasional* phænomenon. Let the range of observation be confined to a sea *always* vexed with storm. It would then, I suppose, be seen that the casting up of the weeds and of the zoophytes ought to be regarded as independent events. Now, to speak more generally, there are conditions common to all phænomena,—conditions which, it is felt, do not affect their mutual independence. I apprehend therefore that the solution indicates, that when a particular condition has prevailed through the whole of our *recorded experience*, it assumes the above character with reference to the class of phænomena over which that experience has extended.

8. PROBLEM IV.—To illustrate in some degree the above observations, let there be given, in addition to the data of the last problem, the absolute probability of the event E, the completed system of data being

$$\text{Prob. } x = a, \quad \text{Prob. } y = b, \quad \text{Prob. } z = c,$$
$$\text{Prob. } xz = ap, \quad \text{Prob. } yz = bq,$$

and let it be required to find Prob. xy.

Assuming, as before, $xz = s$, $yz = t$, $xy = w$, the final logical equation is

$$w = xystz + xy\bar{s}\bar{t}\bar{z} + 0\,(x\bar{y}s\bar{t}z + \bar{x}y\,\bar{}\,tz + x\bar{y}\bar{z}\bar{s}\bar{t} + xy\bar{z}\bar{s}\bar{t}$$
$$+ \bar{x}\bar{y}z\bar{s}\bar{t} + \bar{x}\bar{y}\bar{z}\bar{s}\bar{t}).$$
$$+ \text{ terms whose coefficients are } \frac{1}{0}. \qquad (1)$$

The algebraic system having been formed, the subsequent eliminations may be simplified by the transformations adopted in the previous problem. The final result is

$$\text{Prob. } xy = ab\left\{\frac{pq}{c} + \frac{(1-p)(1-q)}{1-c}\right\}. \qquad (2)$$

The conditions among the constants are

$$c \overset{=}{>} ap, \quad c \overset{=}{>} bq, \quad c \overset{=}{<} 1 - a(1-p), \quad c \overset{=}{<} 1 - b(1-q).$$

Now if $p = 1$, $q = 1$, we find

$$\text{Prob. } xy = \frac{ab}{c},$$

c not admitting of any value less than a or b. It follows hence that if the event E is known to be an *occasional* one, its invariable attendance on the events x and y *increases* the probability of their conjunction in the inverse ratio of its own frequency.

The formula (2) may be verified in a large number of cases. As a particular instance, let $q = c$, we find

$$\text{Prob. } xy = ab. \qquad (3)$$

Now the assumption $q = c$ involves, by Definition (Chap. XVI.) the independence of the events B and E. If then B and E are independent, no relation which may exist between A and E can establish a relation between A and B; wherefore A and B are also independent, as the above equation (3) implies.

It may readily be shown from (2) that the value of Prob. z, which renders Prob. xy a minimum, is

$$\text{Prob. } z = \frac{\sqrt{(pq)}}{\sqrt{(pq)} + \sqrt{(1-p)(1-q)}}.$$

If $p = q$, this gives

$$\text{Prob. } z = p;$$

a result, the correctness of which may be shown by the same considerations which have been applied to (3).

PROBLEM V.—Given the probabilities of any three events, and the probability of their conjunction; required the probability of the conjunction of any two of them.

Suppose the data to be

$$\text{Prob. } x = p, \ \text{Prob. } y = q, \ \text{Prob. } z = r, \ \text{Prob. } xyz = m,$$

and the quæsitum to be Prob. xy.

Assuming $xyz = s$, $xy = t$, we find as the final logical equation,

$$t = xyzs + xy\bar{z}\bar{s} + 0(x\bar{y}\bar{s} + \bar{x}\bar{s}) + \frac{1}{0}(\text{sum of all other constituents});$$

whence, finally,

$$\text{Prob. } xy = \frac{H - \sqrt{(H^2 - 4pq\bar{r}^2 - 4\bar{p}\bar{q}\bar{r}m)}}{2\bar{r}},$$

wherein $\quad \bar{p} = 1 - p$, &c. $\quad H = \bar{p}\bar{q} + (p+q)\bar{r}.$

This admits of verification when $p = 1$, when $q = 1$, when $r = 0$, and therefore $m = 0$, &c.

Had the condition, Prob. $z = r$, been omitted, the solution would still have been definite. We should have had

$$\text{Prob.}\, xy = \frac{pq(1-m) + (1-p)(1-q)m}{1-m};$$

and it may be added, as a final confirmation of their correctness, that the above results become identical when $m = pqr$.

9. The following problem is a generalization of Problem I., and its solution, though necessarily more complex, is obtained by a similar analysis.

PROBLEM VI.—If an event can only happen as a consequence of one or more of certain causes $A_1, A_2, \ldots A_n$, and if generally c_i represent the probability of the cause A_i, and p_i the probability that if the cause A_i exist, the event E will occur, then the series of values of c_i and p_i being given, required the probability of the event E.*

Let the causes $A_1, A_2, \ldots A_n$ be represented by $x_1, x_2, \ldots x_n$, and the event E by z.

Then we have generally,

$$\text{Prob.}\, x_i = c_i, \qquad \text{Prob.}\, x_i z = c_i p_i.$$

Further, the condition that E can only happen in connexion with some one or more of the causes $A_1, A_2, \ldots A_n$ establishes the logical condition,

$$z(1-x_1)(1-x_2)\ldots(1-x_n) = 0. \qquad (1)$$

* It may be proper to remark, that the above problem was proposed to the notice of mathematicians by the author in the Cambridge and Dublin Mathematical Journal, Nov. 1851, accompanied by the subjoined observations:

"The motives which have led me, after much consideration, to adopt, with reference to this question, a course unusual in the present day, and not upon slight grounds to be revived, are the following:—First, I propose the question as a test of the sufficiency of received methods. Secondly, I anticipate that its discussion will in some measure add to our knowledge of an important branch of pure analysis. However, it is upon the former of these grounds alone that I desire to rest my apology.

"While hoping that some may be found who, without departing from the line of their previous studies, may deem this question worthy of their attention, I wholly disclaim the notion of its being offered as a trial of personal skill or knowledge, but desire that it may be viewed solely with reference to those public and scientific ends for the sake of which alone it is proposed."

The author thinks it right to add, that the publication of the above problem led to some interesting private correspondence, but did not elicit a solution.

Now let us assume generally

$$x_i z = t_i,$$

which is reducible to the form

$$x_i z (1 - t_i) + t_i (1 - x_i z) = 0,$$

forming the type of a system of n equations which, together with (1), express the logical conditions of the problem. Adding all these equations together, as after the previous reduction we are permitted to do, we have

$$\Sigma \{x_i z (1 - t_i) + t_i (1 - x_i z)\} + z (1 - x_1)(1 - x_2)\ldots(1 - x_n) = 0, \quad (2)$$

(the summation implied by Σ extending from $i = 1$ to $i = n$), and this single and sufficient logical equation, together with the $2n$ data, represented by the general equations

$$\text{Prob. } x_i = c_i, \quad \text{Prob. } t_i = c_i p_i, \quad (3)$$

constitute the elements from which we are to determine Prob. z.

Let (2) be developed with respect to z. We have

$$[\Sigma\{x_i(1 - t_i) + t_i (1 - x_i)\} + (1 - x_1)(1 - x_2)\ldots(1 - x_n)] z + \Sigma t_i (1 - z) = 0,$$

whence

$$z = \frac{\Sigma t_i}{\Sigma t_i - \Sigma\{x_i(1 - t_i) + t_i(1 - x_i)\} - (1 - x_1)(1 - x_2)\ldots(1 - x_n)}. \quad (4)$$

Now any constituent in the expansion of the second member of the above equation will consist of $2n$ factors, of which n are taken out of the set $x_1, x_2, \ldots x_n, 1 - x_1, 1 - x_2, \ldots 1 - x_n$, and n out of the set $t_1, t_2, \ldots t_n, 1 - t_1, 1 - t_2, \ldots 1 - t_n$, no such combination as $x_1(1 - x_1)$, $t_1(1 - t_1)$, being admissible. Let us consider first those constituents of which $(1 - t_1), (1 - t_2)\ldots(1 - t_n)$ forms the t-factor, that is the factor derived from the set $t_1, \ldots 1 - t_1$.

The coefficient of any such constituent will be found by changing $t_1, t_2, \ldots t_n$ respectively into 0 in the second member of (4), and then assigning to $x_1, x_2, \ldots x_n$ their values as dependent upon the nature of the x-factor of the constituent. Now simply substituting for $t_1, t_2, \ldots t_n$ the value 0, the second member becomes

$$\frac{0}{-\Sigma x_i - (1 - x_1)(1 - x_2)\ldots(1 - x_n)},$$

Z

and this vanishes whatever values, 0, 1, we subsequently assign to $x_1, x_2, \ldots x_n$. For if those values are not all equal to 0, the term Σx_i does not vanish, and if they are all equal to 0, the term $-(1-x_1)\ldots(1-x_n)$ becomes -1, so that in either case the denominator does not vanish, and therefore the fraction does. Hence the coefficients of all constituents of which $(1-t_1)\ldots(1-t_n)$ is a factor will be 0, and as the sum of all possible x-constituents is unity, there will be an aggregate term $0(1-t_1)\ldots(1-t_n)$ in the development of z.

Consider, in the next place, any constituent of which the t-factor is $t_1 t_2 \ldots t_r (1-t_{r+1})\ldots(1-t_n)$, r being equal to or greater than unity. Making in the second member of (4), $t_1 = 1, \ldots t_r = 1$, $t_{r+1} = 0, \ldots t_n = 0$, we get the expression

$$\frac{r}{x_1 \ldots + x_r - x_{r+1} \ldots - x_n - (1-x_1)(1-x_2)\ldots(1-x_n)}.$$

Now the only admissible values of the symbols being 0 and 1, it is evident that the above expression will be equal to 1 when $x_1 = 1 \ldots x_r = 1$, $x_{r+1} = 0, \ldots x_n = 0$, and that for all other combinations of value that expression will assume a value greater than unity. Hence the coefficient 1 will be applied to all constituents of the final development which are of the form

$$x_1 \ldots x_r (1-x_{r+1})\ldots(1-x_n)\, t_1 \ldots t_r (1-t_{r+1})\ldots(1-t_n),$$

the x-factor being similar to the t-factor, while other constituents included under the present case will have the virtual coefficient $\frac{1}{0}$. Also, it is manifest that this reasoning is independent of the particular arrangement and succession of the individual symbols.

Hence the complete expansion of z will be of the form

$$z = \Sigma(XT) + 0(1-t_1)(1-t_2)\ldots(1-t_n)$$
$$+ \text{ constituents whose coefficients are } \frac{1}{0}, \quad (5)$$

where T represents any t-constituent except $(1-t_1)\ldots(1-t_n)$, and X the corresponding or similar constituent of $x_1 \ldots x_n$.

For instance, if $n = 2$, we shall have

$$\Sigma\,(XT) = x_1\,x_2\,t_1\,t_2 + x_1\,\bar{x}_2\,t_1\,\bar{t}_2 + \bar{x}_1\,x_2\,\bar{t}_1\,t_2,$$

$\bar{x}_1$, $\bar{x}_2$, &c. standing for $1 - x_1$, $1 - x_2$, &c.; whence

$$\begin{aligned} z = {} & x_1\,x_2\,t_1\,t_2 + x_1\,\bar{x}_2\,t_1\,\bar{t}_2 + \bar{x}_1\,x_2\,\bar{t}_1\,t_2 \\ & + 0\,(x_1\,x_2\,\bar{t}_1\,\bar{t}_2 + x_1\,\bar{x}_2\,\bar{t}_1\,\bar{t}_2 + \bar{x}_1\,x_2\,\bar{t}_1\,\bar{t}_2 + \bar{x}_1\,\bar{x}_2\,\bar{t}_1\,\bar{t}_2) \qquad (6) \\ & + \text{constituents whose coefficients are } \frac{1}{0}. \end{aligned}$$

This result agrees, difference of notation being allowed for, with the developed form of z in Problem I. of this chapter, as it evidently ought to do.

10. To avoid complexity, I purpose to deduce from the above equation (6) the necessary conditions for the determination of Prob. z for the particular case in which $n = 2$, in such a form as may enable us, by pursuing in thought the same line of investigation, to assign the corresponding conditions for the more general case in which n possesses any integral value whatever.

Supposing then $n = 2$, we have

$$\begin{aligned} V = {} & x_1\,x_2\,t_1\,t_2 + x_1\,\bar{x}_2\,t_1\bar{t}_2 + \bar{x}_1\,x_2\bar{t}_1\,t_2 + x_1\,x_2\,\bar{t}_1\,\bar{t}_2 + x_1\,\bar{x}_2\bar{t}_1\,\bar{t}_2 \\ & + \bar{x}_1\,x_2\,\bar{t}_1\,\bar{t}_2 + \bar{x}_1\,\bar{x}_2\,\bar{t}_1\,\bar{t}_2. \end{aligned}$$

$$\text{Prob. } z = \frac{x_1\,x_2\,t_1\,t_2 + x_1\,\bar{x}_2\,t_1\,\bar{t}_2 + \bar{x}_1\,x_2\,\bar{t}_1 t_2}{V},$$

the conditions for the determination of x_1, t_1, &c., being

$$\begin{aligned} & \frac{x_1\,x_2\,t_1\,t_2 + x_1\,\bar{x}_2\,t_1\,\bar{t}_2 + x_1\,x_2\,\bar{t}_1\,\bar{t}_2 + x_1\,\bar{x}_2\,\bar{t}_1\,\bar{t}_2}{c_1} \\ & = \frac{x_1\,x_2\,t_1\,t_2 + \bar{x}_1\,x_2\,\bar{t}_1\,t_2 + x_1\,x_2\,\bar{t}_1\,\bar{t}_2 + \bar{x}_1\,x_2\,\bar{t}_1\,\bar{t}_2}{c_2} \\ & = \frac{x_1\,x_2\,t_1\,t_2 + x_1\,\bar{x}_2\,t_1\,\bar{t}_2}{c_1\,p_1} = \frac{x_1\,x_2\,t_1\,t_2 + \bar{x}_1\,x_2\,\bar{t}_1\,t_2}{c_2\,p_2} = V. \end{aligned}$$

Divide the members of this system of equations by $\bar{x}_1\,\bar{x}_2\,\bar{t}_1\,\bar{t}_2$, and the numerator and denominator of Prob. z by the same quantity, and in the results assume

$$\frac{x_1\,t_1}{\bar{x}_1\,\bar{t}_1} = m_1, \quad \frac{x_2\,t_2}{\bar{x}_2\,\bar{t}_2} = m_2, \quad \frac{x_1}{\bar{x}_1} = n_1, \quad \frac{x_2}{\bar{x}_2} = n_2; \qquad (7)$$

we find

$$\text{Prob. } z = \frac{m_1 m_2 + m_1 + m_2}{m_1 m_2 + m_1 + m_2 + n_1 n_2 + n_1 + n_2 + 1},$$

$$\text{and } \frac{m_1 m_2 + m_1 + n_1 n_2 + n_1}{c_1} = \frac{m_1 m_2 + m_2 + n_1 n_2 + n_2}{c_2}$$

$$= \frac{m_1 m_2 + m_1}{c_1 p_1} = \frac{m_1 m_2 + m_2}{c_2 p_2} = m_1 m_2 + m_1 + m_2 + n_1 n_2 + n_1 + n_2 + 1, \quad (8)$$

whence, if we assume,

$$(m_1 + 1)(m_2 + 1) = M, \quad (n_1 + 1)(n_2 + 1) = N, \quad (9)$$

we have, after a slight reduction,

$$\text{Prob. } z = \frac{M - 1}{M + N - 1},$$

$$\frac{n_1(n_2 + 1)}{c_1(1 - p_1)} = \frac{n_2(n_1 + 1)}{c_2(1 - p_2)} = \frac{m_1(m_2 + 1)}{c_1 p_1} = \frac{m_2(m_1 + 1)}{c_2 p_2} = M + N - 1;$$

or,

$$\frac{m_1 M}{(m_1 + 1) c_1 p_1} = \frac{m_2 M}{(m_2 + 1) c_2 p_2} = \frac{n_1 N}{(n_1 + 1) c_1 (1 - p_1)}$$

$$= \frac{n_2 N}{(n_2 + 1) c_2 (1 - p_2)} = M + N - 1.$$

Now let a similar series of transformations and reductions be performed in thought upon the final logical equation (5). We shall obtain for the determination of Prob. z the following expression:

$$\text{Prob. } z = \frac{M - 1}{M + N - 1}, \quad (10)$$

wherein

$$M = (m_1 + 1)(m_2 + 1) \ldots (m_n + 1),$$

$$N = (n_1 + 1)(n_2 + 1) \ldots (n_n + 1),$$

$m_1, \ldots m_n, n_1, \ldots n_n$, being given by the system of equations,

$$\frac{m_1 M}{(m_1 + 1) c_1 p_1} = \frac{m_2 M}{(m_2 + 1) c_2 p_2} \ldots = \frac{m_n M}{(m_n + 1) c_n p_n}$$

$$= \frac{n_1 N}{(n_1 + 1) c_1 (1 - p_1)} \ldots = \frac{n_n N}{(n_n + 1) c_n (1 - p_n)} = M + N - 1. \quad (11)$$

Still further to simplify the results, assume

$$\frac{M+N-1}{M}=\frac{1}{\mu}, \quad \frac{M+N-1}{N}=\frac{1}{\nu};$$

whence

$$M=\frac{\mu}{\mu+\nu-1}, \quad N=\frac{\nu}{\mu+\nu-1}.$$

We find

$$\frac{m_1}{(m_1+1)\,c_1p_1}=\frac{m_2}{(m_2+1)\,c_2p_2}\,..=\frac{m_n}{(m_n+1)\,c_np_n}=\frac{1}{\mu},$$

$$\frac{n_1}{(n_1+1)\,c_1(1-p_1)}=\frac{n_2}{(n_2+1)\,c_2(1-p_2)}\,..=\frac{n_n}{(n_n+1)\,c_n(1-p_n)}=\frac{1}{\nu};$$

whence

$$m_1=\frac{c_1p_1}{\mu-c_1p_1},\;..\,m_n=\frac{c_np_n}{\mu-c_np_n};$$

and finally,

$$m_1+1=\frac{\mu}{\mu-c_1p_1},\;..\,m_n+1=\frac{\mu}{\mu-c_np_n},$$

$$n_1+1=\frac{\nu}{\nu-c_1(1-p_1)},\;..\,n_n+1=\frac{\nu}{\nu-c_n(1-p_n)}.$$

Substitute these values with those of M and N in (9), and we have

$$\frac{\mu^n}{(\mu-c_1p_1)(\mu-c_2p_2)\,..\,(\mu-c_np_n)}=\frac{\mu}{\mu+\nu-1},$$

$$\frac{\nu^n}{\{\nu-c_1(1-p_1)\}\{\nu-c_2(1-p_2)\}\,..\,\{\nu-c_n(1-p_n)\}}=\frac{\nu}{\mu+\nu-1},$$

which may be reduced to the symmetrical form

$$\mu+\nu-1=\frac{(\mu-c_1p_1)\,..\,(\mu-c_np_n)}{\mu^{n-1}}$$

$$=\frac{\{\nu-c_1(1-p_1)\}\,..\,\{\nu-c_n(1-p_n)\}}{\nu^{n-1}}. \qquad (12)$$

Finally,

$$\text{Prob. } z=\frac{M-1}{M+N-1}=1-\nu. \qquad (13)$$

Let us then assume $1-\nu=u$, we have then

$$\mu-u=\frac{(\mu-c_1p_1)\,..\,(\mu-c_np_n)}{\mu^{n-1}}$$

$$=\frac{\{1-c_1(1-p_1)-u\}\,..\,\{1-c_n(1-p_n)-u\}}{(1-u)^{n-1}}.$$

If we make for simplicity

$$c_1 p_1 = a_1, \quad c_n p_n = a_n, \quad 1 - c_1(1 - p_1) = b_1, \text{ \&c.},$$

the above equations may be written as follows:

$$\mu - u = \frac{(\mu - a_1) \ldots (\mu - a_n)}{\mu^{n-1}}, \tag{14}$$

wherein

$$\mu = u + \frac{(b_1 - u) \ldots (b_n - u)}{(1 - u)^{n-1}}. \tag{15}$$

This value of μ substituted in (14) will give an equation involving only u, the solution of which will determine Prob. z, since by (13) Prob. $z = u$. It remains to assign the limits of u.

11. Now the very same analysis by which the limits were determined in the particular case in which $n = 2$, (XIX. 12) conducts us in the present case to the following result. The quantity u, in order that it may represent the value of Prob. z, must must have for its inferior limits the quantities $a_1, a_2, \ldots a_n$, and for its superior limits the quantities $b_1, b_2, \ldots b_n$, $a_1 + a_2 \ldots + a_n$. We may hence infer, *à priori*, that there will always exist one root, and only one root, of the equation (14) satisfying these conditions. I deem it sufficient, for practical verification, to show that there will exist one, and only one, root of the equation (14), between the limits $a_1, a_2, \ldots a_n$, and $b_1, b_2, \ldots b_n$.

First, let us consider the nature of the changes to which μ is subject in (15), as u varies from a_1, which we will suppose the greatest of its minor limits, to b_1, which we will suppose the least of its major limits. When $u = a_1$, it is evident that μ is positive and greater than a_1. When $u = b_1$, we have $\mu = b_1$, which is also positive. Between the limits $u = a_1$, $u = b_1$, it may be shown that μ increases with u. Thus we have

$$\frac{d\mu}{du} = 1 - \frac{(b_2 - u) \ldots (b_n - u)}{(1 - u)^{n-1}} - \frac{(b_1 - u)(b_3 - u) \ldots (b_n - u)}{(1 - u)^{n-1}} \ldots$$

$$+ (n - 1) \frac{(b_1 - u)(b_2 - u) \ldots (b_n - u)}{(1 - u)^n}. \tag{16}$$

Now let

$$\frac{b_1 - u}{1 - u} = x_1 \ldots \frac{b_n - u}{1 - u} = x_n.$$

Evidently $x_1, x_2, \ldots x_n$, will be proper fractions, and we have

$$\frac{d\mu}{du} = 1 - x_2 x_3 \ldots x_n - x_1 x_3 \ldots x_n \ldots - x_1 x_2 \ldots x_{n-1} + (n-1)\, x_1 x_2 \ldots x_n$$
$$= 1 - (1 - x_1)\, x_2 x_3 \ldots x_n - x_1 (1 - x_2)\, x_3 \ldots x_n \ldots$$
$$- x_1 x_2 \ldots x_{n-1} (1 - x_n) - x_1 x_2 \ldots x_n.$$

Now the negative terms in the second member are (if we may borrow the language of the logical developments) *constituents* formed from the fractional quantities $x_1, x_2, \ldots x_n$. Their sum cannot therefore exceed unity; whence $\frac{d\mu}{du}$ is positive, and μ increases with u between the limits specified.

Now let (14) be written in the form

$$\frac{(\mu - a_1) \ldots (\mu - a_n)}{\mu^{n-1}} - (\mu - u) = 0, \qquad (17)$$

and assume $u = a_1$. The first member becomes

$$(\mu - a_1) \left\{ \frac{(\mu - a_2) \ldots (\mu - a_n)}{\mu^{n-1}} - 1 \right\}, \qquad (18)$$

and this expression is negative in value. For, making the same assumption in (15), we find

$$\mu - a_1 = \frac{(b_1 - u) \ldots (b_n - u)}{(1 - u)^{n-1}} = \text{a positive quantity.}$$

At the same time we have

$$\frac{(\mu - a_2) \ldots (\mu - a_n)}{\mu^{n-1}} = \frac{\mu - a_2}{\mu} \ldots \frac{\mu - a_n}{\mu},$$

and since the factors of the second member are positive fractions, that member is less than unity, whence (18) is negative. Wherefore *the assumption* $u = a_1$ *makes the first member of* (17) *negative.*

Secondly, let $u = b_1$, then by (15) $\mu = u = b_1$, and *the first member of* (17) *becomes positive.*

Lastly, between the limits $u = a_1$ and $u = b_1$, the first member of (17) continuously increases. For the first term of that expression written under the form

$$(\mu - a_1) \frac{\mu - a_1}{\mu} \ldots \frac{\mu - a_n}{\mu}$$

increases, since μ increases, and, with it, every factor contained. Again, the negative term $\mu - u$ diminishes with the increase of u, as appears from its value deduced from (15), viz.,

$$\frac{(b_1 - u) \ldots (b_n - u)}{(1 - u)^{n-1}}.$$

Hence then, between the limits $u = a_1$, $u = b_1$, the first member of (17) continuously increases, changing in so doing from a negative to a positive value. Wherefore, between the limits assigned, there exists one value of u, and only one, by which the said equation is satisfied.

12. Collecting these results together, we arrive at the following solution of the general problem.

The probability of the event E will be that value of u deduced from the equation

$$\mu - u = \frac{(\mu - c_1 p_1) \ldots (\mu - c_n p_n)}{\mu^{n-1}}, \tag{19}$$

wherein

$$\mu = u + \frac{\{1 - c_1 (1 - p_1) - u\} \ldots \{1 - c_n (1 - p_n) - u\}}{(1 - u)^{n-1}},$$

which (value) lies between the two sets of quantities,

$c_1 p_1$, $c_2 p_2$, .. $c_n p_n$ and $1 - c_1 (1 - p_1)$, $1 - c_2 (1 - p_2)$ .. $1 - c_n (1 - p_n)$,

the former set being its inferior, the latter its superior, limits.

And it may further be inferred in the general case, as it has been proved in the particular case of $n = 2$, that the value of u, determined as above, will not exceed the quantity

$$c_1 p_1 + c_2 p_2 \ldots + c_n p_n.$$

13. Particular verifications are subjoined.

1st. Let $p_1 = 1$, $p_2 = 1$, .. $p_n = 1$. This is to suppose it certain, that if any one of the events A_1, A_2 .. A_n, happen, the event E will happen. In this case, then, the probability of the occurrence of E will simply be the probability that the events or causes A_1, A_2 .. A_n do not all fail of occurring, and its expression will therefore be $1 - (1 - c_1)(1 - c_2) \ldots (1 - c_n)$.

Now the general solution (19) gives

$$\mu - u = \frac{(\mu - c_1) \ldots (\mu - c_n)}{\mu^{n-1}},$$

wherein

$$\mu = u + \frac{(1 - u)^n}{(1 - u)^{n-1}} = 1.$$

Hence,

$$1 - u = (1 - c_1) \ldots (1 - c_n),$$

$$\therefore u = 1 - (1 - c_1) \ldots (1 - c_n),$$

equivalent to the *à priori* determination above.

2nd. Let $p_1 = 0$, $p_2 = 0$, $p_n = 0$, then (19) gives

$$\mu - u = \mu,$$

$$\therefore u = 0,$$

as it evidently ought to be.

3rd. Let c_1, $c_2 \ldots c_n$ be small quantities, so that their squares and products may be neglected. Then developing the second members of the equation (19),

$$\mu - u = \frac{\mu^n - (c_1 p_1 + c_2 p_2 \ldots + c_n p_n)\,\mu^{n-1}}{\mu^{n-1}}$$

$$= \mu - (c_1 p_1 + c_2 p_2 \ldots + c_n p_n),$$

$$\therefore u = c_1 p_1 + c_2 p_2 \ldots + c_n p_n.$$

Now this is what the solution would be were the causes A_1, $A_2 \ldots A_n$ mutually exclusive. But the smaller the probabilities of those causes, the more do they approach the condition of being mutually exclusive, since the smaller is the probability of any concurrence among them. Hence the result above obtained will undoubtedly be the limiting form of the expression for the probability of E.

4th. In the particular case of $n = 2$, we may readily eliminate μ from the general solution. The result is

$$\frac{(u - c_1 p_1)(u - c_2 p_2)}{c_1 p_1 + c_2 p_2 - u} = \frac{\{1 - c_1(1 - p_1) - u\}\{1 - c_2(1 - p_2) - u\}}{1 - u},$$

which agrees with the particular solution before obtained for this case, Problem I.

Though by the system (19), the solution is in general made to depend upon the solution of an equation of a high order, its practical difficulty will not be great. For the conditions relating to the limits enable us to select at once a near value of u, and the forms of the system (19) are suitable for the processes of successive approximation.

14. PROBLEM 7.—The data being the same as in the last problem, required the probability, that if any definite and given combination of the causes $A_1, A_2, \ldots A_n$, present itself, the event E will be realized.

The cases $A_1, A_2, \ldots A_n$, being represented as before by $x_1, x_2, \ldots x_n$ respectively, let the definite combination of them, referred to in the statement of the problem, be represented by the $\phi(x_1, x_2 \ldots x_n)$ so that the actual occurrence of that combination will be expressed by the logical equation,

$$\phi(x_1, x_2, \ldots x_n) = 1.$$

The data are

$$\begin{aligned} &\text{Prob. } x_1 = c_1, \ \ldots \ \text{Prob. } x_n = c_n, \\ &\text{Prob. } x_1 z = c_1 p_1, \ \text{Prob. } x_n z = c_n p_n; \end{aligned} \qquad (1)$$

and the object of investigation is

$$\frac{\text{Prob. } \phi(x_1, x_2 \ldots x_n)\, z}{\text{Prob. } \phi(x_1, x_2 \ldots x_n)}. \qquad (2)$$

We shall first seek the value of the numerator.

Let us assume,

$$x_1 z = t_1 \ldots x_n z = t_n, \qquad (3)$$

$$\phi(x_1, x_2 \ldots x_n)\, z = w. \qquad (4)$$

Or, if for simplicity, we represent $\phi(x_1, x_2 \ldots x_n)$ by ϕ, the last equation will be

$$\phi z = w, \qquad (5)$$

to which must be added the equation

$$\bar{x}_1 \bar{x}_2 \ldots \bar{x}_n z = 0. \qquad (6)$$

Now any equation $x_r z = t_r$ of the system (3) may be reduced to the form

$$x_r z \bar{t}_r + t_r (1 - x_r z) = 0.$$

Similarly reducing (5), and adding the different results together, we obtain the logical equation

$$\Sigma\{x_r z\bar{t}_r + t_r(1 - x_r z)\} + \bar{x}_1 \ldots \bar{x}_n z + \phi z\bar{w} + w(1 - \phi z) = 0, \quad (7)$$

from which z being eliminated, w must be determined as a developed logical function of $x_1, \ldots x_n, t_1, \ldots t_n$.

Now making successively $z = 1$, $z = 0$ in the above equation, and multiplying the results together, we have

$$\{\Sigma(x_r\bar{t}_r + \bar{x}_r t_r) + \bar{x}_1 \ldots \bar{x}_n + \phi\bar{w} + w\bar{\phi}\} \times (\Sigma t_r + w) = 0.$$

Developing this equation with reference to w, and replacing in the result $\Sigma t_r + 1$ by 1, in accordance with Prop. I. Chap. IX., we have

$$Ew + E'(1 - w) = 0;$$

wherein

$$E = \Sigma(x_r\bar{t}_r + t_r\bar{x}_r) + \bar{x}_1 \ldots \bar{x}_n + \bar{\phi},$$

$$E' = \Sigma t_r\{\Sigma(x_r\bar{t}_r + t_r\bar{x}_r) + \bar{x}_1 \ldots \bar{x}_n + \phi\}.$$

And hence

$$w = \frac{E'}{E' - E}. \quad (8)$$

The second member of this equation we must now develop with respect to the double series of symbols $x_1, x_2, \ldots x_n, t_1, t_2, \ldots t_n$. In effecting this object, it will be most convenient to arrange the constituents of the resulting development in three distinct classes, and to determine the coefficients proper to those classes separately.

First, let us consider those constituents of which $\bar{t}_1 \ldots \bar{t}_n$ is a factor. Making $t_1 = 0 \ldots t_n = 0$, we find

$$E' = 0, \ E = \Sigma x_r + \bar{x}_1 \ldots \bar{x}_n + \bar{\phi}.$$

It is evident, that whatever values (0, 1) are given to the x-symbols, E does not vanish. Hence the coefficients of all constituents involving $\bar{t}_1 \ldots \bar{t}_n$ are 0.

Consider secondly, those constituents which do not involve the factor $\bar{t}_1 \ldots \bar{t}_n$, and which are symmetrical with reference to the two sets of symbols $x_1 \ldots x_n$ and $t_1 \ldots t_n$. By symmetrical constituents is here meant those which would remain unchanged if x_1 were converted into t_1, x_2 into t_2, &c., and *vice versâ*. The constituents $x_1 \ldots x_n \, t_1 \ldots t_n$, $\bar{x}_1 \ldots \bar{x}_n \, \bar{t}_1 \ldots \bar{t}_n$, &c., are in this sense symmetrical.

For all symmetrical constituents it is evident that

$$\Sigma\,(x_r\,\bar{t}_r + t_r\,\bar{x}_r)$$

vanishes. For those which do not involve $\bar{t}_1 \ldots \bar{t}_n$, it is further evident that $\bar{x}_1 \ldots \bar{x}_n$ also vanishes, whence

$$E = \bar{\phi} \qquad E' = \Sigma t_r\,(\phi),$$

$$w = \frac{\Sigma t_r(\phi)}{\Sigma t_r(\phi) - \bar{\phi}}.$$

For those constituents of which the x-factor is found in ϕ the second member of the above equation becomes 1; for those of which the x-factor is found in $\bar{\phi}$ it becomes 0. Hence *the coefficients of symmetrical constituents not involving* $\bar{t}_1 \ldots \bar{t}_n$, *of which the x-factor is found in ϕ will be* 1; *of those of which the x-factor is not found in ϕ it will be* 0.

Consider lastly, those constituents which are unsymmetrical with reference to the two sets of symbols, and which at the same time do not involve $\bar{t}_1 \ldots \bar{t}_n$.

Here it is evident, that neither E nor E' can vanish, whence the numerator of the fractional value of w in (8) must exceed the denominator. That value cannot therefore be represented by 1, 0, or $\frac{0}{0}$. It must then, in the logical development, be represented by $\frac{1}{0}$. Such then will be the coefficient of this class of constituents.

15. Hence the final logical equation by which w is expressed as a developed logical function of $x_1, \ldots x_n,\ t_1, \ldots t_n$, will be of the form

$$w = \Sigma_1\,(XT) + 0\,\{\Sigma_2(XT) + \bar{t}_1 \ldots \bar{t}_n\} + \frac{1}{0}\ \text{(sum of other constituents)}, \qquad (9)$$

wherein $\Sigma_1\,(XT)$ represents the sum of all symmetrical constituents of which the factor X is found in ϕ, and $\Sigma_2\,(XT)$, the sum of all symmetrical constituents of which the factor X is not found in ϕ,—the constituent $\bar{x}_1 \ldots \bar{x}_n\,\bar{t}_1 \ldots \bar{t}_n$, should it appear, being in either case rejected.

Passing from Logic to Algebra, it may be observed, that

here and in all similar instances, the function V, by the aid of which the algebraic system of equations for the determination of the values of $x_1, \ldots x_n, t_1, \ldots t_n$ is formed, is independent of the nature of any function ϕ involved, not in the expression of the *data*, but in that of the *quæsitum* of the problem proposed. Thus we have in the present example,

$$\text{Prob. } w = \frac{\Sigma_1 (XT)}{V},$$

wherein

$$V = \Sigma_1 (XT) + \Sigma_2 (XT) + t_1 \ldots \bar{t}_n$$
$$= \Sigma (XT) + \bar{t}_1 \ldots t_n. \qquad (10)$$

Here $\Sigma (XT)$ represents the sum of all symmetrical constituents of the x and t symbols, except the constituent $\bar{x}_1 \ldots \bar{x}_n, \bar{t}_1 \ldots \bar{t}_n$. This value of V is the same as that virtually employed in the solution of the preceding problem, and hence we may avail ourselves of the results there obtained.

If then, as in the solution referred to, we assume

$$\frac{x_1 t_1}{\bar{x}_1 \bar{t}_1} = m_1, \quad \frac{x_n t_n}{\bar{x}_n \bar{t}_n} = m_n, \quad \frac{x_1}{\bar{x}_1} = n_1, \text{ \&c.,}$$

we shall obtain a result which may be thus written:

$$\text{Prob. } w = \frac{M_1}{M + N - 1}, \qquad (11)$$

M_1 being formed by rejecting from the function ϕ the constituent $x_1 \ldots \bar{x}_n$, if it is there found, dividing the result by the same constituent $\bar{x}_1 \ldots \bar{x}_n$, and then changing $\frac{x_1}{\bar{x}_1}$ into m_1, $\frac{x_2}{\bar{x}_2}$ into m_2, and so on. The values of M and N are the same as in the preceding problem. Reverting to these and to the corresponding values of m_1, m_2, &c., we find

$$\text{Prob. } w = M_1 (\mu + \nu - 1),$$

the general values of m_r, n_r being

$$m_r = \frac{c_r p_r}{\mu - c_r p_r}, \qquad n_r = \frac{c_r (1 - p_r)}{\mu - c_r (1 - p_r)},$$

and μ and ν being given by the solution of the system of equations,

$$\mu + \nu - 1 = \frac{(\mu - c_1 p_1) \ldots (\mu - c_n p_n)}{\mu^{n-1}} = \frac{\{\nu - c_1(1-p_1)\} \ldots \{\nu - c_n(1-p_n)\}}{\nu^{n-1}}.$$

The above value of Prob. w will be the numerator of the fraction (2). It now remains to determine its denominator.

For this purpose assume

$$\phi(x_1, x_2 \ldots x_n) = v,$$

or

$$\phi = v;$$

whence

$$\phi \bar{v} + v \bar{\phi} = 0.$$

Substituting the first member of this equation in (7) in place of the corresponding form $\phi z \bar{w} + w(1 - \phi z)$ we obtain as the primary logical equation,

$$\Sigma \{x_r z \bar{t}_r + t_r (1 - x_r z)\} + \bar{x}_1 \ldots \bar{x}_n z + \phi \bar{v} + v \bar{\phi} = 0,$$

whence eliminating z, and reducing by Prop. II. Chap. IX.,

$$\phi \bar{v} + v \bar{\phi} + \Sigma t_r \{\Sigma (x_r \bar{t}_r + t_r \bar{x}_r) + \bar{x}_1 \ldots \bar{x}_n\} = 0.$$

Hence

$$v = \frac{\phi + \Sigma t_r \{\Sigma (x_r \bar{t}_r + t_r \bar{x}_r) + \bar{x}_1 \ldots \bar{x}_n\}}{2\phi - 1},$$

and developing as before,

$$v = \Sigma_1 (XT) + \bar{t}_1 \ldots \bar{t}_n \Sigma_1 (X) + 0 \{\Sigma_2 (XT) + \bar{t}_1 \ldots \bar{t}_n \Sigma_2 (X)\}$$
$$+ \frac{1}{0} \text{(sum of other constituents).} \qquad (12)$$

Here $\Sigma_1 (X)$ indicates the sum of all constituents found in ϕ, $\Sigma_2 (X)$ the sum of all constituents not found in ϕ. The expressions are indeed used in place of ϕ and $1 - \phi$ to preserve symmetry.

It follows hence that $\Sigma_1 (X) + \Sigma_2 (X) = 1$, and that, as before, $\Sigma_1 (XT) + \Sigma_2 (XT) = \Sigma (XT)$. Hence V will have the same value as before, and we shall have

$$\text{Prob. } v = \frac{\Sigma_1 (XT) + \bar{t}_1 \ldots \bar{t}_n \Sigma_1 (X)}{V},$$

Or transforming, as in the previous case,

$$\text{Prob. } v = \frac{M_1 + N_1}{M + N - 1}, \qquad (13)$$

wherein N_1 is formed by dividing ϕ by $\bar{x}_1 \dots \bar{x}_n$, and changing in the result $\frac{x_1}{\bar{x}_1}$ into n_1, $\frac{x_2}{\bar{x}_2}$ into n_2, &c.

Now the final solution of the problem proposed will be given by assigning their determined values to the terms of the fraction

$$\frac{\text{Prob.}\ \phi_1(x_1, \dots x_n)\, z}{\text{Prob.}\ \phi\,(x_1, \dots x_n)}, \text{ or } \frac{\text{Prob.}\ w}{\text{Prob.}\ v}.$$

Hence, therefore, by (11) and (13) we have

$$\text{Prob. sought} = \frac{M_1}{M_1 + N_1}.$$

A very slight attention to the mode of formation of the functions M_1 and N_1 will show that the process may be greatly simplified. We may, indeed, exhibit the solution of the general problem in the form of a rule, as follows:

Reject from the function $\phi\,(x_1, x_2 \dots x_n)$ the constituent $\bar{x}_1 \dots \bar{x}_n$ if it is therein contained, suppress in all the remaining constituents the factors $\bar{x}_1$, $\bar{x}_2$, &c., and change generally in the result x_r into $\frac{c_r p_r}{\mu - c_r p_r}$. Call this result M_1.

Again, replace in the function $\phi\,(x_1, x_2 \dots x_n)$ the constituent $\bar{x}_1 \dots \bar{x}_n$ if it is therein found, by unity; suppress in all the remaining constituents the factors $\bar{x}_1$, $\bar{x}_2$, &c., and change generally in the result x_r into $\frac{c_r (1 - p_r)}{\nu - c_r(1 - p_r)}$.

Then the solution required will be expressed by the formula

$$\frac{M_1}{M_1 + N_1}, \qquad (14)$$

μ and ν being determined by the solution of the system of equations

$$\mu + \nu - 1 = \frac{(\mu - c_1 p_1) \dots (\mu - c_n p_n)}{\mu^{n-1}}$$
$$= \frac{\{\nu - c_1(1 - p_1)\} \dots \{\nu - c_n(1 - p_n)\}}{\nu^{n-1}}. \qquad (15)$$

It may be added, that the limits of μ and ν are the same as in the previous problem. This might be inferred from the general principle of continuity; but conditions of limitation, which are probably sufficient, may also be established by other considerations.

Thus from the demonstration of the general method in probabilities, Chap. XVII. Prop. IV., it appears that the quantities $x_1, .. x_n, t_1, .. t_n$, in the primary system of algebraic equations, must be *positive proper fractions.* Now

$$\frac{x_r}{1 - x_r} = n_r = \frac{c_r (1 - p_r)}{\nu - c_r (1 - p_r)}.$$

Hence generally n_r must be a positive quantity, and therefore we must have

$$\nu \overset{=}{>} c_r (1 - p_r).$$

In like manner since we have

$$\frac{x_r t_r}{(1 - x_r)(1 - t_r)} = m_r = \frac{c_r p_r}{\mu - c_r p_r},$$

we must have generally

$$\mu \overset{=}{>} c_r p_r.$$

16. It is probable that the two classes of conditions thus represented are together sufficient to determine generally which of the roots of the equations determining μ and ν are to be taken. Let us take in particular the case in which $n = 2$. Here we have

$$\mu + \nu - 1 = \frac{(\mu - c_1 p_1)(\mu - c_2 p_2)}{\mu} = \mu - (c_1 p_1 + c_2 p_2) + \frac{c_1 p_1 c_2 p_2}{\mu},$$

$$\therefore \nu = 1 - c_1 p_1 - c_2 p_2 + \frac{c_1 p_1 c_2 p_2}{\mu} = 1 - c_1 p_1 - \frac{(\mu - c_1 p_1) c_2 p_2}{\mu}.$$

Whence, since $\mu \overset{=}{>} c_1 p_1$ we have generally

$$\nu \overset{=}{<} 1 - c_1 p_1.$$

In like manner we have

$$\nu \overset{=}{<} 1 - c_2 p_2, \quad \mu \overset{=}{<} 1 - c_1 (1 - p_1), \quad \mu \overset{=}{<} 1 - c_2 (1 - p_2).$$

Now it has already been shown that there will exist but one value of μ satisfying the whole of the above conditions relative to that quantity, viz.

$$\mu \overset{=}{>} c_r p_r, \quad \mu \overset{=}{<} 1 - c_r (1 - p_r),$$

whence the solution for this case, at least, is determinate. And I apprehend that the same method is generally applicable and sufficient. But this is a question upon which a further degree of light is desirable.

To verify the above results, suppose $\phi(x_1, .. x_n) = 1$, which is virtually the case considered in the previous problem. Now the development of 1 gives all possible constituents of the symbols $x_1, .. x_n$. Proceeding then according to the Rule, we find

$$M_1 = \frac{\mu^n}{(\mu - c_1 p_1)..(\mu - c_n p_n)} - 1 = \frac{\mu}{\mu + \nu - 1} - 1 \text{ by (15).}$$

$$N_1 = \frac{\nu^n}{\{\nu - c_1(1 - p_1)\}..\{\nu - c_n(1 - p_n)\}} - 1 = \frac{\nu}{\mu + \nu - 1} - 1.$$

Substituting in (14) we find

$$\text{Prob. } z = 1 - \nu,$$

which agrees with the previous solution.

Again, let $\phi(x_1, .. x_n) = x_1$, which, after development and suppression of the factors $\bar{x}_2, .. \bar{x}_n$, gives $x_1(x_2 + 1)..(x_n + 1)$, whence we find

$$M_1 = \frac{c_1 p_1 \mu^{n-1}}{(\mu - c_1 p_1)..(\mu - c_n p_n)} = \frac{c_1 p_1}{\mu + \nu - 1} \text{ by (15).}$$

$$N_1 = \frac{c_1(1 - p_1)\nu^{n-1}}{\{\nu - c_1(1 - p_1)\}..\{\nu - c_n(1 - p_n)\}} = \frac{c_1(1 - p_1)}{\mu + \nu - 1}.$$

Substituting, we have

Probability that if the event A_1 occur, E will occur $= p_1$.

And this result is verified by the data. Similar verifications might easily be added.

Let us examine the case in which

$$\phi(x_1, .. x_n) = x_1 \bar{x}_2 .. \bar{x}_n + x_2 \bar{x}_1 \bar{x}_3 .. \bar{x}_n .. + x_n \bar{x}_1 .. \bar{x}_{n-1}.$$

Here we find

$$M_1 = \frac{c_1 p_1}{\mu - c_1 p_1} .. + \frac{c_n p_n}{\mu - c_n p_n},$$

$$N_1 = \frac{c_1(1 - p_1)}{\nu - c_1(1 - p_1)} .. + \frac{c_n(1 - p_n)}{\nu - c_n(1 - p_n)};$$

whence we have the following result—

Probability that if some one alone of the causes $A_1, A_2 .. A_n$ present itself, the event E will follow. $\Bigg\} = \dfrac{\Sigma \dfrac{c_r p_r}{\mu - c_r p_r}}{\Sigma \dfrac{c_r p_r}{\mu - c_r p_r} + \Sigma \dfrac{c_r (1 - p_r)}{\nu - c_r (1 - p_r)}}$

Let it be observed that this case is quite different from the well-known one in which the mutually exclusive character of the causes $A_1, .. A_n$ is one of the elements of the data, expressing a condition under which the very observations by which the probabilities of A_1, A_2, &c. are supposed to have been determined, were made.

Consider, lastly, the case in which $\phi(x_1, .. x_n) = x_1 x_2 .. x_n$. Here

$$M_1 = \frac{c_1 p_1 .. c_n p_n}{(\mu - c_1 p_1) .. (\mu - c_n p_n)} = \frac{c_1 p_1 .. c_n p_n}{\mu^{n-1} (\mu + \nu - 1)},$$

$$N_1 = \frac{c_1 (1 - p_1) .. c_n (1 - p_n)}{\{\nu - c_1 (1 - p_1)\} .. \{\nu - c_n (1 - p_n)\}} = \frac{c_1 (1 - p_1) .. c_n (1 - p_n)}{\nu^{n-1} (\mu + \nu - 1)}.$$

Hence the following result—

Probability that if all the causes $A_1, .. A_n$ conspire, the event E will follow. $\Bigg\} = \dfrac{p_1 .. p_n \nu^{n-1}}{p_1 .. p_n \nu^{n-1} + (1 - p_1) .. (1 - p_n) \mu^{n-1}}.$

This expression assumes, as it ought to do, the value 1 when any one of the quantities $p_1, .. p_n$ is equal to 1.

17. Problem VIII.—Certain causes A_1, $A_2 .. A_n$ being so restricted that they cannot all fail, but still can only occur in certain definite combinations denoted by the equation

$$\phi(A_1, A_2 .. A_n) = 1,$$

and there being given the separate probabilities $c_1, .. c_n$ of the said causes, and the corresponding probabilities $p_1, .. p_n$ that an event E will follow if those respective causes are realized, required the probability of the event E.

This problem differs from the one last considered in several particulars, but chiefly in this, that the restriction denoted by the equation $\phi(A_1, .. A_n) = 1$, forms one of the data, and is supposed to be furnished by or to be accordant with the very experience from which the knowledge of the numerical elements of the problem is derived.

Representing the events $A_1, \ldots A_n$ by $x_1, \ldots x_n$ respectively, and the event E by z, we have—

$$\text{Prob.}\, x_r = c_r, \quad \text{Prob.}\, x_r z = c_r p_r. \tag{1}$$

Let us assume, generally,

$$x_r z = t_r,$$

then combining the system of equations thus indicated with the equations

$$\bar{x}_1 \ldots \bar{x}_n = 0, \quad \phi(x_1, \ldots x_n) = 1, \quad \text{or } \phi = 1,$$

furnished in the data, we ultimately find, as the developed expression of z,

$$z = \Sigma(XT) + 0\bar{t}_1 \bar{t}_2 \ldots \bar{t}_n \Sigma(X), \tag{2}$$

where X represents in succession each constituent found in ϕ, and T a similar series of constituents of the symbols $t_1, \ldots t_n$; $\Sigma(XT)$ including only *symmetrical* constituents with reference to the two sets of symbols.

The method of reduction to be employed in the present case is so similar to the one already exemplified in former problems, that I shall merely exhibit the results to which it leads. We find

$$\text{Prob.}\, z = \frac{M}{M+N}, \tag{3}$$

with the relations

$$\frac{M_1}{c_1 p_1} \ldots = \frac{M_n}{c_n p_n} = \frac{N_1}{c_1(1-p_1)} = \frac{N_n}{c_n(1-p_n)} = M + N. \tag{4}$$

Wherein M is formed by suppressing in $\phi(x_1, \ldots x_n)$ all the factors $\bar{x}_1, \ldots \bar{x}_n$, and changing in the result x_1 into m_1, x_n into m_n, while N is formed by substituting in M, n_1 for m_1, &c.; moreover M_1 consists of that portion of M of which m_1 is a factor, N_1 of that portion of N of which n_1 is a factor; and so on.

Let us take, in illustration, the particular case in which the causes $A_1 \ldots A_n$ are mutually exclusive. Here we have

$$\phi(x_1, \ldots x_n) = x_1 \bar{x}_2 \ldots \bar{x}_n \ldots + x_n \bar{x}_1 \ldots \bar{x}_{n-1}.$$

Whence

$$M = m_1 + m_2 \ldots + m_n,$$
$$N = n_1 + n_2 \ldots + n_n,$$
$$M_1 = m_1, \quad N_1 = n_1, \text{ \&c.}$$

Substituting, we have

$$\frac{m_1}{c_1 p_1} \ldots = \frac{m_n}{c_n p_n} = \frac{n_1}{c_1(1-p_1)} \ldots = \frac{n_n}{c_n(1-p_n)} = M + N.$$

Hence we find

$$\frac{m_1 + m_2 \ldots + m_n}{c_1 p_1 + c_2 p_2 \ldots + c_n p_n} = M + N,$$

or

$$\frac{M}{c_1 p_1 \ldots + c_n p_n} = M + N.$$

Hence, by (3),

$$\text{Prob. } z = c_1 p_1 \ldots + c_n p_n,$$

a known result.

There are other particular cases in which the system (4) admits of ready solution. It is, however, obvious that in most instances it would lead to results of great complexity. Nor does it seem probable that the existence of a functional relation among causes, such as is assumed in the data of the general problem, will often be presented in actual experience; if we except only the particular cases above discussed.

Had the general problem been modified by the restriction that the event E cannot occur, all the causes $A_1 \ldots A_n$ being absent, instead of the restriction that the said causes cannot all fail, the remaining condition denoted by the equation $\phi(A_1, \ldots A_n) = 1$ being retained, we should have found for the final logical equation

$$z = \Sigma_1(XT) + 0\,\Sigma(X),$$

$\Sigma(X)$ being, as before, equal to $\phi(x_1, \ldots x_n)$, but $\Sigma_1(XT)$ formed by rejecting from ϕ the particular constituent $\bar{x}_1 \ldots \bar{x}_n$ if therein contained, and then multiplying each x-constituent of the result by the corresponding t-constituent. It is obvious that in the particular case in which the causes are mutually exclusive the value of Prob. z hence deduced will be the same as before.

18. Problem IX.—Assuming the data of any of the pre-

vious problems, let it be required to determine the probability that if the event E present itself, it will be associated with the particular cause A_r; in other words, to determine the *à posteriori* probability of the cause A_r when the event E has been observed to occur.

In this case we must seek the value of the fraction

$$\frac{\text{Prob. } x_r z}{\text{Prob. } z}, \text{ or } \frac{c_r p_r}{\text{Prob. } z}, \text{ by the data.} \qquad (1)$$

As in the previous problems, the value of Prob. z has been assigned upon different hypotheses relative to the connexion or want of connexion of the causes, it is evident that in all those cases the present problem is susceptible of a determinate solution by simply substituting in (1) the value of that element thus determined.

If the *à priori* probabilities of the causes are equal, we have $c_1 = c_2 \ldots = c_r$. Hence for the different causes the value (1) will vary directly as the quantity p_r. Wherefore *whatever the nature of the connexion among the causes*, the *à posteriori* probability of each cause will be proportional to the probability of the observed event E when that cause is known to exist. The particular case of this theorem, which presents itself when the causes are mutually exclusive, is well known. We have then

$$\frac{\text{Prob. } x_r z}{\text{Prob. } z} = \frac{c_r p_r}{\Sigma c_r p_r} = \frac{p_r}{p_1 + p_2 \ldots + p_n},$$

the values of $c_1, \ldots c_n$ being equal.

Although, for the demonstration of these and similar theorems in the particular case in which the causes are mutually exclusive, it is not necessary to introduce the functional symbol ϕ, which is, indeed, to claim for ourselves the choice of all possible and conceivable hypotheses of the connexion of the causes, yet, under every form, the solution by the method of this work of problems, in which the number of the data is indefinitely great, must always partake of a somewhat complex character. Whether the systematic evolution which it presents, first, of the logical, secondly, of the numerical relations of a problem, furnishes any compensation for the length and occasional tediousness of its processes, I do not presume to inquire. Its chief value undoubtedly consists in its power,—in the mastery which it gives us over questions which would apparently baffle the unassisted strength of human reason. For this cause it has not been deemed superfluous to exhibit in this chapter its application to problems, some of which may possibly be regarded as repulsive, from their difficulty, without being recommended by any prospect of immediate utility. Of the ulterior value of such speculations it is, I conceive, impossible for us, at present, to form any decided judgment.

19. The following problem is of a much easier description than the previous ones.

PROBLEM X.—*The probability of the occurrence of a certain natural phænomenon under given circumstances is* p. *Observation has also recorded a probability* a *of the existence of a permanent cause of that phænomenon, i.e. of a cause which would always produce the event under the circumstances supposed. What is the probability that if the phænomenon is observed to occur* n *times in succession under the given circumstances, it will occur the* $n+1^{th}$ *time? What also is the probability, after such observation, of the existence of the permanent cause referred to?*

FIRST CASE.—Let t represent the existence of a permanent cause, and $x_1, x_2 \ldots x_{n+1}$ the successive occurrences of the natural phænomenon.

If the permanent cause exist, the events $x_1, x_2 \ldots x_{n+1}$ are necessary consequences. Hence

$$t = vx_1, \quad t = vx_2, \text{ \&c.},$$

and eliminating the indefinite symbols,

$$t(1-x_1)=0, \quad t(1-x_2)=0, \quad t(1-x_{n+1})=0.$$

Now we are to seek the probability that if the combination $x_1 x_2 \ldots x_n$ happen, the event x_{n+1} will happen, i. e. we are to seek the value of the fraction

$$\frac{\text{Prob. } x_1 x_2 \ldots x_{n+1}}{\text{Prob. } x_1 x_2 \ldots x_n}.$$

We will first seek the value of Prob. $x_1 x_2 \ldots x_n$.

Represent the combination $x_1 x_2 .. x_n$ by w, then we have the following logical equations:

$$t(1 - x_1) = 0, \quad t(1 - x_2) = 0 \; .. \; t(1 - x_n) = 0,$$
$$x_1 x_2 .. x_n = w.$$

Reducing the last to the form

$$(x_1 x_2 .. x_n)(1 - w) + w(1 - x_1 x_2 .. x_n) = 0,$$

and adding it to the former ones, we have

$$\Sigma t(1 - x_i) + x_1 x_2 .. x_n (1 - w) + w(1 - x_1 x_2 .. x_n) = 0, \quad (1)$$

wherein Σ extends to all values of i from 1 to n, for the one logical equation of the data. With this we must connect the numerical conditions,

$$\text{Prob.}\, x_1 = \text{Prob.}\, x_2 .. = \text{Prob.}\, x_n = p, \quad \text{Prob.}\, t = a;$$

and our object is to find Prob. w.

From (1) we have

$$w = \frac{\Sigma t(1 - x_i) + x_1 x_2 .. x_n}{2 x_1 x_2 .. x_n - 1}$$

$$= \frac{\Sigma(1 - x_i) + x_1 x_2 .. x_n}{2 x_1 x_2 .. x_n - 1} t + \frac{x_1 x_2 .. x_n}{2 x_1 x_2 .. x_n - 1}(1 - t), \quad (2)$$

on developing with respect to t. This result must further be developed with respect to $x_1, x_2, .. x_n$.

Now if we make $x_1 = 1, x_2 = 1, .. x_n = 1$, the coefficients both of t and of $1 - t$ become 1. If we give to the same symbols any other set of values formed by the interchange of 0 and 1, it is evident that the coefficient of t will become negative, while that of $1 - t$ will become 0. Hence the full development (2) will be

$$w = x_1 x_2 .. x_n t + x_1 x_2 .. x_n (1 - t) + 0(1 - x_1 x_2 .. x_n)(1 - t)$$
$$+ \text{ constituents whose coefficients are } \frac{1}{0}, \text{ or equivalent to } \frac{1}{0}.$$

Here we have

$$V = x_1 x_2 .. x_n t + x_1 x_2 .. x_n (1 - t) + (1 - x_1 x_2 .. x_n)(1 - t)$$
$$= x_1 x_2 .. x_n t + 1 - t;$$

whence, passing from Logic to Algebra,

$$\frac{x_1 x_2 \ldots x_n t + x_1(1-t)}{p} = \frac{x_1 x_2 \ldots x_n t + x_2(1-t)}{p} \ldots$$

$$= \frac{x_1 x_2 \ldots x_n t + x_n(1-t)}{p} = \frac{x_1 x_2 \ldots x_n t}{a} = x_1 x_2 \ldots x_n t + 1 - t.$$

$$\text{Prob. } w = \frac{x_1 x_2 \ldots x_n}{x_1 x_2 \ldots x_n t + 1 - t}.$$

From the forms of the above equations it is evident that we have $x_1 = x_2 \ldots = x_n$. Replace then each of these quantities by x, and the system becomes

$$\frac{x^n t + (1-t)x}{p} = \frac{x^n t}{a} = x^n t + 1 - t,$$

$$\text{Prob. } w = \frac{x^n}{x^n t + 1 - t};$$

from which we readily deduce

$$\text{Prob. } w = \text{Prob. } x_1 x_2 \ldots x_n = a + (p-a)\left(\frac{p-a}{1-a}\right)^{n-1}$$

If in this result we change n into $n+1$, we get

$$\text{Prob. } x_1 x_2 \ldots x_{n+1} = a + (p-a)\left(\frac{p-a}{1-a}\right)^n.$$

Hence we find—

$$\frac{\text{Prob. } x_1 x_2 \ldots x_{n+1}}{\text{Prob. } x_1 x_2 \ldots x_n} = \frac{a + (p-a)\left(\frac{p-a}{1-a}\right)^n}{a + (p-a)\left(\frac{p-a}{1-a}\right)^{n-1}} \qquad (3)$$

as the expression of the probability that if the phænomenon be n times repeated, it will also present itself the $n+1^{th}$ time. By the method of Chapter XIX. it is found that a cannot exceed p in value.

The following verifications are obvious :—

1st. If $a = 0$, the expression reduces to p, as it ought to do. For when it is certain that no permanent cause exists, the successive occurrences of the phænomenon are independent.

2nd. If $p = 1$, the expression becomes 1, as it ought to do.

3rd. If $p = a$, the expression becomes 1, unless $a = 0$. If the probability of a phænomenon is equal to the probability that there exists a cause which under given circumstances would always produce it, then the fact that that phænomenon has ever been noticed under those circumstances, renders certain its re-appearance under the same.*

4th. As n increases, the expression approaches in value to unity. This indicates that the probability of the recurrence of the event increases with the frequency of its successive appearances,—a result agreeable to the natural laws of expectation.

SECOND CASE.—We are now to seek the probability *à posteriori* of the existence of a permanent cause of the phænomenon. This requires that we ascertain the value of the fraction

$$\frac{\text{Prob. } tx_1 x_2 \ldots x_n}{\text{Prob. } x_1 x_2 \ldots x_n},$$

the denominator of which has already been determined.

To determine the numerator assume

$$tx_1 x_2 \ldots x_n = w,$$

then proceeding as before, we obtain for the logical development,

$$w = tx_1 x_2 \ldots x_n + 0 (1 - t).$$

Whence, passing from Logic to Algebra, we have at once

$$\text{Prob. } w = a,$$

a result which might have been anticipated. Substituting then for the numerator and denominator of the above fraction their values, we have for the *à posteriori* probability of a permanent cause, the expression

---

* As we can neither re-enter nor recall the state of infancy, we are unable to say how far such results as the above serve to explain the confidence with which young children connect events whose association they have once perceived. But we may conjecture, generally, that the strength of their expectations is due to the necessity of inferring (as a part of their rational nature), and the narrow but impressive experience upon which the faculty is exercised. Hence the reference of every kind of sequence to that of cause and effect. A little friend of the author's, on being put to bed, was heard to ask his brother the pertinent question,—"Why does going to sleep at night make it light in the morning?" The brother, who was a year older, was able to reply, that it would be light in the morning even if little boys did not go to sleep at night.

$$\frac{a}{a + (p - a)\left(\frac{p - a}{1 - a}\right)^{n-1}}$$

It is obvious that the value of this expression increases with the value of n.

I am indebted to a learned correspondent,* whose original contributions to the theory of probabilities have already been referred to, for the following verification of the first of the above results (3).

"The whole *à priori* probability of the event (under the circumstances) being p, and the probability of some cause C which would necessarily produce it, a, let x be the probability that it will happen if no such cause as C exist. Then we have the equation

$$p = a + (1 - a)\,x,$$

whence

$$x = \frac{p - a}{1 - a}.$$

Now the phænomenon observed is the occurrence of the event n times. The *à priori* probability of this would be—

1 supposing C to exist,
x^n supposing C not to exist;

whence the *à posteriori* probability that C exists is

$$\frac{a}{a + (1 - a)\,x^n},$$

that C does not exist is

$$\frac{(1 - a)\,x^n}{a + (1 - a)\,x^n}.$$

Consequently the probability of another occurrence is

$$\frac{a}{a + (1 - a)\,x^n} \times 1 + \frac{(1 - a)\,x^n}{a + (1 - a)\,x^n} \times a,$$

or

$$\frac{a + (1 - a)\,x^{n+1}}{a + (1 - a)\,x^n},$$

---

* Professor Donkin.

which, on replacing n by its value $\frac{p-a}{1-a}$, will be found to agree with (3)."

Similar verifications might, it is probable, also be found for the following results, obtained by the direct application of the general method.

The probability, under the same circumstances, that if, out of n occasions, the event happen r times, and fail $n-r$ times, it will happen on the $n+1^{th}$ time is

$$\frac{a+m(p-a)\left(\frac{p-la}{1-a}\right)^r}{a+m(p-la)\left(\frac{p-la}{1-a}\right)^{r-1}}$$

wherein $m = \frac{n(n-1)\ldots n-r+1}{1.2\ldots r}$ and $l = \frac{r}{n}$.

The probability of a permanent cause (r being less than n) is 0. This is easily verified.

If p be the probability of an event, and c the probability that if it occur it will be due to a permanent cause; the probability after n successive observed occurrences that it will recur on the $n+1^{th}$ similar occasion is

$$\frac{c+(1-c)\,x^n}{c+(1-c)\,x^{n-1}},$$

wherein $x = \frac{p(1-c)}{1-cp}$.

20. It is remarkable that the solutions of the previous problems are void of any arbitrary element. We should scarcely, from the appearance of the data, have anticipated such a circumstance. It is, however, to be observed, that in all those problems the probabilities of the *causes* involved are supposed to be known *à priori*. In the absence of this assumed element of knowledge, it seems probable that arbitrary constants would *necessarily* appear in the final solution. Some confirmation of this remark is afforded by a class of problems to which considerable attention has been directed, and which, in conclusion, I shall briefly consider.

It has been observed that there exists in the heavens a large number of double stars of extreme closeness. Either these apparent instances of connexion have some physical ground or they have not. If they have not, we may regard the phænomenon of a double star as the accidental result of a "random distribution" of stars over the celestial vault, i. e. of a distribution which would render it just as probable that either member of the binary system should appear in one spot as in another. If this hypothesis be assumed, and if the number of stars of a requisite brightness be known, we can determine what is the probability that two of them should be found within such limits of mutual distance as to constitute the observed phænomenon. Thus Mitchell,* estimating that there are 230 stars in the heavens equal in brightness to β Capricorni, determines that it is 80 to 1 against such a combination being presented were those stars distributed at random. The probability, when such a combination has been observed, that there exists between its members a physical ground of connexion, is then required.

Again, the sum of the inclinations of the orbits of the ten known planets to the plane of the ecliptic in the year 1801 was 91° 4187, according to the French measures. Were all inclinations equally probable, Laplace† determines, that there would be only the excessively small probability .00000011235 that the mean of the inclinations should fall within the limit thus assigned. And he hence concludes, that there is a very high probability in favour of a disposing cause, by which the inclinations of the planetary orbits have been confined within such narrow bounds. Professor De Morgan,‡ taking the sum of the inclinations at 92°, gives to the above probability the value .00000012, and infers that "it is 1 : .00000012, that there was a necessary cause in the formation of the solar system for the inclinations being what they are." An equally determinate conclusion has been drawn from observed coincidences between the direction of

* Phil. Transactions, An. 1767.

† Théorie Analytique des Probabilités, p. 276.

‡ Encyclopædia Metropolitana. Art. Probabilities.

circular polarization in rock-crystal, and that of certain oblique faces in its crystalline structure.*

These problems are all of a similar character. A certain hypothesis is framed, of the various possible consequences of which we are able to assign the probabilities with perfect rigour. Now some actual result of observation being found among those consequences, and its *hypothetical* probability being therefore known, it is required thence to determine the probability of the hypothesis assumed, or its contrary. In Mitchell's problem, the hypothesis is that of a "random distribution of the stars,"—the possible and observed consequence, the appearance of a close double star. The very small probability of such a result is held to imply that the probability of the hypothesis is equally small, or, at least, of the same order of smallness. And hence the high and, and as some think, *determinate* probability of a disposing cause in the stellar arrangements is inferred. Similar remarks apply to the other examples adduced.

21. The general problem, in whatsoever form it may be presented, admits only of an *indefinite* solution. Let x represent the proposed hypothesis, y a phænomenon which might occur as one of its possible consequences, and whose calculated probability, on the assumption of the truth of the hypothesis, is p, and let it be required to determine the probability that if the phænomenon y is observed, the hypothesis x is true. The very data of this problem cannot be expressed without the introduction of an arbitrary element. We can only write

$$\text{Prob. } x = a, \qquad \text{Prob. } xy = ap; \tag{1}$$

a being perfectly arbitrary, except that it must fall within the limits 0 and 1 inclusive. If then P represent the conditional probability sought, we have

$$P = \frac{\text{Prob. } xy}{\text{Prob. } y} = \frac{ap}{\text{Prob. } y}. \tag{2}$$

It remains then to determine Prob. y.

* Edinburgh Review, No. 185, p. 32. This article, though not entirely free from error, is well worthy of attention.

Let $xy = t$, then

$$y = \frac{t}{x} = tx + \frac{1}{0}t(1-x) + 0(1-t)x + \frac{0}{0}(1-t)(1-x). \quad (3)$$

Hence observing that Prob. $x = a$, Prob. $t = ap$, and passing from Logic to Algebra, we have

$$\text{Prob. } y = \frac{tx + c(1-t)x}{tx + 1 - t},$$

with the relations

$$\frac{tx + (1-t)x}{a} = \frac{tx}{ap} = tx + 1 - t.$$

Hence we readily find

$$\text{Prob. } y = ap + c(1-a). \quad (4)$$

Now recurring to (3), we find that c is the probability, that if the event $(1-t)(1-x)$ occur, the event y will occur. But

$$(1-t)(1-x) = (1-xy)(1-x) = 1-x.$$

Hence c is the probability *that if the event x do not occur, the event y will occur.*

Substituting the value of Prob. y in (2), we have the following theorem:

The calculated probability of any phænomenon y, upon an assumed physical hypothesis x, being p, the à posteriori probability P of the physical hypothesis, when the phænomenon has been observed, is expressed by the equation

$$P = \frac{ap}{ap + c(1-a)}, \quad (5)$$

where a and c are arbitrary constants, the former representing the à priori probability of the hypothesis, the latter the probability that if the hypothesis were false, the event y would present itself.

The principal conclusion deducible from the above theorem is that, other things being the same, the value of P increases and diminishes simultaneously with that of p. Hence the greater or less the probability of the phænomenon when the *hypothesis* is *assumed*, the greater or less is the probability of the hypothesis when the *phænomenon* has been *observed*. When p is very small, then generally P also is small, unless either a is large or c small.

Hence, secondly, if the probability of the phænomenon is very small when the hypothesis is assumed, the probability of the hypothesis is very small when the phænomenon is observed, unless either the *à priori* probability a of the hypothesis is large, or the probability of the phænomenon upon any other hypothesis small.

The formula (5) admits of exact verification in various cases, as when $c = 0$, or $a = 1$, or $a = 0$. But it is evident that it does not, unless there be means for determining the values of a and c, yield a *definite* value of P. Any solutions which profess to accomplish this object, either are erroneous in principle, or involve a tacit assumption respecting the above arbitrary elements. Mr. De Morgan's solution of Laplace's problem concerning the existence of a determining cause of the narrow limits within which the inclinations of the planetary orbits to the plane of the ecliptic are confined, appears to me to be of the latter description. Having found a probability $p = .00000012$, that the sum of the inclinations would be less than 92° were all degrees of inclination equally probable in each orbit, this able writer remarks: "If there be a reason for the inclinations being as described, the probability of the event is 1. Consequently, it is $1 : .00000012$ (i. e. $1 : p$), that there was a necessary cause in the formation of the solar system for the inclinations being what they are." Now this result is what the equation (5) would really give, if, assigning to p the above value, we should assume $c = 1$, $a = \frac{1}{2}$. For we should thus find,

$$P = \frac{\frac{1}{2}p}{\frac{1}{2}p + \frac{1}{2}} = \frac{p}{1+p}.$$

$$\therefore\ 1 - P : P :: 1 : p. \qquad (6)$$

But P representing the probability, *à posteriori*, that all inclinations are equally probable, $1 - P$ is the probability, *à posteriori*, that such is not the case, or, adopting Mr. De Morgan's alternative, that a determining cause exists. The equation (6), therefore, agrees with Mr. De Morgan's result.

22. Are we, however, justified in assigning to a and c particular values? I am strongly disposed to think that we are not.

The question is of less importance in the special instance than in its ulterior bearings. In the received applications of the theory of probabilities, arbitrary constants do not explicitly appear; but in the above, and in many other instances sanctioned by the highest authorities, some virtual determination of them has been attempted. And this circumstance has given to the results of the theory, especially in reference to questions of causation, a character of definite precision, which, while on the one hand it has seemed to exalt the dominion and extend the province of numbers, even beyond the measure of their ancient claim to rule the world;* on the other hand has called forth vigorous protests against their intrusion into realms in which conjecture is the only basis of inference. The very fact of the appearance of arbitrary constants in the solutions of problems like the above, treated by the method of this work, seems to imply, that definite solution is impossible, and to mark the point where inquiry ought to stop. We possess indeed the means of interpreting those constants, but the experience which is thus indicated is as much beyond our reach as the experience which would preclude the necessity of any attempt at solution whatever.

Another difficulty attendant upon these questions, and inherent, perhaps, in the very constitution of our faculties, is that of precisely defining what is meant by Order. The manifestations of that principle, except in very complex instances, we have no difficulty in detecting, nor do we hesitate to impute to it an almost necessary foundation in causes operating under Law. But to assign to it a standard of *numerical* value would be a vain, not to say a presumptuous, endeavour. Yet must the attempt be made, before we can aspire to weigh with accuracy the probabilities of different constitutions of the universe, so as to determine the elements upon which alone a definite solution of the problems in question can be established.

23. The most usual mode of endeavouring to evade the *necessary* arbitrariness of the solution of problems in the theory of

---

* Mundum regunt numeri.

† See an interesting paper by Prof. Forbes in the Philosophical Magazine, Dec. 1850; also Mill's Logic, chap. xviii.

probabilities which rest upon insufficient data, is to assign to some element whose real probability is unknown all possible degrees of probability; to suppose that these degrees of probability are themselves equally probable; and, regarding them as so many distinct causes of the phænomenon observed, to apply the theorems which represent the case of an effect due to some one of a number of equally probable but mutually exclusive causes (Problem 9). For instance, the rising of the sun after a certain interval of darkness having been observed m times in succession, the probability of its again rising under the same circumstances is determined, on received principles, in the following manner. Let p be any unknown probability between 0 and 1, and c (infinitesimal and constant) the probability, that the probability of the sun's rising after an interval of darkness lies between the limits p and $p + dp$. Then the probability that the sun will rise m times in succession is

$$c\int_0^1 p^m dp;$$

and the probability that he will do this, and will rise again, or, which is the same thing, that he will rise $m + 1$ times in succession, is

$$c\int_0^1 p^{m+1} dp,$$

Hence the probability that if he rise m times in succession, he will rise the $m + 1^{th}$ time, is

$$\frac{c\int_0^1 p^{m+1} dp}{c\int_0^1 p^m dp} = \frac{m+1}{m+2},$$

the known and generally received solution.

The above solution is usually founded upon a supposed analogy of the problem with that of the drawing of balls from an urn containing a mixture of black and white balls, between which all possible numerical ratios are assumed to be equally probable. And it is remarkable, that there are two or three distinct hypotheses which lead to the same final result. For instance, if the balls are finite in number, and those which are drawn are not replaced, or if they are infinite in number, whether those drawn are replaced or not, then, supposing that m successive drawings have yielded only white balls, the probability of the issue of a white ball at the $m + 1^{th}$ drawing is

$$\frac{m+1}{m+2}.*$$

It has been said, that the principle involved in the above and in similar applications is that of the equal distribution of our knowledge, or rather of our ignorance—the assigning to different states of things of which we know nothing, and upon the very ground that we know nothing, equal degrees of probability. I apprehend, however, that this is an arbitrary method of procedure. Instances may occur, and one such has been adduced, in which different hypotheses lead to the same final conclusion. But those instances are exceptional. With reference to the particular problem in question, it is shown in the memoir cited, that there is one hypothesis, viz., when the balls are finite in number and not replaced, which leads to a different conclusion, and it is easy to see that there are other hypotheses, as strictly involving the principle of the "equal distribution of knowledge or ignorance," which would also conduct to conflicting results.

24. For instance, let the case of sunrise be represented by the drawing of a white ball from a bag containing an infinite number of balls, which are all either black or white, and let the assumed principle be, that *all possible constitutions of the system of balls are equally probable*. By a constitution of the system, I mean an arrangement which assigns to every ball in the system a determinate colour, either black or white. Let us thence seek the probability, that if m white balls are drawn in m drawings, a white ball will be drawn in the $m + 1^{th}$ drawing.

First, suppose the number of the balls to be μ, and let the symbols $x_1, x_2, \ldots x_\mu$ be appropriated to them in the following manner. Let x_i denote that event which consists in the ith ball of the system being white, the proposition declaratory of such a state of things being $x_i = 1$. In like manner the compound

---

* ee a memoir by Bishop Terrot, Edinburgh Phil. Trans. vol. xx. Part iv.

symbol $1 - x_i$ will represent the circumstance of the ith ball being black. It is evident that the several constituents formed of the entire set of symbols $x_1, x_2, \ldots x_\mu$ will represent in like manner the several possible constitutions of the system of balls with respect to blackness and whiteness, and the number of such constitutions being 2^μ, the probability of each will, in accordance with the hypothesis, be $\frac{1}{2^\mu}$. This is the value which we should find if we substituted in the expression of any constituent for each of the symbols $x_1, x_2, \ldots x_\mu$, the value $\frac{1}{2}$. Hence, then, the probability of any event which can be expressed as a series of constituents of the above description, will be found by substituting in such expression the value $\frac{1}{2}$ for each of the above symbols.

Now the larger μ is, the less probable it is that any ball which has been drawn and replaced will be drawn again. As μ approaches to infinity, this probability approaches to 0. And this being the case, the state of the balls actually drawn can be expressed as a logical function of m of the symbols $x_1, \ldots x_2 \ldots x_\mu$, and therefore, by development, as a series of constituents of the said m symbols. Hence, therefore, its probability will be fonnd by substituting for each of the symbols, whether in the undeveloped or the developed form, the value $\frac{1}{2}$. But this is the very substitution which it would be necessary, and which it would suffice, to make, if the probability of a white ball at each drawing were known, *à priori*, to be $\frac{1}{2}$.

It follows, therefore, that if the number of balls be infinite, and all constitutions of the system equally probable, the probability of drawing m white balls in succession will be $\frac{1}{2^m}$, and the probability of drawing $m + 1$ white balls in succession $\frac{1}{2^{m+1}}$; whence the probability that after m white balls have been drawn, the next drawing will furnish a white one, will be $\frac{1}{2}$. In other words, past experience does not in this case affect future expectation.

25. It may be satisfactory to verify this result by ordinary methods. To accomplish this, we shall seek—

First: The probability of drawing r white balls, and $p-r$ black balls, in p trials, out of a bag containing μ balls, every ball being replaced after drawing, and all constitutions of the systems being equally probable, *à priori*.

Secondly: The value which this probability assumes when μ becomes infinite.

Thirdly: The probability hence derived, that if m white balls are drawn in succession, the $m+1^{th}$ ball drawn will be white also.

The probability that r white balls and $p-r$ black ones will be drawn in p trials out of an urn containing μ balls, each ball being replaced after trial, and all constitutions of the system as above defined being equally probable, is equal to the sum of the probabilities of the same result upon the separate hypotheses of there being no white balls, 1 white ball,—lastly μ white balls in the urn. Therefore, it is the sum of the probabilities of this result on the hypothesis of there being n white balls, n varying from 0 to μ.

Now supposing that there are n white balls, the probability of drawing a white ball in a single drawing is $\frac{n}{\mu}$, and the probability of drawing r white balls and $p-r$ black ones in a particular order in p drawings, is

$$\left(\frac{n}{\mu}\right)^r\left(1-\frac{n}{\mu}\right)^{p-r}.$$

But there being as many such orders as there are combinations of r things in p things, the total probability of drawing r white balls in p drawings out of the system of μ balls of which n are white, is

$$\frac{p(p-1)..(p-r+1)}{1.2..r}\left(\frac{n}{\mu}\right)^r\left(1-\frac{n}{\mu}\right)^{p-r}. \qquad (1)$$

Again, the number of constitutions of the system of μ balls, which admit of exactly n balls being white, is

$$\frac{\mu(\mu-1)\,..\,(\mu-n+1)}{1\,.\,2\,..\,n},$$

and the number of possible constitutions of the system is 2^μ.

Hence the probability that exactly n balls are white is

$$\frac{\mu(\mu-1)\,..\,(\mu-n+1)}{1\,.\,2\,..\,n2^\mu}.$$

Multiplying (1) by this expression, and taking the sum of the products from $n=0$ to $n=\mu$, we have

$$\frac{p(p-1)\,..\,p-r+1}{1\,.\,2\,..\,r}\,\Sigma_{n=0}^{n=\mu}\frac{\mu(\mu-1)\,..\,\mu-n+1}{1\,.\,2\,..\,n\,2^\mu}\left(\frac{n}{\mu}\right)^r\left(1-\frac{n}{\mu}\right)^{p=r}, \quad (2)$$

for the expression of the total probability, that out of a system of μ balls of which all constitutions are equally probable, r white balls will issue in p drawings. Now

$$\Sigma_{n=0}^{n=\mu}\frac{\mu(\mu-1)\,..\,(\mu-n+1)}{1\,.\,2\,..\,n\,.\,2^\mu}\left(\frac{n}{\mu}\right)^r\left(1-\frac{n}{\mu}\right)^{p-r}$$

$$=\Sigma_{n=0}^{n=\mu}\frac{\mu(\mu-1)\,..\,(\mu-n+1)}{1\,.\,2\,..\,n\,2^\mu}\left(\frac{n}{\mu}\right)^r\left(1-\frac{n}{\mu}\right)^{p-r}\epsilon^{n\theta}\,..\,(\theta=0)$$

$$=\frac{1}{2^\mu}\left(\frac{D}{\mu}\right)^r\left(1-\frac{D}{\mu}\right)^{p-r}\Sigma_{n=0}^{n=\mu}\frac{\mu(\mu-1)\,..(\mu-n+1)}{1\,.\,2\,..\,n}\,\epsilon^{n\theta}$$

$$=\frac{1}{2^\mu}\left(\frac{D}{\mu}\right)^r\left(1-\frac{D}{\mu}\right)^{p-r}(1+\epsilon^\theta)^\mu, \quad (3)$$

D standing for the symbol $\frac{d}{d\theta}$, so that $\phi(D)\,\epsilon_{n\theta}=\phi(n)\,\epsilon^{n\theta}$. But by a known theorem,

$$t^m=1+\Delta 0^m t+\frac{\Delta^2 0^m}{1\,.\,2}t(t-1)+\frac{\Delta^3 0^m}{1\,.\,2\,.\,3}t(t-1)(t-2).$$

$$\therefore\ D^m(1+\epsilon^\theta)^\mu=\{1+\Delta 0^m D+\frac{\Delta^2 0^m}{1\,.\,2}D(D-1)+\&c.\}\,(1+\epsilon^\theta)^\mu.$$

In the second member let $\epsilon^\theta=x$, then

$$D^m(1+\epsilon^\theta)^\mu=(1+\Delta 0^m x\frac{d}{dx}+\frac{\Delta^2 0^m}{1\,.\,2}x^2\frac{d^2}{dx^2}+\&c.)\,(1+x)^\mu,$$

since

$$D(D-1)\,..\,(D-i+1)=x^i\left(\frac{d}{dx}\right)^i.$$

In the second member of the above equation, performing the differentiations and making $x = 1$ (since $\theta = 0$), we get

$$D^m (1 + \epsilon^\theta)^\mu = \mu (\Delta 0^m) 2^{\mu-1} + \frac{\mu(\mu-1)}{1.2} (\Delta^2 0^m) 2^{\mu-2} + \&c.$$

The last term of the second member of this equation will be

$$\frac{\mu(\mu-1)\,..\,(\mu-m+1)\,\Delta^m 0^m}{1.2\,..\,m} 2^{\mu-m} = \mu(\mu-1)\,..\,(\mu-m+1)\,2^{\mu-m};$$

since $\Delta^m 0^m = 1.2\,..\,m$. When μ is a large quantity this term exceeds all the others in value, and as μ approaches to infinity tends to become infinitely great in comparison with them. And as moreover it assumes the form $\mu^m 2^{\mu-m}$, we have, on passing to the limit,

$$D^m (1 + \epsilon^\theta)^\mu = \mu^m 2^{\mu-m} = \left(\frac{\mu}{2}\right)^m 2^\mu.$$

Hence if $\phi(D)$ represent any function of the symbol D, which is capable of being expanded in a series of ascending powers of D, we have

$$\phi(D)(1 + \epsilon^\theta)^\mu = \phi\left(\frac{\mu}{2}\right) 2^\mu, \qquad (4)$$

if $\theta = 0$ and $\mu = \infty$. Strictly speaking, this implies that the ratio of the two members of the above equation approaches a state of equality, as μ increases towards infinity, θ being equal to 0.

By means of this theorem, the last member of (3) reduces to the form

$$\frac{1}{2^\mu}\left(\frac{1}{2}\right)^r \left(1 - \frac{1}{2}\right)^{p-r} 2^\mu = \left(\frac{1}{2}\right)^p.$$

Hence (2) gives

$$\frac{p(p-1)\,..\,(p-r+1)}{1.2\,..\,r}\left(\frac{1}{2}\right)^p,$$

as the expression for the probability that from an urn containing an infinite number of black and white balls, all constitutions of the system being equally probable, r white balls will issue in p drawings.

Hence, making $p = m$, $r = m$, the probability that in m drawings all the balls will be white is $\left(\frac{1}{2}\right)^m$, and the probability that this will be the case, and that moreover the $m + 1^{th}$ drawing will yield a white ball is $\left(\frac{1}{2}\right)^{m+1}$, whence the probability, that if the first m drawings yield white balls only, the $m + 1^{th}$ drawing will also yield a white ball, is

$$\left(\frac{1}{2}\right)^{m+1} \div \left(\frac{1}{2}\right)^{m} = \frac{1}{2};$$

and generally, any proposed result will have the same probability as if it were an even chance whether each particular drawing yielded a white or a black ball. This agrees with the conclusion before obtained.

26. These results only illustrate the fact, that when the defect of data is supplied by hypothesis, the solutions will, in general, vary with the nature of the hypotheses assumed; so that the question still remains, only more definite in form, whether the principles of the theory of probabilities serve to guide us in the election of such hypotheses. I have already expressed my conviction that they do not—a conviction strengthened by other reasons than those above stated. Thus, a definite solution of a problem having been found by the method of this work, an equally definite solution is sometimes attainable by the same method when one of the data, suppose Prob. $x = p_1$ is omitted. But I have not been able to discover any mode of deducing the second solution from the first by *integration*, with respect to p supposed variable within limits determined by Chap. XIX. This deduction would, however, I conceive, be possible, were the principle adverted to in Art. 23 valid. Still it is with diffidence that I express my dissent on these points from mathematicians generally, and more especially from one who, of English writers, has most fully entered into the spirit and the methods of Laplace; and I venture to hope, that a question, second to none other in the Theory of Probabilities in importance, will receive the careful attention which it deserves.

CHAPTER XXI.

PARTICULAR APPLICATION OF THE PREVIOUS GENERAL METHOD TO THE QUESTION OF THE PROBABILITY OF JUDGMENTS.

1. ON the presumption that the general method of this treatise for the solution of questions in the theory of probabilities, has been sufficiently elucidated in the previous chapters, it is proposed here to enter upon one of its practical applications selected out of the wide field of social statistics, viz., the estimation of the probability of judgments. Perhaps this application, if weighed by its immediate results, is not the best that could have been chosen. One of the first conclusions to which it leads is that of the *necessary* insufficiency of any data that experience alone can furnish, for the accomplishment of the most important object of the inquiry. But in setting clearly before us the *necessity* of hypotheses as supplementary to the data of experience, and in enabling us to deduce with rigour the consequences of *any* hypothesis which may be assumed, the method accomplishes all that properly lies within its scope. And it may be remarked, that in questions which relate to the conduct of our own species, hypotheses are more justifiable than in questions such as those referred to in the concluding sections of the previous chapter. Our general experience of human nature comes in aid of the scantiness and imperfection of statistical records.

2. The elements involved in problems relating to criminal assize are the following:—

1st. The probability that a particular member of the jury will form a correct opinion upon the case.

2nd. The probability that the accused party is guilty.

3rd. The probability that he will be condemned, or that he will be acquitted.

4th. The probability that his condemnation or acquittal will be just.

5th. The constitution of the jury.

6th. The data furnished by experience, such as the relative numbers of cases in which unanimous decisions have been arrived at, or particular majorities obtained; the number of cases in which decisions have been reversed by superior courts, &c.

Again, the class of questions under consideration may be regarded as either direct or inverse. The direct questions of probability are those in which the probability of correct decision for each member of the tribunal, or of guilt for the accused party, are supposed to be known *à priori*, and in which the probability of a decision of a particular kind, or with a definite majority, is sought. Inverse problems are those in which, from the data furnished by experience, it is required to determine some element which, though it stand to those data in the relation of cause to effect, cannot directly be made the subject of observation; as when from the records of the decisions of courts it is required to determine the probability that a member of a court will judge correctly. To this species of problems, the most difficult and the most important of the whole series, attention will chiefly be directed here.

3. There is no difficulty in solving the direct problems referred to in the above enumeration. Suppose there is but one juryman. Let k be the probability that the accused person is guilty; x the probability that the juryman will form a correct opinion; X the probability that the accused person will be condemned: then—

kx = probability that the accused party is guilty, and that the juryman judges him to be guilty.

$(1-k)(1-x)$ = probability that the accused person is innocent, and that the juryman pronounces him guilty.

Now these being the only cases in which a verdict of condemnation can be given, and being moreover mutually exclusive, we have

$$X = kx + (1-k)(1-x). \qquad (1)$$

In like manner, if there be n jurymen whose separate probabilities of correct judgment are $x_1, x_2 \ldots x_n$, the probability of an unanimous verdict of condemnation will be

$$X = kx_1 x_2 \ldots x_n + (1-k)(1-x_1)(1-x_2)\ldots(1-x_n).$$

Whence, if the several probabilities x_1, x_2 .. x_n are equal, and are each represented by x, we have

$$X = kx^n + (1-k)(1-x)^n. \qquad (2)$$

The probability in the latter case, that the accused person is guilty, will be

$$\frac{kx^n}{kx^n + (1-k)(1-x)^n}.$$

All these results assume, that the events whose probabilities are denoted by k, x_1, x_2, &c., are independent, an assumption which, however, so far as we are concerned, is involved in the fact that those events are the only ones of which the probabilities are given.

The probability of condemnation by a given number of voices may be found on the same principles. If a jury is composed of three persons, whose several probabilities of correct decision are x, x', x'', the probability X_2 that the accused person will be declared guilty by two of them will be

$$X_2 = k\{xx'(1-x'') + xx''(1-x') + x'x''(1-x)\}$$
$$+ (1-k)\{(1-x)(1-x')x'' + (1-x)(1-x'')x' + (1-x')(1-x'')x\},$$

which if $x = x' = x''$ reduces to

$$3kx^2(1-x) + 3(1-k)x(1-x)^2.$$

And by the same mode of reasoning, it will appear that if X_i represent the probability that the accused person will be declared guilty by i voices out of a jury consisting of n persons, whose separate probabilities of correct judgment are equal, and represented by x, then

$$X_i = \frac{n(n-1)..(n-i+1)}{1.2..i}\{kx^i(1-x)^{n-i} + (1-k)x^{n-i}(1-x)^i\}. \quad (3)$$

If the probability of condemnation by a determinate majority a is required, we have simply

$$i - a = n - i,$$

whence

$$i = \frac{n+a}{2},$$

which must be substituted in the above formula. Of course a admits only of such values as make i an integer. If n is even, those values are 0, 2, 4, &c.; if odd, 1, 3, 5, &c., as is otherwise obvious.

The probability of a condemnation by a majority of at least a given number of voices m, will be found by adding together the following several probabilities determined as above, viz.:

1st. The probability of a condemnation by an exact majority m;

2nd. The probability of condemnation by the next greater majority $m + 2$;

and so on; the last element of the series being the probability of unanimous condemnation. Thus the probability of condemnation by a majority of 4 at least out of 12 jurors, would be

$$X_8 + X_9 \ldots + X_{12},$$

the values of the above terms being given by (3) after making therein $n = 12$.

4. When, instead of a jury, we are considering the case of a simple deliberative assembly consisting of n persons, whose separate probabilities of correct judgment are denoted by x, the above formulæ are replaced by others, made somewhat more simple by the omission of the quantity k.

The probability of unanimous decision is

$$X = x^n + (1 - x)^n.$$

The probability of an agreement of i voices out of the whole number is

$$X_i = \frac{n(n-1)\ldots(n-i+1)}{1 \cdot 2 \ldots i}\{x^i(1-x)^{n-i} + x^{n-i}(1-x)^i\}. \quad (4)$$

Of this class of investigations it is unnecessary to give any further account. They have been pursued to a considerable extent by Condorcet, Laplace, Poisson, and other writers, who have investigated in particular the modes of calculation and reduction which are necessary to be employed when n and i are large numbers. It is apparent that the whole inquiry is of a very speculative character. The values of x and k cannot be determined by direct observation. We can only presume that they must both in general exceed the value $\frac{1}{2}$; that the former, x, must increase with the progress of public intelligence; while the latter, k, must depend much upon those preliminary steps in the administration of the law by which persons suspected of crime are brought before the tribunal of their country. It has been remarked by Poisson, that in periods of revolution, as during the Reign of Terror in France, the value of k may fall, if account be taken of political offences, far below the limit $\frac{1}{2}$. The history of Europe in days nearer to our own would probably confirm this observation, and would show that it is not from the wild license of democracy alone, that the accusation of innocence is to be apprehended.

Laplace makes the assumption, that all values of x from

$$x = \frac{1}{2}, \text{ to } x = 1,$$

are equally probable. He thus excludes the supposition that a juryman is more likely to be deceived than not, but assumes that within the limits to which the probabilities of individual correctness of judgment are confined, we have no reason to give preference to one value of x over another. This hypothesis is entirely arbitrary, and it would be unavailing here to examine into its consequences.

Poisson seems first to have endeavoured to deduce the values of x and k, inferentially, from experience. In the six years from 1825 to 1830 inclusively, the number of individuals accused of crimes against the person before the tribunals of France was 11016, and the number of persons condemned was 5286. The juries consisted each of 12 persons, and the decision was pronounced by a simple majority. Assuming the above numbers to be sufficiently large for the estimation of probabilities, there would therefore be a probability measured by the fraction $\frac{5286}{11016}$, or .4782 that an accused person would be condemned by a simple majority. We should have the equation

$$X_7 + X_8 \ldots + X_{12} = .4782, \tag{5}$$

the general expression for X_i being given by (3) after making therein $n = 12$. In the year 1831 the law, having received alteration, required a majority of at least four persons for condemnation, and the number of persons tried for crimes against the person during that year being 2046, and the number condemned 743, the probability of the condemnation of an individual by the above majority was $\frac{743}{2046}$, or .3631. Hence we should have

$$X_8' + X_9 \ldots + X_{12} = .3631. \qquad (6)$$

Assuming that the values of k and x were the same for the year 1831 as for the previous six years, the two equations (5) and (6) enable us to determine approximately their values. Poisson thus found,

$$k = .5354, \quad x = .6786.$$

For crimes against property during the same periods, he found by a similar analysis,

$$k = .6744, \quad x = .7771.$$

The solution of the system (5) (6) conducts in each case to two values of k, and to two values of x, the one value in each pair being greater, and the other less, than $\frac{1}{2}$. It was assumed, that in each case the larger value should be preferred, it being conceived more probable that a party accused should be guilty than innocent, and more probable that a juryman should form a correct than an erroneous opinion upon the evidence.

5. The data employed by Poisson, especially those which were furnished by the year 1831, are evidently too imperfect to permit us to attach much confidence to the above determinations of x and k; and it is chiefly for the sake of the method that they are here introduced. It would have been possible to record during the six years, 1825–30, or during any similar period, the number of condemnations pronounced with each possible majority of voices. The values of the several elements X_8, X_9, .. X_{12}, were there no reasons of policy to forbid, might have been accurately ascertained. Here then the conception of the general problem, of which Poisson's is a particular case, arises. How shall we, from this apparently supernumerary system of data, determine the values of x and k? If the hypothesis, adopted by Poisson and all other writers on the subject, of the absolute independence of the events whose probabilities are denoted by x and k be retained, we should be led to form a system of five equations of the type (3), and either select from these that particular pair of equations which might appear to be most advantageous, or combine together the equations of the system by the method of least squares. There might exist a doubt as to whether the latter method would be strictly applicable in such cases, especially if the values of x and k afforded by different selected pairs of the given equations were very different from each other. M. Cournot has considered a somewhat similar problem, in which, from the records of individual votes in a court consisting of four judges, it is proposed to investigate the separate probabilities of a correct verdict from each judge. For the determination of the elements x, x', x'', x''', he obtains eight equations, which he divides into two sets of four equations, and he remarks, that should any considerable discrepancy exist between the values of x, x', x'', x''', determined from those sets, it might be regarded as an indication that the hypothesis of the independence of the opinions of the judges was, in the particular case, untenable. The principle of this mode of investigation has been adverted to in (XVIII. 4).

6. I proceed to apply to the class of problems above indicated, the method of this treatise, and shall inquire, first, whether the records of courts and deliberative assemblies, *alone*, can furnish any information respecting the probabilities of correct judgment for their individual members, and, it appearing that they cannot, secondly, what kind and amount of necessary hypothesis will best comport with the actual data.

PROPOSITION I.

From the mere records of the decisions of a court or deliberative assembly, it is not possible to deduce any definite conclusion respecting the correctness of the individual judgments of its members.

Though this Proposition may appear to express but the conviction of unassisted good sense, it will not be without interest to show that it admits of rigorous demonstration.

Let us suppose the case of a deliberative assembly consisting of n members, no hypothesis whatever being made respecting the dependence or independence of their judgments. Let the logical symbols $x_1, x_2, \ldots x_n$ be employed according to the following definition, viz.: Let the generic symbol x_i denote that event which consists in the uttering of a correct opinion by the ith member, A_i of the court. We shall consider the values of Prob. x_1, Prob. x_2, . . Prob. x_n, as the *quæsita* of a problem, the expression of whose possible data we must in the next place investigate.

Now those data are the probabilities of events capable of being expressed by definite logical functions of the symbols x_1, $x_2, \ldots x_n$. Let $X_1, X_2, \ldots X_m$ represent the functions in question, and let the actual system of data be

$$\text{Prob. } X_1 = a_1, \quad \text{Prob. } X_2 = a_2 \quad \text{Prob. } X_m = a_m.$$

Then from the very nature of the case it may be shown that $X_1, X_2, \ldots X_m$, are functions which remain unchanged if $x_1, x_2, \ldots x_n$ are therein changed into $1 - x_1, 1 - x_2, \ldots 1 - x_n$ respectively. Thus, if it were recorded that in a certain proportion of instances the votes given were unanimous, the event whose probability, supposing the instances sufficiently numerous, is thence determined, is expressed by the logical function

$$x_1 x_2 \ldots x_n + (1 - x_1)(1 - x_2) \ldots (1 - x_n),$$

a function which satisfies the above condition. Again, let it be recorded, that in a certain proportion of instances, the vote of an individual, suppose A_1, differs from that of all the other members of the court. The event, whose probability is thus given, will be expressed by the function

$$x_1 (1 - x_2) \ldots (1 - x_n) + (1 - x_1) x_2 \ldots x_n;$$

also satisfying the above conditions. Thus, as agreement in opinion may be an agreement in either truth or error; and as, when opinions are divided, either party may be right or wrong; it is manifest that the expression of any particular state, whether of agreement or difference of sentiment in the assembly, will depend upon a logical function of the symbols $x_1, x_2, \ldots x_n$, which similarly involves the privative symbols $1 - x_1$, $1 - x_2$, .. $1 - x_n$. But in the records of assemblies, it is not presumed to declare which set of opinions is right or wrong. Hence the functions $X_1, X_2, .. X_m$ must be solely of the kind above described.

7. Now in proceeding, according to the general method, to determine the value of Prob. x_1, we should first equate the functions $X_1, .. X_m$ to a new set of symbols $t_1, .. t_m$. From the equations

$$X_1 = t_1,\ X_2 = t_2,\ ..\ X_m = t_m,$$

thus formed, we should eliminate the symbols $x_2, x_3, .. x_n$, and then determine x_1 as a developed logical function of the symbols $t_1, t_2, .. t_m$, expressive of events whose probabilities are given. Let the result of the above elimination be

$$Ex_1 + E'(1 - x_1) = 0; \tag{1}$$

E and E' being function of $t_1, t_2, .. t_m$. Then

$$x_1 = \frac{E'}{E' - E}. \tag{2}$$

Now the functions $X_1, X_2, .. X_m$ are symmetrical with reference to the symbols $x_1, .. x_n$ and $1 - x_1, .. 1 - x_n$. It is evident, therefore, that in the equation E' must be identical with E. Hence (2) gives

$$x = \frac{E}{0},$$

and it is evident, that the only coefficients which can appear in the development of the second member of the above equation are $\frac{0}{0}$ and $\frac{1}{0}$. The former will present itself whenever the values assigned to $t_1, .. t_m$ in determining the coefficient of a constituent, are such as to make $E = 0$, the latter, or an equivalent result, in every other case. Hence we may represent the development under the form

$$x_1 = \frac{0}{0}C + \frac{1}{0}D, \tag{3}$$

C and D being constituents, or aggregates of constituents, of the symbols $t_1, t_2, .. t_m$.

Passing then from Logic to Algebra, we have

$$\text{Prob. } x_1 = \frac{cC}{C} = c,$$

the function V of the general Rule (XVII. 17) reducing in the present case to C. The value of Prob. x_1 is therefore wholly arbitrary, if we except the condition that it must not transcend the limits 0 and 1. The individual values of Prob. x_2, .. Prob. x_m, are in like manner arbitrary. It does not hence follow, that these arbitrary values are not connected with each other by necessary conditions dependent upon the data. The investigation of such conditions would, however, properly fall under the methods of Chap. XIX.

If, reverting to the final logical equation, we seek the interpretation of c, we obtain but a restatement of the original problem. For since C and D together include all possible constituents of t_1, t_2, .. t_m, we have

$$C + D = 1;$$

and since D is affected by the coefficient $\frac{1}{0}$, it is evident that on substituting therein for t_1, t_2, .. t_m, their expressions in terms of x_1, x_2, .. x_n, we should have $D = 0$. Hence the same substitution would give $C = 1$. Now by the rule, c is the probability that if the event denoted by C take place, the event x_1 will take place. Hence C being equal to 1, and, therefore, embracing all possible contingencies, c must be interpreted as the absolute probability of the occurrence of the event x_1.

It may be interesting to determine in a particular case the actual form of the final logical equation. Suppose, then, that the elements from which the data are derived are the records of events distinct and mutually exclusive. For instance, let the numerical data a_1, a_2, .. a_m, be the respective probabilities of distinct and definite majorities. Then the logical functions X_1, X_2, .. X_m being mutually exclusive, must satisfy the conditions

$$X_1 X_2 = 0, \; .. \; X_1 X_m = 0, \quad X_2 X_m = 0, \text{ \&c.}$$

Whence we have,

$$t_1 t_2 = 0, \quad t_1 t_m = 0, \text{ \&c.}$$

Under these circumstances it may easily be shown, that the developed logical value of x_1 will be

$$x_1 = \frac{0}{0}(\bar{t}_1\bar{t}_2 \,..\, \bar{t}_m + t_1\bar{t}_2 \,..\, \bar{t}_m \,..+ t_m\bar{t}_1 \,..\, \bar{t}_{m-1})$$

$$+ \text{ constitutents whose coefficients are } \frac{1}{0}.$$

In the above equation $\bar{t}_1$ stands for $1 - t_1$, &c.

These investigations are equally applicable to the case in which the probabilities of the verdicts of a jury, so far as agreement and disagreement of opinion are concerned, form the data of a problem. Let the logical symbol w denote that event or state of things which consists in the guilt of the accused person. Then the functions $X_1, X_2 \,..\, X_m$ of the present problem are such, that no change would therein ensue from simultaneously converting $w, x_1, x_2 \,..\, x_n$ into $\bar{w}, \bar{x}_1, \bar{x}_2, \,..\, \bar{x}_n$ respectively. Hence the final logical value of w, as well as those of $x_1, x_2, \,..\, x_n$ will be exhibited under the same form (3), and a like general conclusion thence deduced.

It is therefore established, that from mere statistical documents nothing can be inferred respecting either the individual correctness of opinion of a judge or counsellor, the guilt of an individual, or the merits of a disputed question. If the determination of such elements as the above can be reduced within the province of science at all, it must be by virtue either of some assumed criterion of truth furnishing us with new data, or of some hypothesis relative to the connexion or the independence of individual judgments, which may warrant a new form of the investigation. In the examination of the results of different hypotheses, the following general Proposition will be of importance.

PROPOSITION II.

8. *Given the probabilities of the n simple events* $x_1, x_2, \,..\, x_n$, *viz.:—*

$$\text{Prob. } x_1 = c_1, \quad \text{Prob. } x_2 = c_2, \,..\, \text{Prob. } x_n = c_n; \qquad (1)$$

also the probabilities of the $m-1$ *compound events* $X_1, X_2, \,..\, X_{m-1}$, *viz.:—*

$$\text{Prob. } X_1 = a_1, \quad \text{Prob. } X_2 = a_2, \,..\, \text{Prob. } X_{m-1} = a_{m-1}; \quad (2)$$

the latter events $X_1 \,..\, X_{m-1}$ being distinct and mutually exclusive; required the probability of any other compound event X.

In this proposition it is supposed, that X_1, X_2, .. X_{m-1}, as well as X, are functions of the symbols x_1, x_2, .. x_n alone. Moreover, the events X_1, X_2, .. X_{m-1}, being mutually exclusive, we have

$$X_1 X_2 = 0, \;..\; X_1 X_{m-1} = 0, \quad X_2 X_3 = 0, \text{ \&c.}; \qquad (3)$$

the product of any two members of the system vanishing. Now assume

$$X_1 = t_1, \quad X_{m-1} = t_{m-1}, \quad X = t. \qquad (4)$$

Then t must be determined as a logical function of x_1, .. x_n, t_1, .. t_{m-1}.

Now by (3),

$$t_1 t_2 = 0, \quad t_1 t_{m-1} = 0, \quad t_2 t_3 = 0, \text{ \&c.}; \qquad (5)$$

all binary products of t_1, .. t_{m-1}, vanishing. The developed expression for t can, therefore, only involve in the list of constituents which have 1, 0, or $\frac{0}{0}$ for their coefficients, such as contain some one of the following factors, viz.:—

$$\bar{t}_1 \bar{t}_2 \,..\, \bar{t}_{m-1}, \quad t_1 \bar{t}_2 \,..\, \bar{t}_{m-1}, \;..\; \bar{t}_1 \,..\, \bar{t}_{m-2} t_{m-1}; \qquad (6)$$

$\bar{t}_1$ standing for $1 - t_1$, &c. It remains to assign that portion of each constituent which involves the symbols $x_1 \,..\, x_n$; together with the corresponding coefficients.

Since $X_i = t_i$ (i being any integer between 1 and $m - 1$ inclusive), it is evident that

$$X_i \bar{t}_1 \,..\, \bar{t}_{m-1} = 0,$$

from the very constitution of the functions. Any constituent included in the first member of the above equation would, therefore, have $\frac{1}{0}$ for its coefficient.

Now let

$$X_m = 1 - X_1 \,..\, - X_{m-1}; \qquad (7)$$

and it is evident that such constituents as involve $\bar{t}_1 \,..\, \bar{t}_{m-1}$, as a factor, and yet have coefficients of the form 1, 0, or $\frac{0}{0}$, must be included in the expression

$$X_m \bar{t}_1 \,.\, \bar{t}_{m-1}.$$

Now X_m may be resolved into two portions, viz., XX_m and $(1 - X)\, X_m$, the former being the sum of those constituents of X_m which are found in X, the latter of those which are not found in X. It is evident that in the developed expression of t, which is equivalent to X, the coefficients of the constituents in the former portion XX_m will be 1, while those of the latter portion $(1 - X)\, X_m$ will be 0. Hence the elements we have now considered will contribute to the development of t the terms

$$XX_m \bar{t}_1 \,..\, \bar{t}_{m-1} + 0\,(1 - X)\, X_m \bar{t}_1 \,..\, \bar{t}_{m-1}.$$

Again, since $X_1 = t_1$, while $X_2\, t_1 = t_2\, t_1 = 0$, &c., it is evident that the only constituents involving $t_1 \bar{t}_2 \,..\, \bar{t}_{m-1}$ as a factor which have coefficients of the form 1, 0, or $\frac{0}{0}$, will be included in the expression

$$X_1 t_1 \bar{t}_2 \,..\, t_{m-1};$$

and reasoning as before, we see that this will contribute to the development of t the terms

$$XX_1 t_1 \bar{t}_2 \,..\, \bar{t}_{m-1} + 0\,(1 - X)\, X_1 t_1 \bar{t}_2 \,..\, \bar{t}_{m-1}.$$

Proceeding thus with the remaining terms of (6), we deduce for the final expression of t,

$$\begin{aligned} t = XX_m \bar{t}_1 \,..\, \bar{t}_{m-1} + XX_1 t_1 \bar{t}_2 \,..\, \bar{t}_{m-1} \,..+ XX_{m-1} \bar{t}_1 \,..\, \bar{t}_{m-2}\, t_{m-1} \\ + 0\,(1 - X)\, X_m \bar{t}_1 \,..\, \bar{t}_{m-1} + 0\,(1 - X)\, X_1 t_1 \bar{t}_2 \,..\, \bar{t}_{m-1} + \text{\&c.} \qquad (8) \\ + \text{ terms whose coefficients are } \frac{1}{0}. \end{aligned}$$

In this expression it is to be noted that XX_m denotes the sum of those constituents which are common to X and X_m, that sum being actually given by multiplying X and X_m together, according to the rules of the calculus of Logic.

In passing from Logic to Algebra, we shall represent by (XX_m) what the above product becomes, when, after effecting the multiplication, or selecting the common constituents, we give to the symbols $x_1, \,..\, x_n$, a quantitative meaning.

With this understanding we shall have, by the general Rule (XVII. 17),

$$\text{Prob. } t = \frac{(XX_m)\bar{t}_1..\bar{t}_{m-1}+(XX_1)\,t_1\,\bar{t}_2..\bar{t}_{m-1}..+(XX_{m-1})\bar{t}_1..\bar{t}_{m-2}t_{m-1}}{V}, \quad (9)$$

$$V = X_m\,\bar{t}_1\,..\,\bar{t}_{m-1} + X_1\,t_1\,\bar{t}_2\,..\,\bar{t}_{m-1}\,..+X_{m-1}\,\bar{t}_1\,..\,\bar{t}_{m-2}\,t_{m-1} \quad (10)$$

whence the relations determining $x_1, .. x_n, t_1, .. t_{m-1}$ will be of the following type (i varying from 1 to n),

$$\frac{(x_i\,X_m)\,\bar{t}_1\,..\,\bar{t}_{m-1} + (x_i\,X_1)\,t_1\,\bar{t}_2\,..\,\bar{t}_{m-1}\,..+(x_i\,X_{m-1})\,\bar{t}_1\,..\,\bar{t}_{m-2}\,t_{m-1}}{c_i}$$

$$= \frac{X_1\,t_1\,\bar{t}_2\,..\,\bar{t}_{m-1}}{a_1}\,..=\frac{X_{m-1}\,\bar{t}_1\,..\,\bar{t}_{m-2}\,t_{m-1}}{a_{m-1}} = V. \quad (11)$$

From the above system we shall next eliminate the symbols $t_1, .. t_{m-1}$.

We have

$$t_1\,\bar{t}_2\,..\,\bar{t}_{m-1} = \frac{a_1 V}{X_1}, \qquad \bar{t}_1\,..\,\bar{t}_{m-2}\,t_{m-1} = \frac{a_{m-1} V}{X_{m-1}}. \quad (12)$$

Substituting these values in (10), we find

$$V = X_m\,\bar{t}_1\,..\,\bar{t}_{m-1} + a_1 V\,..+a_{m-1} V.$$

Hence,

$$\bar{t}_1\,..\,\bar{t}_{m-1} = \frac{(1 - a_1\,..-a_{m-1})\,V}{X_m}.$$

Now let

$$a_m = 1 - a_1\,..-a_{m-1}, \quad (13)$$

then we have

$$t_1\,..\,t_{m-1} = \frac{a_m V}{X_m}. \quad (14)$$

Now reducing, by means of (12) and (14), the equation (9), and the equation formed by equating the first line of (11) to the symbol V; writing also Prob. X for Prob. t, we have

$$\text{Prob. } X = \frac{a_1\,(XX_1)}{X_1} + \frac{a_2\,(XX_2)}{X_2}\,..+\frac{a_m\,(XX_m)}{X_m}, \quad (15)$$

$$\frac{a_1\,(x_i X_1)}{X_1} + \frac{a_2\,(x_i X_2)}{X_2}\,..+\frac{a_m\,(x_i X_m)}{X_m} = c_i; \quad (16)$$

wherein X_m and a_m are given by (7) and (13).

These equations involve the direct solution of the problem under consideration. In (16) we have the type of n equations (formed by giving to i the values 1, 2, .. n successively), from which the values of x_1, x_2, .. x_n, will be found, and those values substituted in (15) give the value of Prob. X as a function of the constants a_1, c_1, &c.

One conclusion deserving of notice, which is deducible from the above solution, is, that if the probabilities of the compound events X_1, .. X_{m-1}, are the same as they would be were the events x_1, .. x_n entirely independent, and with given probabilities c_1, .. c_n, then the probability of the event X will be the same as if calculated upon the same hypothesis of the absolute independence of the events x_1, .. x_n. For upon the hypothesis supposed, the assumption $x_1 = c_1$, $x_n = c_n$, in the quantitative system would give $X_1 = a_1$, $X_m = a_m$, whence (15) and (16) would give

$$\text{Prob. } X = (XX_1) + (XX_2) \ldots + (XX_m), \tag{17}$$

$$(x_i X_1) + (x_i X_2) \ldots + (x_i X_m) = c_i. \tag{18}$$

But since $X_1 + X_2 \ldots + X_m = 1$, it is evident that the second member of (17) will be formed by taking all the constituents that are contained in X, and giving them an algebraic significance. And a similar remark applies to (18). Whence those equations respectively give

$$\text{Prob. } X \text{ (logical)} = X \text{ (algebraic)},$$

$$x_i = c_i.$$

Wherefore, if $X = \phi(x_1, x_2, \ldots x_n)$, we have

$$\text{Prob. } X = \phi(c_1, c_2, \ldots c_n),$$

which is the result in question.

Hence too it would follow, that if the quantities c_1, .. c_n were indeterminate, and no hypothesis were made as to the possession of a mean common value, the system (15) (16) would be satisfied by giving to those quantities any such values, x_1, x_2, .. x_n, as would satisfy the equations

$$X_1 = a_1 \ldots X_{m-1} = a_{m-1}, \quad X = a,$$

supposing the value of the element a, like the values of a_1,.. a_{m-1}, to be given by experience.

9. Before applying the general solution (15) (16), to the question of the probability of judgments, it will be convenient to make the following transformation. Let the data be

$$x_1 = c_1 \ldots . x_n = c_n,$$

namely,

$$\text{Prob. } X_1 = a_1 \ldots . \text{ Prob. } X_{m-2} = a_{m-2};$$

and let it be required to determine Prob. X_{m-1}, the unknown value of which we will represent by a_{m-1}. Then in (15) and (16) we must change

$$X \text{ into } X_{m-1}, \qquad \text{Prob. } X \text{ into } a_{m-1},$$

$$X_{m-1} \text{ into } X_{m-2}, \qquad a_{m-1} \text{ into } a_{m-2}.$$

$$X_m \text{ into } X_{m-1} + X_m, \qquad a_m \text{ into } a_{m-1} + a_m;$$

with these transformations, and observing that $(X_{m-1} X_r) = 0$, except when $r = m - 1$, and that it is then equal to X_{m-1}, the equations (15) (16) give

$$a_{m-1} = \frac{(a_{m-1} + a_m) X_{m-1}}{X_{m-1} + X_m}, \tag{19}$$

$$c_i = \frac{a_1(x_i X_1)}{X_1} \ldots + \frac{a_{m-2}(x_i X_{m-2})}{X_{m-2}} + \frac{(a_{m-1} + a_m)(x_i X_{m-1} + x_i X_m)}{X_{m-1} + X_m}. \tag{20}$$

Now from (19) we find

$$\frac{X_{m-1}}{a_{m-1}} = \frac{X_m}{a_m} = \frac{X_{m-1} + X_m}{a_{m-1} + a_m},$$

by virtue of which the last term of (20) may be reduced to the form

$$\frac{a_{m-1}(x_i X_{m-1})}{X_{m-1}} + \frac{a_m(x_i X_m)}{X_m}.$$

With these reductions the system (17) and (18) may be replaced by the following symmetrical one, viz.:

$$\frac{X_{m-1}}{a_{m-1}} = \frac{X_m}{a_m}, \tag{21}$$

$$\frac{a_1(x_i X_1)}{X_1} + \frac{a_2(x_i X_2)}{X_2} \ldots + \frac{a_m(x_i X_m)}{X_m} = c_i. \tag{22}$$

These equations, in connexion with (7) and (13), enable us to determine a_{m-1} as a function of $c_1 \ldots c_n$, $a_1 \ldots a_{m-2}$, the numerical data supposed to be furnished by experience. We now proceed to their application.

PROPOSITION III.

10. *Given any system of probabilities drawn from recorded instances of unanimity, or of assigned numerical majority in the decisions of a deliberative assembly; required, upon a certain determinate hypothesis, the mean probability of correct judgment for a member of the assembly.*

In what way the probabilities of unanimous decision and of specific numerical majorities may be determined from experience, has been intimated in a former part of this chapter. Adopting the notation of Prop. I. we shall represent the events whose probabilities are given by the functions $X_1, X_2, \ldots X_{m-1}$. It has appeared from the very nature of the case that these events are mutually exclusive, and that the functions by which they are represented are symmetrical with reference to the symbols $x_1, x_2, \ldots x_n$. Those symbols we continue to use in the same sense as in Prop. I., viz., by x_i we understand that event which consists in the formation of a correct opinion by the ith member of the assembly.

Now the immediate data of experience are—

$$\text{Prob. } X_1 = a_1, \quad \text{Prob. } X_2 = a_2, \ldots \text{Prob. } X_{m-2} = a_{m-2}, \quad (1)$$

$$\text{Prob. } X_{m-1} = a_{m-1}. \quad (2)$$

$X_1 \ldots X_{m-1}$ being functions of the logical symbols $x_1, \ldots x_n$ to the probabilities of the events denoted by which, we shall assign the indeterminate value c. Thus we shall have

$$\text{Prob. } x_1 = \text{Prob. } x_2 \ldots = \text{Prob. } x_n = c. \quad (3)$$

Now it has been seen, Prop. I., that the immediate data (1) (2), unassisted by any hypothesis, merely conduct us to a re-statement of the problem. On the other hand, it is manifest that if, adopting the methods of Laplace and Poisson, we employ the system (3) alone as the data for the application of the method of this work, finally comparing the results obtained with the experimental system (1) (2), we are relying *wholly* upon a doubtful hypothesis,—the independence of individual judgments. But though we ought not wholly to rely upon this hypothesis, we cannot wholly dispense with it, or with some equivalent substitute. Let us then examine the consequences of a *limited* independence of the individual judgments; the conditions of limitation being furnished by the apparently superfluous data. From the system (1) (3) let us, by the method of this work, determine Prob. X_{m-1}, and, comparing the result with (2), determine c. Even here an arbitrary power of selection is claimed. But it is manifest from Prop. I. that something of this kind is unavoidable, if we would obtain a definite solution at all. As to the principle of selection, I apprehend that the equation (2) reserved for final comparison should be that which, from the magnitude of its numerical element a_{m-1}, is esteemed the most important of the primary series furnished by experience.

Now, from the mutually exclusive character of the events denoted by the functions X_1, X_2, .. X_{m-1}, the concluding equations of the previous proposition become applicable. On account of the symmetry of the same functions, and the reduction of the system of values denoted by c_i to a single value c, the equations represented by (22) become identical, the values of $x_1, x_2, .. x_n$ become equal, and may be replaced by a single value x, and we have simply,

$$\frac{X_{m-1}}{a_{m-1}} = \frac{X_m}{a_m}, \qquad (4)$$

$$\frac{a_1(xX_1)}{X_1} + \frac{a_2(xX_2)}{X_2} .. + \frac{a_m(xX_m)}{X_m} = c. \qquad (5)$$

The following is the nature of the solution thus indicated:

The functions $X_1, .. X_{m-1}$, and the values $a_1 .. a_{m-1}$, being given in the data, we have first,

$$X_m = 1 - X_1 .. - X_{m-1},$$

$$a_m = 1 - a_1 .. - a_{m-1}.$$

From each of the functions X_1, X_2, .. X_m thus given or determined, we must select those constituents which contain a particular symbol, as x_1, for a factor. This will determine the functions (xX_1), (xX_2), &c., and then in all the functions we must change $x_1, x_2, .. x_n$ individually to x. Or we may regard any algebraic function X_i in the system (4) (5) as expressing the probability of the event denoted by the logical function X_i, on the supposition that the logical symbols $x_1, x_2, .. x_n$ denote independent events whose common probability is x. On the same supposition (xX_i) would denote the probability of the concurrence of any particular event of the series $x_1, x_2, .. x_n$ with X_i. The forms of X_i, (xX_i), &c. being determined, the equation (4) gives the value of x, and this, substituted in (5), determines the value of the element c required. Of the two values which its solution will offer, one being greater, and the other less, than $\frac{1}{2}$, the greater one must be chosen, whensoever, upon general considerations, it is thought more probable that a member of the assembly will judge correctly, than that he will judge incorrectly.

Here then, upon the assumed principle that the largest of the values a_{m-1} shall be reserved for final comparison in the equation (2), we possess a definite solution of the problem proposed. And the same form of solution remains applicable should any other equation of the system, upon any other ground, as that of superior accuracy, be similarly reserved in the place of (2).

11. Let us examine to what extent the above reservation has influenced the final solution. It is evident that the equation (5) is quite independent of the choice in question. So is likewise the second member of (4). Had we reserved the function X_1, instead of X_{m-1}, the equation for the determination of x would have been

$$\frac{X_1}{a_1} = \frac{X_m}{a_m}; \tag{6}$$

but the value of x thence determined would still have to be substituted in the same final equation (5). We know that were the events $x_1, x_2, .. x_n$ really independent, the equations (4), (6), and all others of which they are types, would prove equivalent, and that the value of x furnished by any one of them would be the true value of c. This affords a means of verifying (5). For if that equation be correct, it ought, under the above circumstances, to be satisfied by the assumption $c = x$. In other words, the equation

$$\frac{a_1 (xX_1)}{X_1} + \frac{a_2 (xX_2)}{X_2} .. + \frac{a_m (xX_m)}{X_m} = x \tag{7}$$

ought, on solution, to give the same value of x as the equation (4) or (6). Now this will be the case. For since, by hypothesis,

$$\frac{X_1}{a_1} = \frac{X_2}{a_2} \ldots = \frac{X_m}{a_m},$$

we have, by a known theorem,

$$\frac{X_1}{a_1} = \frac{X_2}{a_2} \ldots = \frac{X_m}{a_m} = \frac{X_1 + X_2 \ldots + X_m}{a_1 + a_2 \ldots + a_m} = 1.$$

Hence (7) becomes on substituting a_1 for X_1, &c.

$$(xX_1) + (xX_2) \ldots + (xX_m) = x$$

a mere identity.

Whenever, therefore, the events $x_1, x_2, \ldots x_n$ are really independent, the system (4) (5) is a correct one, and is independent of the arbitrariness of the first step of the process by which it was obtained. When the said events are not independent, the final system of equations will possess, leaving in abeyance the principle of selection above stated, an arbitrary element. But from the persistent form of the equation (5) it may be inferred that the solution is arbitrary in a less degree than the solutions to which the hypothesis of the absolute independence of the individual judgments would conduct us. The discussion of the limits of the value of c, as dependent upon the limits of the value of x, would determine such points.

These considerations suggest to us the question whether the equation (7), which is symmetrical with reference to the functions $X_1, X_2, \ldots X_m$, free from any arbitrary elements, and rigorously exact when the events $x_1, x_2, \ldots x_n$ are really independent, might not be accepted as a mean general solution of the problem. The proper mode of determining this point would, I conceive, be to ascertain whether the value of x which it would afford would, in general, fall within the limits of the value of c, as determined by the systems of equations of which the system (4), (5), presents the type. It seems probable that under ordinary circumstances this would be the case. Independently of such considerations, however, we may regard (7) as itself the expression of a certain principle of solution, viz., that regarding $X_1, X_2, \ldots X_m$ as exclusive *causes* of the event whose probability is x, we accept the probabilities of those causes $a_1, \ldots a_m$ from experience, but form the conditional probabilities of the event as dependent upon such causes,

$$\frac{(xX_1)}{X_1}, \quad \frac{(xX_2)}{X_2}, \text{ \&c. (XVII. Prop. I.)}$$

on the hypothesis of the independence of individual judgments, and so deduce the equation (7). I conceive this, however, to be a less rigorous, though possibly, in practice a more convenient mode of procedure than that adopted in the general solution.

12. It now only remains to assign the particular forms which the algebraic functions X_i, (xX_i), &c. in the above equations assume when the logical function X_i represents that event which consists in r members of the assembly voting one way, and $n - r$ members the other way. It is evident that in this case the algebraic function X_i expresses what the probability of the supposed event would be were the events $x_1, x_2, \ldots x_n$ independent, and their common probability measured by x. Hence we should have, by Art. 3,

$$X_i = \frac{n(n-1)\ldots(n-r+1)}{1 \cdot 2 \ldots r}\{x^r + (1-x)^{n-r}\}.$$

Under the same circumstances (xX_i) would represent the probability of the compound event, which consists in a particular member of the assembly forming a correct judgment, conjointly with the general state of voting recorded above. It would, therefore, be the probability that a particular member votes correctly, while of the remaining $n - 1$ members, $r - 1$ vote correctly; or that the same member votes correctly, while of the remaining $n - 1$ members r vote incorrectly. Hence

$$(xX_i) = \frac{(n-1)(n-2)\ldots(n-r+1)}{1 \cdot 2 \ldots r-1} x^r + \frac{(n-1)(n-2)\ldots(n-r)}{1 \cdot 2 \ldots r} x^{n-r}.$$

PROPOSITION IV.

13. *Given any system of probabilities drawn from recorded instances of unanimity, or of assigned numerical majority in the decisions of a criminal court of justice, required upon hypotheses similar to those of the last proposition, the mean probability c of*

correct judgment for a member of the court, and the general probability k of guilt in an accused person.

The solution of this problem differs in but a slight degree from that of the last, and may be referred to the same general formulæ, (4) and (5), or (7). It is to be observed, that as there are two elements, c and k, to be determined, it is necessary to reserve two of the functions $X_1, X_2, .. X_{m-1}$, let us suppose X_1 and X_{m-1}, for final comparison, employing either the remaining $m-3$ functions in the expression of the data, or the two respective sets $X_2, X_3, .. X_{m-1}$, and $X_1, X_2, ... X_{m-2}$. In either case it is supposed that there must be at least two original independent data. If the equation (7) be alone employed, it would in the present instance furnish two equations, which may thus be written:

$$\frac{a_1(xX_1)}{X_1} + \frac{a_2(xX_2)}{X_2} .. + \frac{a_m(xX_m)}{X_m} = x, \tag{1}$$

$$\frac{a_1(kX_1)}{X_1} + \frac{a_2(kX_2)}{X_2} .. + \frac{a_m(kX_m)}{X_m} = k. \tag{2}$$

These equations are to be employed in the following manner:—Let $x_1, x_2, .. x_n$ represent those events which consist in the formation of a correct opinion by the members of the court respectively. Let also w represent that event which consists in the guilt of the accused member. By the aid of these symbols we can logically express the functions $X_1, X_2, .. X_{m-1}$, whose probabilities are given, as also the function X_m. Then from the function X_1 select those constituents which contain, as a factor, any particular symbol of the set $x_1, x_2, .. x_n$, and also those constituents which contain as a factor w. In both results change $x_1, x_2, .. x_n$ severally into x, and w into k. The above results will give (xX_1) and (kX_1). Effecting the same transformations throughout, the system (1), (2) will, upon the particular hypothesis involved, determine x and k.

14. We may collect from the above investigations the following facts and conclusions:

1st. That from the mere records of agreement and disagreement in the opinions of any body of men, no definite numerical conclusions can be drawn respecting either the probability of correct judgment in an individual member of the body, or the merit of the questions submitted to its consideration.

2nd. That such conclusions may be drawn upon various distinct hypotheses, as—1st, Upon the usual hypothesis of the absolute independence of individual judgments; 2ndly, upon certain definite modifications of that hypothesis warranted by the actual data; 3rdly, upon a distinct principle of solution suggested by the appearance of a common form in the solutions obtained by the modifications above adverted to.

Lastly. That whatever of doubt may attach to the final results, rests not upon the imperfection of the method, which adapts itself equally to all hypotheses, but upon the uncertainty of the hypotheses themselves.

It seems, however, probable that with even the widest limits of hypothesis, consistent with the taking into account of *all* the data of experience, the deviation of the results obtained would be but slight, and that their mean values might be determined with great confidence by the methods of Prop. III. Of those methods I should be disposed to give the preference to the first. Such a principle of mean solution having been agreed upon, other considerations seem to indicate that the values of c and k for tribunals and assemblies possessing a definite constitution, and governed in their deliberations by fixed rules, would remain nearly constant, subject, however, to a small secular variation, dependent upon the progress of knowledge and of justice among mankind. There exist at present few, if any, data proper for their determination.

CHAPTER XXII.

ON THE NATURE OF SCIENCE, AND THE CONSTITUTION OF THE INTELLECT.

1. WHAT I mean by the constitution of a system is the aggregate of those causes and tendencies which produce its observed character, when operating, without interference, under those conditions to which the system is conceived to be adapted. Our judgment of such adaptation must be founded upon a study of the circumstances in which the system attains its freest action, produces its most harmonious results, or fulfils in some other way the apparent design of its construction. There are cases in which we know distinctly the causes upon which the operation of a system depends, as well as its conditions and its end. This is the most perfect kind of knowledge relatively to the subject under consideration. There are also cases in which we know only imperfectly or partially the causes which are at work, but are able, nevertheless, to determine to some extent the laws of their action, and, beyond this, to discover general tendencies, and to infer ulterior purpose. It has thus, I think rightly, been concluded that there is a moral faculty in our nature, not because we can understand the special instruments by which it works, as we connect the organ with the faculty of sight, nor upon the ground that men agree in the adoption of universal rules of conduct; but because while, in some form or other, the sentiment of moral approbation or disapprobation manifests itself in all, it tends, wherever human progress is observable, wherever society is not either stationary or hastening to decay, to attach itself to certain classes of actions, consentaneously, and after a manner indicative both of permanency and of law. Always and everywhere the manifestation of Order affords a presumption, not measurable indeed, but real (XX. 22), of the fulfilment of an end or purpose, and the existence of a ground of orderly causation.

2. The particular question of the constitution of the intellect has, it is almost needless to say, attracted the efforts of speculative ingenuity in every age. For it not only addresses itself to that desire of knowledge which the greatest masters of ancient thought believed to be innate in our species, but it adds to the ordinary strength of this motive the inducement of a human and personal interest. A genuine devotion to truth is, indeed, seldom partial in its aims, but while it prompts to expatiate over the fair fields of outward observation, forbids to neglect the study of our own faculties. Even in ages the most devoted to material interests, some portion of the current of thought has been reflected inwards, and the desire to comprehend that by which all else is comprehended has only been baffled in order to be renewed.

It is probable that this *pertinacity* of effort would not have been maintained among sincere inquirers after truth, had the conviction been general that such speculations are hopelessly barren. We may conceive that it has been felt that if something of error and uncertainty, always incidental to a state of partial information, must ever be attached to the results of such inquiries, a residue of positive knowledge may yet remain; that the contradictions which are met with are more often verbal than real; above all, that even *probable* conclusions derive here an interest and a value from their subject, which render them not unworthy to claim regard beside the more definite and more splendid results of physical science. Such considerations seem to be perfectly legitimate. Insoluble as many of the problems connected with the inquiry into the nature and constitution of the mind must be presumed to be, there are not wanting others upon which a limited but not doubtful knowledge, others upon which the conclusions of a highly probable analogy, are attainable. As the realms of day and night are not strictly conterminous, but are separated by a crepuscular zone, through which the light of the one fades gradually off into the darkness of the other, so it may be said that every region of positive knowledge lies surrounded by a debateable and speculative territory, over which it in some degree extends its influence and its light. Thus there may be questions relating to the constitution of the intellect which, though they do not admit, in the present state of knowledge, of an absolute decision, may receive so much of reflected information as to render their probable solution not difficult; and there may also be questions relating to the nature of science, and even to particular truths and doctrines of science, upon which they who accept the general principles of this work cannot but be led to entertain positive opinions, differing, it may be, from those which are usually received in the present day.* In what follows I shall recapitulate some of the more definite conclusions established in the former parts of this treatise, and shall then indicate one or two trains of thought, connected with the general objects above adverted to, which they seem to me calculated to suggest.

3. Among those conclusions, relating to the intellectual constitution, which may be considered as belonging to the realm of positive knowledge, we may reckon the scientific laws of thought and reasoning, which have formed the basis of the general methods of this treatise, together with the principles, Chap. v., by which their application has been determined. The resolution of the domain of thought into two spheres, distinct but coexistent (IV. XI.); the subjection of the intellectual operations within those spheres to a common system of laws (XI.); the general mathematical character of those laws, and their actual expression (II. III.); the extent of their affinity with the laws of thought in the domain of number, and the point of their divergence therefrom; the dominant character of the two limiting conceptions of universe and eternity among all the subjects of thought with which Logic is concerned; the relation of those conceptions to the fundamental conception of unity in the science of number,—these, with many similar results, are not to be ranked as merely

---

* The following illustration may suffice:—

It is maintained by some of the highest modern authorities in grammar that conjunctions connect propositions only. Now, without inquiring directly whether this opinion is sound or not, it is obvious that it cannot consistently be held by any who admit the scientific principles of this treatise; for to such it would seem to involve a denial, either, 1st, of the possibility of *performing*, or 2ndly, of the possibility of *expressing*, a mental operation, the laws of which, viewed in both these relations, have been investigated and applied in the present work.—(Latham on the English Language; Sir John Stoddart's Universal Grammar, &c.)

probable or analogical conclusions, but are entitled to be regarded as truths of science. Whether they be termed metaphysical or not, is a matter of indifference. The nature of the evidence upon which they rest, though in kind distinct, is not inferior in value to any which can be adduced in support of the general truths of physical science.

Again, it is agreed that there is a certain order observable in the progress of all the exacter forms of knowledge. The study of every department of physical science begins with observation, it advances by the collation of facts to a presumptive acquaintance with their connecting law, the validity of such presumption it tests by new experiments so devised as to augment, if the presumption be well founded, its probability indefinitely; and finally, the law of the phænomenon having been with sufficient confidence determined, the investigation of causes, conducted by the due mixture of hypothesis and deduction, crowns the inquiry. In this advancing order of knowledge, the particular faculties and laws whose nature has been considered in this work bear their part. It is evident, therefore, that if we would impartially investigate either the nature of science, or the intellectual constitution in its relation to science, no part of the two series above presented ought to be regarded as isolated. More especially ought those truths which stand in any kind of *supplemental* relation to each other to be considered in their mutual bearing and connexion.

4. Thus the necessity of an experimental basis for all positive knowledge, viewed in connexion with the existence and the peculiar character of that system of mental laws, and principles, and operations, to which attention has been directed, tends to throw light upon some important questions by which the world of speculative thought is still in a great measure divided. How, from the particular facts which experience presents, do we arrive at the general propositions of science? What is the nature of these propositions? Are they solely the collections of experience, or does the mind supply some connecting principle of its own? In a word, what is the nature of scientific truth, and what are the grounds of that confidence with which it claims to be received?

That to such questions as the above, no single and general answer can be given, must be evident. There are cases in which they do not even need discussion. Instances are familiar, in which general propositions merely express *per enumerationem simplicem*, a fact established by actual observation in all the cases to which the proposition applies. The astronomer asserts upon this ground, that all the known planets move from west to east round the sun. But there are also cases in which general propositions are assumed from observation of their truth in particular instances, and extension of that truth to instances unobserved. No principle of merely deductive reasoning can warrant such a procedure. When from a large number of observations on the planet Mars, Kepler inferred that it revolved in an ellipse, the conclusion was larger than his premises, or indeed than any premises which mere observation could give. What other element, then, is necessary to give even a prospective validity to such generalizations as this? It is the ability inherent in our nature to appreciate Order, and the concurrent presumption, however founded, that the phænomena of Nature are connected by a principle of Order. Without these, the general truths of physical science could never have been ascertained. Grant that the procedure thus established can only conduct us to probable or to approximate results; it only follows, that the larger number of the generalizations of physical science possess but a probable or approximate truth. The security of the tenure of knowledge consists in this, that wheresoever such conclusions do truly represent the constitution of Nature, our confidence in their truth receives indefinite confirmation, and soon becomes undistinguishable from certainty. The existence of that principle above represented as the basis of inductive reasoning enables us to solve the much disputed question as to the necessity of general propositions in reasoning. The logician affirms, that it is impossible to deduce any conclusion from particular premises. Modern writers of high repute have contended, that all reasoning is from particular to particular truths. They instance, that in concluding from the possession of a property by certain members of a class, its possession by some other member, it is not necessary to establish the intermediate *general* conclusion which affirms its possession by *all* the members of the class in common. Now whether it is so or not, that principle of order or analogy upon which the reasoning is conducted must either be stated or apprehended as a general truth, to give validity to the final conclusion. In this form, at least, the necessity of general propositions as the basis of inference is confirmed,—a necessity which, however, I conceive to be involved in the very existence, and still more in the peculiar *nature*, of those faculties whose laws have been investigated in this work. For if the process of reasoning be carefully analyzed, it will appear that abstraction is made of all peculiarities of the individual to which the conclusion refers, and the attention confined to those properties by which its membership of the class is defined.

5. But besides the general propositions which are derived by induction from the collated facts of experience, there exist others belonging to the domain of what is termed *necessary* truth. Such are the general propositions of Arithmetic, as well as the propositions expressing the laws of thought upon which the general methods of this treatise are founded; and these propositions are not only capable of being rigorously verified in particular instances, but are made manifest in all their generality from the study of particular instances. Again, there exist general propositions expressive of necessary truths, but incapable, from the imperfection of the senses, of being exactly verified. Some, if not all, of the propositions of Geometry are of this nature; but it is not in the region of Geometry alone that such propositions are found. The question concerning their nature and origin is a very ancient one, and as it is more intimately connected with the inquiry into the constitution of the intellect than any other to which allusion has been made, it will not be irrelevant to consider it here. Among the opinions which have most widely prevailed upon the subject are the following. It has been maintained, that propositions of the class referred to exist in the mind independently of experience, and that those conceptions which are the subjects of them are the imprints of eternal archetypes. With such archetypes, conceived, however, to possess a reality of which all the objects of sense are but a faint shadow or dim suggestion, Plato furnished his ideal world. It

has, on the other hand, been variously contended, that the subjects of such propositions are copies of individual objects of experience; that they are mere names; that they are individual objects of experience themselves; and that the propositions which relate to them are, on account of the imperfection of those objects, bnt partially true; lastly, that they are intellectual products formed by abstraction from the sensible perceptions of individual things, but so formed as to become, what the individual things never can be, subjects of science, i. e. subjects concerning which exact and general propositions may be affirmed. And there exist, perhaps, yet other views, in some of which the sensible, in others the intellectual or ideal, element predominates.

Now if the last of the views above adverted to be taken (for it is not proposed to consider either the purely ideal or the purely nominalist view) and if it be inquired what, in the sense above stated, are the proper objects of science, objects in relation to which its propositions are true without any mixture of error, it is conceived that but one answer can be given. It is, that neither do individual objects of experience, nor with all probability do the mental images which they suggest, possess any strict claim to this title. It seems to be certain, that neither in nature nor in art do we meet with anything absolutely agreeing with the geometrical definition of a straight line, or of a triangle, or of a circle, though the deviation therefrom may be inappreciable by sense; and it may be conceived as at least doubtful, whether we can form a perfect mental image, or conception, with which the agreement shall be more exact. But it is not doubtful that such conceptions, however imperfect, do point to something beyond themselves, in the gradual approach towards which all imperfection tends to disappear. Although the perfect triangle, or square, or circle, exists not in nature, eludes all our powers of *representative* conception, and is presented to us in thought only, as the limit of an indefinite process of abstraction, yet, by a wonderful faculty of the understanding, it may be made the subject of propositions which are *absolutely* true. The domain of reason is thus revealed to us as larger than that of imagination. Should any, indeed, think that we are able to picture to ourselves, with rigid accuracy, the scientific elements of form, direction, magnitude, &c., these things, as actually conceived, will, in the view of such persons, be the proper objects of science. But if, as seems to me the more just opinion, an incurable imperfection attaches to all our attempts to realize with precision these elements, then we can only affirm, that the more external objects do approach in reality, or the conceptions of fancy by abstraction, to certain *limiting* states, never, it may be, actually attained, the more do the general propositions of science concerning those things or conceptions approach to absolute truth, the actual deviation therefrom tending to disappear. To some extent, the same observations are applicable also to the physical sciences. What have been termed the "fundamental ideas" of those sciences as force, polarity, crystallization, &c.,* are neither, as I conceive, intellectual products independent of experience, nor mere copies of external things; but while, on the one hand, they have a necessary antecedent in experience, on the other hand they require for their formation the exercise of the power of abstraction, in obedience to some general faculty or disposition of our nature, which ever prompts us to the research, and qualifies us for the appreciation, of order.† Thus we study approximately the effects of gravitation on the motions of the heavenly bodies, by a reference to the *limiting* supposition, that the planets are perfect

---

* Whewell's Philosophy of the Inductive Sciences, pp. 71, 77, 213.

† Of the idea of order it has been profoundly said, that it carries within itself its own justification or its own control, the very trustworthiness of our faculties being judged by the conformity of their results to an order which satisfies the reason. "L'idée de l'ordre a cela de singulier et d'eminent, qu'elle porte en elle même sa justification ou son contrôle. Pour trouver si nos autres facultés nous trompent ou nous ne trompent pas, nous examinons si les notions qu'elles nous donnent s'enchainent on ne s'enchaînent pas suivant un ordre qui satisfasse la raison."—*Cournot, Essai sur les fondements de nos Connaissances.* Admitting this principle as the guide of those powers of abstraction which we undoubtedly possess, it seems unphilosophical to assume that the fundamental ideas of the sciences are not derivable from experience. Doubtless the capacities which have been given to us for the comprehension of the actual world would avail us in a differently constituted scene, if in some form or other the dominion of order was still maintained. It is conceivable that in such a new theatre of speculation, the laws of the intellectual procedure remaining the same, the fundamental ideas of the sciences might be wholly different from those with which we are at present acquainted.

spheres or spheroids. We determine approximately the path of a ray of light through the atmosphere, by a process in which abstraction is made of all disturbing influences of temperature. And such is the order of procedure in all the higher walks of human knowledge. Now what is remarkable in connexion with these processes of the intellect is the disposition, and the corresponding ability, to ascend from the imperfect representations of sense and the diversities of individual experience, to the perception of general, and it may be of immutable truths. Whereever this disposition and this ability unite, each series of connected facts in nature may furnish the intimations of an order more exact than that which it directly manifests. For it may serve as ground and occasion for the exercise of those powers, whose office it is to apprehend the general truths which are indeed exemplified, but never with perfect fidelity, in a world of changeful phænomena.

6. The truth that the ultimate laws of thought are mathematical in their form, viewed in connexion with the fact of the possibility of error, establishes a ground for some remarkable conclusions. If we directed our attention to the scientific truth alone, we might be led to infer an almost exact parallelism between the intellectual operations and the movements of external nature. Suppose any one conversant with physical science, but unaccustomed to reflect upon the nature of his own faculties, to have been informed, that it had been proved, that the laws of those faculties were mathematical; it is probable that after the first feelings of incredulity had subsided, the impression would arise, that the order of thought must, *therefore*, be as necessary as that of the material universe. We know that in the realm of natural science, the absolute connexion between the initial and final elements of a problem, exhibited in the mathematical form, fitly symbolizes that physical necessity which binds together effect and cause. The necessary sequence of states and conditions in the inorganic world, and the necessary connexion of premises and conclusion in the processes of exact demonstration thereto applied, seem to be co-ordinate. It may possibly be a question, to which of the two series the primary application of the term "necessary" is due; whether to the observed constancy of

Nature, or to the indissoluble connexion of propositions in all valid reasoning upon her works. Historically we should perhaps give the preference to the former, philosophically to the latter view. But the fact of the connexion is indisputable, and the analogy to which it points is obvious.

Were, then, the laws of valid reasoning uniformly obeyed, a very close parallelism would exist between the operations of the intellect and those of external Nature. Subjection to laws mathematical in their form and expression, even the subjection of an absolute obedience, would stamp upon the two series one common character. The reign of necessity over the intellectual and the physical world would be alike complete and universal.

But while the observation of external Nature testifies with ever-strengthening evidence to the fact, that uniformity of operation and unvarying obedience to appointed laws prevail throughout her entire domain, the slightest attention to the processes of the intellectual world reveals to us another state of things. The mathematical laws of reasoning are, properly speaking, the laws of *right* reasoning only, and their actual transgression is a perpetually recurring phænomenon. Error, which has no place in the material system, occupies a large one here. We must accept this as one of those ultimate facts, the origin of which it lies beyond the province of science to determine. We must admit that there exist laws which even the rigour of their mathematical forms does not preserve from violation. We must ascribe to them an authority the essence of which does not consist in power, a supremacy which the analogy of the inviolable order of the natural world in no way assists us to comprehend.

As the distinction thus pointed out is *real*, it remains unaffected by any peculiarity in our views respecting other portions of the mental constitution. If we regard the intellect as free, and this is apparently the view most in accordance with the general spirit of these speculations, its freedom must be viewed as opposed to the dominion of necessity, not to the existence of a certain just supremacy of truth. The laws of correct inference may be violated, but they do not the less truly *exist* on this account. Equally do they remain unaffected in character and authority if the hypothesis of necessity in its extreme form be adopted. Let it be granted that the laws of valid reasoning, such as they are determined to be in this work, or, to speak more generally, such as they would finally appear in the conclusions of an exhaustive analysis, form but a *part* of the system of laws by which the actual processes of reasoning, whether right or wrong, are governed. Let it be granted that if that system were known to us in its completeness, we should perceive that the whole intellectual procedure was *necessary*, even as the movements of the inorganic world are necessary. And let it finally, as a consequence of this hypothesis, be granted that the phænomena of incorrect reasoning or error, wheresoever presented, are due to the interference of other laws with those laws of which *right* reasoning is the product. Still it would remain that there exist among the intellectual laws a number marked out from the rest by this special character, viz., that every movement of the intellectual system which is accomplished solely under their direction is *right*, that every interference therewith by other laws is not interference only, but *violation*. It cannot but be felt that this circumstance would give to the laws in question a character of distinction and of predominance. They would but the more evidently seem to indicate a final purpose which is not always fulfilled, to possess an authority inherent and just, but not always commanding obedience.

Now a little consideration will show that there is nothing analogous to this in the government of the world by natural law. The realm of inorganic Nature admits neither of preference nor of distinctions. We cannot separate any portion of her laws from the rest, and pronounce them alone worthy of obedience,—alone charged with the fulfilment of her highest purpose. On the contrary, all her laws seem to stand co-ordinate, and the larger our acquaintance with them, the more necessary does their united action seem to the harmony and, so far as we can comprehend it, to the general *design* of the system. How often the most signal departures from apparent order in the inorganic world, such as the perturbations of the planetary system, the interruption of the process of crystallization by the intrusion of a foreign force, and others of a like nature, either merge into the conception of some more exalted scheme of order, or lose to a more attentive and instructed gaze their abnormal aspect, it is needless to remark. One explanation only of these facts can be given, viz., that the distinction between *true* and *false*, between *correct* and *incorrect*, exists in the processes of the intellect, but not in the region of a physical necessity. As we advance from the lower stages of organic being to the higher grade of conscious intelligence, this contrast gradually dawns upon us. Wherever the phænomena of life are manifested, the dominion of rigid law in some degree yields to that mysterious principle of activity. Thus, although the structure of the animal tribes is conformable to certain general types, yet are those types sometimes, perhaps, in relation to the highest standards of beauty and proportion, always, imperfectly realized. The two alternatives, between which Art in the present day fluctuates, are the exact imitation of individual forms, and the endeavour, by abstraction from all such, to arrive at the conception of an ideal grace and expression, never, it may be, perfectly manifested in forms of earthly mould. Again, those teleological adaptations by which, without the organic type being sacrificed, species become fitted to new conditions or abodes, are but slowly accomplished,—accomplished, however, not, apparently, by the fateful power of external circumstances, but by the calling forth of an energy from within. Life in all its forms may thus be contrasted with the passive fixity of inorganic nature. But inasmuch as the perfection of the types in which it is corporeally manifested is in some measure of an ideal character, inasmuch as we cannot precisely define the highest *suggested* excellency of form and of adaptation, the contrast is less marked here than that which exists between the intellectual processes and those of the purely material world. For the definite and technical character of the mathematical laws by which both are governed, places in stronger light the fundamental difference between the kind of authority which, in their capacity of government, they respectively exercise.

7. There is yet another instance connected with the general objects of this chapter, in which the collation of truths or facts, drawn from different sources, suggests an instructive train of reflection. It consists in the comparison of the laws of thought, in their scientific expression, with the actual forms which physical

speculation in early ages, and metaphysical speculation in all ages, have tended to assume. There are two illustrations of this remark, to which, in particular, I wish to direct attention here.

1st. It has been shown (III. 13) that there is a scientific connexion between the conceptions of unity in Number, and the universe in Logic. They occupy in their respective systems the same relative place, and are subject to the same formal laws. Now to the Greek mind, in that early stage of activity,—a stage not less marked, perhaps not less *necessary*, in the progression of the human intellect, than the era of Bacon or of Newton, — when the great problems of Nature began to unfold themselves, while the means of observation were as yet wanting, and its necessity not understood, the terms "Universe" and "The One" seem to have been regarded as almost identical. To assign the nature of that unity of which all existence was thought to be a manifestation, was the first aim of philosophy.* Thales sought for this fundamental unity in water. Anaximenes and Diogenes conceived it to be air. Hippasus of Metapontum, and Heraclitus the Ephesian, pronounced that it was fire. Less definite or less confident in his views, Parmenides simply declared that all existing things were One; Melissus that the Universe was infinite, unsusceptible of change or motion, One, like to itself, and that motion was not, but seemed to be.† In a spirit which, to the reflective mind of Aristotle, appeared sober when contrasted with the rashness of previous speculation, Anaxagoras of Clazomenæ, following, perhaps, the steps of his fellow-citizen, Hermotimus, sought in Intelligence the cause of the world and of its order.‡ The pantheistic tendency which pervaded many of these speculations is manifest in the language of Xenophanes, the founder of the Eleatic school, who, "surveying the expanse of

---

* See various passages in Aristotle's Metaphysics, Book I.

† Ἐδόκει δὲ αὐτῷ τὸ πᾶν ἄπειρον εἶναι, καὶ ἀναλλοίωτον, καὶ ἀκίνητον, καὶ ἓν, ὅμοιον ἑαυτῷ και πλῆρες. κινησίν τε μὴ εἶναι δοκεῖν δὲ εἶναι.—*Diog. Laert.* IX. cap. 4.

‡ Νοῦν δή τις εἰπὼν ἐνεῖναι, καθάπερ ἐν τοῖς ζῴοις, καὶ ἐν τῇ φύσει, τὸν αἴτιον τοῦ κόσμου καὶ τῆς τάξεως πάσης οἷον νήφων ἐφάνη παρ' εἰκῇ λέγοντας τοὺς πρότερον. Φανερῶς μὲν οὖν Ἀναξαγόραν ἴσμεν ἁψάμενον τούτων τῶν λόγων, αἰτίαν δ' ἔχει πρότερον Ἑρμότιμος ὁ Κλαζομένιος εἰπεῖν.—*Arist. Met.* I. 3.

heaven, declared that the One was God."* Perhaps there are few, if any, of the forms in which unity can be conceived, in the abstract as numerical or rational, in the concrete as a passive substance, or a central and living principle, of which we do not meet with applications in these ancient doctrines. The writings of Aristotle, to which I have chiefly referred, abound with allusions of this nature, though of the larger number of those who once addicted themselves to such speculations, it is probable that the very names have perished. Strange, but suggestive truth, that while Nature in all but the aspect of the heavens must have appeared as little else than a scene of unexplained disorder, while the popular belief was distracted amid the multiplicity of its gods, —the conception of a primal unity, if only in a rude, material form, should have struck deepest root; surviving in many a thoughtful breast the chills of a lifelong disappointment, and an endless search !†

2ndly. In equally intimate alliance with that law of thought which is expressed by an equation of the second degree, and which has been termed in this treatise the law of duality, stands the tendency of ancient thought to those forms of philosophical speculation which are known under the name of dualism. The theory of Empedocles,‡ which explained the apparent contradictions of nature by referring them to the two opposing principles

---

* Ξενοφάνης δὲ... εἰς τὸν ὅλον οὐρανὸν ἀποβλέψας, τὸ ἓν εἶναι φησι τὸν θεόν.—*Ib.*

† The following lines, preserved by Sextus Empiricus, and ascribed to Timon the Sillograph, are not devoid of pathos :—

> ὡς καὶ ἐγὼν ὄφελον πυκινοῦ νόου ἀντιβολῆσαι
> ἀμφοτερόβλεπτος (δολίῃ δ' ὁδῷ ἐξεπατήθην,
> πρεσβυγενὴς ἔτ ἐὼν) καὶ ἀναμφήριστος ἁπάσης
> σκεπτοσύνης· ὅππῃ γὰρ ἐμὸν νόον εἰρύσαιμι,
> εἰς ἓν τ' αὐτό τε πᾶν ἀνέλυετο.

I quote them from Ritter, and venture to give the following version :—

> Be mine, to partial views no more confin'd
> Or sceptic doubts, the truth-illumin'd mind!
> For, long deceiv'd, yet still on Truth intent,
> Life's waning years in wand'rings wild are spent.
> Still restless thought the same high quest essays,
> And still the One, the All, eludes my gaze.

‡ Arist. Met. I. 4. 6.

of "strife" and "friendship;" and the theory of Leucippus,* which resolved all existence into the two elements of a *plenum* and a vacuum, are of this nature. The famous comparison of the universe to a lyre or a bow,† its "recurrent harmony" being the product of opposite states of tension, betrays the same origin. In the system of Pythagoras, which seems to have been a combination of dualism with other elements derived from the study of numbers, and of their relations, ten fundamental antitheses are recognised: finite and infinite, even and odd, unity and multitude, right and left, male and female, rest and motion, straight and curved, light and darkness, good and evil, the square and the oblong. In that of Alcmæon the same fundamental dualism is accepted, but without the definite and numerical limitation with which it is connected in the Pythagorean system. The grand development of this idea is, however, met with in that ancient Manichæan doctrine, which not only formed the basis of the religious system of Persia, but spread widely through other regions of the East, and became memorable in the history of the Christian Church. The origin of dualism as a speculative opinion, not yet connected with the personification of the Evil Principle, but naturally succeeding those doctrines which had assumed the primal unity of Nature, is thus stated by Aristotle:—"Since there manifestly existed in Nature things opposite to the good, and not only order and beauty, but also disorder and deformity; and since the evil things did manifestly preponderate in number over the good, and the deformed over the beautiful, some one else at length introduced strife and friendship as the respective causes of these diverse phænomena."‡ And in Greece, indeed, it seems to have been chiefly as a philosophical opinion, or as an adjunct to philosophical speculation, that the dualistic theory obtained ground.§ The moral application of the doctrine most in

---

* Arist. Met. I. 4, 9.

† *παλίντροπος ἁρμονίη ὅκως περ τόξου καὶ λύρης.—Heraclitus*, quoted in *Origenis Philosophumena*, IX. 9. Also *Plutarch*, *De Iside et Osiride*.

‡ *'Επεὶ δὲ καὶ τἀναντία τοῖς ἀγαθοῖς ἐνόντα ἐφαίνετο ἐν τῇ φύσει, καὶ οὐ μόνον τάξις καὶ τὸ καλὸν ἀλλὰ καὶ ἀταξία καὶ τὸ αἰσχρόν, καὶ πλείω τὰ κακὰ τῶν ἀγαθῶν καί τὰ φαῦλα τῶν καλῶν, οὕτως ἄλλος τις φιλίαν εἰσήνεγκε καὶ νεῖκος, ἑκάτερον ἑκατέρων αἴτιον τούτων.—Arist. Metaphysica*, I. 4.

§ Witness Aristotle's well-known derivation of the elements from the quali-

accordance with the Greek mind is preserved in the great Platonic antithesis of "being and non-being,"—the connexion of the former with whatsoever is good and true, with the eternal ideas, and the archetypal world: of the latter with evil, with error, with the perishable phænomena of the present scene. The two forms of speculation which we have considered were here blended together; nor was it during the youth and maturity of Greek philosophy alone that the tendencies of thought above described were manifested. Ages of imitation caught up and adopted as their own the same spirit. Especially wherever the genius of Plato exercised sway was this influence felt. The unity of all real being, its identity with truth and goodness considered as to their essence; the illusion, the profound unreality, of all merely phænomenal existence; such were the views,—such the dispositions of thought, which it chiefly tended to foster. Hence that strong tendency to mysticism which, when the days of renown, whether on the field of intellectual or on that of social enterprise, had ended in Greece, became prevalent in her schools of philosophy, and reached their culminating point among the Alexandrian Platonists. The supposititious treatises of Dionysius the Areopagite served to convey the same influence, much modified by its contact with Aristotelian doctrines, to the scholastic disputants of the middle ages. It can furnish no just ground of controversy to say, that the tone of thought thus encouraged was as little consistent with genuine devotion as with a sober philosophy. That kindly influence of human affections, that homely intercourse with the common things of life, which form so large a part of the true, because intended, discipline of our nature, would be ill replaced by the contemplation even of the highest object of thought, viewed by an excessive abstraction as something concerning which not a single intelligible proposition could either be affirmed or denied.* I would but slightly allude to those connected speculations on the Divine Nature which ascribed

---

ties "warm," and "dry," and their contraries. It is characteristic that Plato connects their generation with mathematical principles.—*Timæus*, cap. xi.

* *Αὐτὸς καὶ ὑπὲρ θέσιν ἐστὶ καὶ ἀφαίρεσιν.*—*Dion. Areop. De Divinis Nominibus*, cap. II.

to it the perfect union of opposite qualities,* or to the remarkable treatises of Anselm, designed to establish a theory of the universe upon the analogies of thought and being.† The primal unity is there represented as having its abode in the one eternal Truth. The conformity of Nature to her laws, the obedience of moral agents to the dictates of rectitude, are the same Truth seen in action; the world itself being but an expression of the self-reflecting thought of its Author.‡ Still more marked was the revival of the older forms of speculation during the sixteenth and seventeenth centuries. The friends and associates of Lorenzo the Magnificent, the recluses known in England as the Cambridge Platonists, together with many meditative spirits scattered through Europe, devoted themselves anew, either to the task of solving the ancient problem, De Uno, Vero, Bono, or to that of proving that all such inquiries are futile and vain.§ The logical elements which underlie all these speculations, and from which they appear to borrow at least their form, it would be easy to trace in the outlines of more modern systems,—more especially in that association of the doctrine of the absolute unity with the distinction of the *ego* and the *non-ego* as the type of Nature, which forms the basis of the philosophy of Hegel. The attempts of speculative minds to ascend to some high pinnacle of truth, from which they might survey the entire framework and con-

---

* See especially the lofty strain of Hildebert beginning "Alpha et Ω magne Deus." (Trench's Sacred Latin Poetry.) The principle upon which all these speculations rest is thus stated in the treatise referred to in the last note. *Οὐδὲν οὖν ἄτοπον, ἐξ ἀμυδρῶν εἰκόνων επὶ τὸ πάντων αἴτιον ἀναβάντας, ὑπερκοσμιόις ὀφθαλμοῖς θεωρῆσαι πάντα ἐν τῷ παντῶν ἀιτίῳ, καὶ τὰ ἀλλήλοις ἐναντία μονοειδῶς καὶ ἡνωμένως.*—*De Divinis Nominibus*, cap. v. And the kind of knowledge which it is thus sought to attain is described as a "darkness beyond light," *ὑπερφῶτος γνόφος*. (*De Mystica Theologia*, cap. I.) Milton has a similar thought—

"Dark with excessive bright Thy skirts appear."

Par. Lost, Book III.

Contrast with these the nobler simplicity of 1 John, i. 5.

† Monologium, Prosologium, and De Veritate.

‡ "Idcirco cum ipse summus spiritus dicit seipsum dicit omnia quæ facta sunt."—*Monolog.* cap. XXIII.

§ See dissertations in Spinoza, Picus of Mirandula, H. More, &c. Modern discussions of this nature are chiefly in connexion with æsthetics, the ground of the application being contained in the formula of Augustine: "Omnis porro pulchritudinis forma, unitas est."

nexion of things *in the order of deductive thought*, have differed less in the forms of theory which they have produced, than through the nature of the interpretations which have been assigned to those forms.* And herein lies the real question as to the influence of philosophical systems upon the disposition and the life. For though it is of slight moment that men should agree in tracing back all the forms and conditions of being to a primal unity, it is otherwise as concerns their conceptions of what that unity is, and what are the kinds of relation, beside that of mere causality, which it sustains to themselves. Herein too may be felt the powerlessness of mere Logic, the insufficiency of the profoundest knowledge of the laws of the understanding, to resolve those problems which lie nearer to our hearts, as progressive years strip away from our life the illusions of its golden dawn.

8. If the extremely arbitrary character of human opinion be considered, it will not be expected, nor is it here maintained, that the above are the only forms in which speculative men have shaped their conjectural solutions of the problem of existence. Under particular influences other forms of doctrine have arisen, not unfrequently, however, masking those portrayed above.†

---

* For instance, the learned mysticism of Gioberti, widely as it differs in its spirit and its conclusions from the pantheism of Hegel (both being, perhaps, equally remote from truth), resembles it in applying both to *thought* and to *being* the principles of unity and duality. It is asked:—"Or non è egli chiaro che ogni discorso si riduce in fine in fine alle idee di Dio, del mondo, e della creazione, l'ultima delle quali è il legame delle due prime?" And this question being affirmatively answered in the formula, "l'Ente crea le esistenze," it is said of that formula,—"Essa abbraccia la realtà universale nella dualità del necessario e del contingente, esprime il vincolo di questi due ordini, e collocandolo nella creazion sostanziale, riduce la dualità reale a un principio unico, all unità primordiale dell' Ente non astratto, complessivo, e generico, ma concreto, individuato, assoluto, e creatore."—*Del Bello e del Buono*, pp. 30, 31.

† Evidence in support of this statement will be found in the remarkable treatise recently published under the title (the correctness of which seems doubtful) of *Origenis Philosophumena*. The early corruptions of Christianity of which it contains the record, though many of them, as is evident from their Ophite character, derived from the very dregs of paganism, manifest certain persistent forms of philosophical speculation. For the most part they either belong to the dualistic scheme, or recognise three principles, primary or derived, between two of which the dualistic relation may be traced.—*Orig. Phil.*, pp. 135, 139, 150, 235, 253, 264.

But the wide prevalence of the particular theories which we have considered, together with their manifest analogy with the expressed laws of thought, may justly be conceived to indicate a connexion between the two systems. As all other mental acts and procedures are beset by their peculiar fallacies, so the operation of that law of thought termed in this work the law of duality may have its own peculiar tendency to error, exalting mere want of agreement into contrariety, and thus form a world which we necessarily view as formed of parts supplemental to each other, framing the conception of a world fundamentally divided by opposing powers. Such, with some large but hasty inductions from phænomena, may have been the origin of dualism,—independently of the question whether dualism is in any form a true theory or not. Here, however, it is of more importance to consider in detail the bearing of these ancient forms of speculation, as revived in the present day, upon the progress of real knowledge; and upon this point I desire, in pursuance of what has been said in the previous section, to add the following remarks:

1st. All sound philosophy gives its verdict against such speculations, if regarded as a means of determining the actual constitution of things. It may be that the progress of natural knowledge tends towards the recognition of some central Unity in Nature. Of such unity as consists in the mutual relation of the parts of a system there can be little doubt, and able men have speculated, not without grounds, on a more intimate correlation of physical forces than the mere idea of a system would lead us to conjecture. Further, it may be that in the bosom of that supposed unity are involved some general principles of division and re-union, the sources, under the Supreme Will, of much of the *related* variety of Nature. The instances of sex and polarity have been adduced in support of such a view. As a supposition, I will venture to add, that it is not very improbable that, in some such way as this, the constitution of things without may correspond to that of the mind within. But such correspondence, if it shall ever be proved to exist, will appear as the last induction from human knowledge, not as the first principle of scientific inquiry. The natural order of discovery is from the particular to the universal, and it may confidently be affirmed that we have not yet advanced sufficiently far on this track to enable us to determine what are the ultimate forms into which all the special differences of Nature shall merge, and from which they shall receive their explanation.

2ndly. Were this correspondence between the forms of thought and the actual constitution of Nature proved to exist, whatsoever connexion or relation it might be supposed to establish between the two systems, it would in no degree affect the question of their mutual independence. It would in no sense lead to the consequence that the one system is the mere *product* of the other. A too great addiction to metaphysical speculations seems, in some instances, to have produced a tendency toward this species of illusion. Thus, among the many attempts which have been made to explain the existence of evil, it has been sought to assign to the fact a merely *relative* character,—to found it upon a species of logical opposition to the equally relative element of good. It suffices to say, that the assumption is purely gratuitous. What evil may be in the eyes of Infinite wisdom and purity, we can at the best but dimly conjecture; but to us, in all its forms, whether of pain or defect, or moral transgression, or retributory wo, it can wear but one aspect,—that of a sad and stern reality, against which, upon somewhat more than the highest order of prudential considerations, the whole preventive force of our nature may be exerted. Now what has been said upon the particular question just considered, is equally applicable to many other of the debated points of philosophy; such, for instance, as the external reality of space and time. We have no warrant for resolving these into mere forms of the understanding, though they unquestionably determine the present sphere of our knowledge. And, to speak more generally, there is no warrant for the extremely *subjective* tendency of much modern speculation. Whenever, in the view of the intellect, different hypotheses are equally consistent with an observed fact, the instinctive testimony of consciousness as to their relative value must be allowed to possess *authority*.

3rdly. If the study of the laws of thought avails us neither to determine the actual constitution of things, nor to explain the facts involved in that constitution which have perplexed the wise and saddened the thoughtful in all ages,—still less does it enable us to rise above the present conditions of our being, or lend its sanction to the doctrine which affirms the possibility of an *intuitive* knowledge of the infinite, and the unconditioned,—whether such knowledge be sought for in the realm of Nature, or above that realm. We can never be said to *comprehend* that which is represented to thought as the limit of an indefinite process of abstraction. A progression *ad infinitum* is impossible to finite powers. But though we cannot comprehend the infinite, there may be even scientific grounds for believing that human nature is constituted in some relation to the infinite. We cannot perfectly express the laws of thought, or establish in the most general sense the methods of which they form the basis, without at least the implication of elements which ordinary language expresses by the terms "Universe" and "Eternity." As in the pure abstractions of Geometry, so in the domain of Logic it is seen, that the empire of Truth is, in a certain sense, larger than that of Imagination. And as there are many special departments of knowledge which can only be completely surveyed from an external point, so the theory of the intellectual processes, as applied only to finite objects, seems to involve the recognition of a sphere of thought from which all limits are withdrawn. If then, on the one hand, we cannot discover in the laws of thought and their analogies a sufficient basis of proof for the conclusions of a too daring mysticism; on the other hand we should err in regarding them as wholly unsuggestive. As parts of our intellectual nature, it seems not improbable that they should manifest their presence otherwise than by merely prescribing the conditions of formal inference. Whatever grounds we have for connecting them with the peculiar tendencies of physical speculation among the Ionian and Italic philosophers, the same grounds exist for associating them with a disposition of thought at once more common and more legitimate. To no casual influences, at least, ought we to attribute that meditative spirit which then most delights to commune with the external magnificence of Nature, when most impressed with the consciousness of sempiternal verities,—which reads in the nocturnal heavens a bright manifestation of order; or feels in some wild scene among the hills, the intimations of more than that abstract eternity which had rolled away ere yet their dark foundations were laid.*

9. Refraining from the further prosecution of a train of thought which to some may appear to be of too speculative a character, let us briefly review the positive results to which we have been led. It has appeared that there exist in our nature faculties which enable us to ascend from the particular facts of experience to the general propositions which form the basis of Science; as well as faculties whose office it is to deduce from general propositions accepted as true the particular conclusions which they involve. It has been seen, that those faculties are subject in their operations to laws capable of precise scientific expression, but invested with an authority which, as contrasted with the authority of the laws of nature, is distinct, *sui generis*, and underived. Further, there has appeared to be a manifest fitness between the intellectual procedure thus made known to us, and the conditions of that system of things by which we are surrounded,—such conditions, I mean, as the existence of species connected by general resemblances, of facts associated under general laws; together with that union of permanency with order, which while it gives stability to acquired knowledge, lays a foundation for the hope of indefinite progression. Human nature, quite independently of its observed or manifested tendencies, is seen to be *constituted* in a certain relation to Truth; and this relation, considered as a subject of speculative knowledge, is as capable of being studied in its details, is, moreover, as worthy of being so studied, as are the several departments of physical science, considered in the same aspect. I would especially direct attention to that view of the constitution of the intellect which represents it as subject to laws determinate in their character, but not operating by the power of necessity; which exhibits it as redeemed from the dominion of fate, without being abandoned to the lawlessness of chance. We cannot embrace this view without accepting at least as *probable* the intimations which, upon the principle of analogy, it seems to furnish respecting another and a higher aspect of our nature,—its subjection in the sphere of duty as well as in that of knowledge to

---

* Psalm xc. 2.

fixed laws whose authority does not consist in power,—its constitution with reference to an ideal standard and a final purpose. It has been thought, indeed, that scientific pursuits foster a disposition either to overlook the specific differences between the moral and the material world, or to regard the former as in no proper sense a subject for exact knowledge. Doubtless all exclusive pursuits tend to produce partial views, and it may be, that a mind long and deeply immersed in the contemplation of scenes over which the dominion of a physical necessity is unquestioned and supreme, may admit with difficulty the possibility of another order of things. But it is because of the *exclusiveness* of this devotion to a particular sphere of knowledge, that the prejudice in question takes possession, if at all, of the mind. The application of scientific methods to the study of the intellectual phænomena, conducted in an impartial spirit of inquiry, and without overlooking those elements of error and disturbance which must be accepted as *facts*, though they cannot be regarded as *laws*, in the constitution of our nature, seems to furnish the materials of a juster analogy.

10. If it be asked to what *practical* end such inquiries as the above point, it may be replied, that there exist various objects, in relation to which the courses of men's actions are mainly determined by their speculative views of human nature. Education, considered in its largest sense, is one of those objects. The ultimate ground of all inquiry into its nature and its methods must be laid in some previous theory of what man is, what are the ends for which his several faculties were designed, what are the motives which have power to influence them to sustained action, and to elicit their most perfect and most stable results. It may be doubted, whether these questions have ever been considered fully, and at the same time impartially, in the relations here suggested. The highest cultivation of taste by the study of the pure models of antiquity, the largest acquaintance with the facts and theories of modern physical science, viewed from this larger aspect of our nature, can only appear as parts of a perfect intellectual discipline. Looking from the same point of view upon the means to be employed, we might be led to inquire, whether that all but exclusive appeal which is made in

the present day to the spirit of emulation or cupidity, does not tend to weaken the influence of those more enduring motives which seem to have been implanted in our nature for the immediate end in view. Upon these, and upon many other questions, the just limits of authority, the reconciliation of freedom of thought with discipline of feelings, habits, manners, and upon the whole *moral* aspect of the question,—what unfixedness of opinion, what diversity of practice, do we meet with! Yet, in the sober view of reason, there is no object within the compass of human endeavours which is of more weight and moment than this, considered, as I have said, in its largest meaning. Now, whatsoever tends to make more exact and definite our view of human nature, in any of its real aspects, tends, in the same proportion, to reduce these questions into narrower compass, and restrict the limits of their possible solution. Thus may even speculative inquiries prove fruitful of the most important principles of action.

11. Perhaps the most obviously legitimate bearing of such speculations would be upon the question of the place of Mathematics in the system of human knowledge, and the nature and office of mathematical studies, as a means of intellectual discipline. No one who has attended to the course of recent discussions can think this question an unimportant one. Those who have maintained that the position of Mathematics is in both respects a fundamental one, have drawn one of their strongest arguments from the actual constitution of things. The material frame is subject in all its parts to the relations of number. All dynamical, chemical, electrical, thermal, actions, seem not only to be measurable in themselves, but to be connected with each other, even to the extent of mutual convertibility, by numerical relations of a perfectly definite kind. But the opinion in question seems to me to rest upon a deeper basis than this. The laws of thought, in all its processes of conception and of reasoning, in all those operations of which language is the expression or the instrument, are of the same kind as are the laws of the acknowledged processes of Mathematics. It is not contended that it is necessary for us to acquaint ourselves with those laws in order to think coherently, or, in the ordinary sense of

the terms, to reason well. Men draw inferences without any consciousness of those elements upon which the entire procedure depends. Still less is it desired to exalt the reasoning faculty over the faculties of observation, of reflection, and of judgment. But upon the very ground that human thought, traced to its ultimate elements, reveals itself in mathematical forms, we have a presumption that the mathematical sciences occupy, by the constitution of our nature, a fundamental place in human knowledge, and that no system of mental culture can be complete or fundamental, which altogether neglects them.

But the very same class of considerations shows with equal force the error of those who regard the study of Mathematics, and of their applications, as a sufficient basis either of knowledge or of discipline. If the constitution of the material frame is mathematical, it is not merely so. If the mind, in its capacity of formal reasoning, obeys, whether consciously or unconsciously, mathematical laws, it claims through its other capacities of sentiment and action, through its perceptions of beauty and of moral fitness, through its deep springs of emotion and affection, to hold relation to a different order of things. There is, moreover, a breadth of intellectual vision, a power of sympathy with truth in all its forms and manifestations, which is not measured by the force and subtlety of the dialectic faculty. Even the revelation of the material universe in its boundless magnitude, and pervading order, and constancy of law, is not necessarily the most fully apprehended by him who has traced with minutest accuracy the steps of the great demonstration. And if we embrace in our survey the interests and duties of life, how little do any processes of mere ratiocination enable us to comprehend the weightier questions which they present! As truly, therefore, as the cultivation of the mathematical or deductive faculty is a part of intellectual discipline, so truly is it only a part. The prejudice which would either banish or make supreme any one department of knowledge or faculty of mind, betrays not only error of judgment, but a defect of that intellectual modesty which is inseparable from a pure devotion to truth. It assumes the office of criticising a constitution of things which no human appointment has established, or can annul. It sets aside the ancient and just conception of truth as one though manifold. Much of this error, as actually existent among us, seems due to the special and isolated character of scientific teaching—which character it, in its turn, tends to foster. The study of philosophy, notwithstanding a few marked instances of exception, has failed to keep pace with the advance of the several departments of knowledge, whose mutual relations it is its province to determine. It is impossible, however, not to contemplate the particular evil in question as part of a larger system, and connect it with the too prevalent view of knowledge as a merely secular thing, and with the undue predominance, already adverted to, of those motives, legitimate within their proper limits, which are founded upon a regard to its secular advantages. In the extreme case it is not difficult to see that the continued operation of such motives, uncontrolled by any higher principles of action, uncorrected by the personal influence of superior minds, must tend to lower the standard of thought in reference to the objects of knowledge, and to render void and ineffectual whatsoever elements of a noble faith may still survive. And ever in proportion as these conditions are realized must the same effects follow. Hence, perhaps, it is that we sometimes find juster conceptions of the unity, the vital connexion, and the subordination to a moral purpose, of the different parts of Truth, among those who acknowledge nothing higher than the changing aspect of collective humanity, than among those who profess an intellectual allegiance to the Father of Lights. But these are questions which cannot further be pursued here. To some they will appear foreign to the professed design of this work. But the consideration of them has arisen naturally, either out of the speculations which that design involved, or in the course of reading and reflection which seemed necessary to its accomplishment.

THE END.

ERRATA.

Page 57, line 11 from bottom, *for* y *read* z.
— 93, — 5, *for* is *read* be.
— 119, — 6, the letter w imperfect, like v.
— ,, — 10, *for* $\bar{w}\bar{x}$ *read* $\bar{w}\bar{z}$.
— 120, — 1, last term, *for* z *read* $\bar{z}$.
— 128, — 4 from bottom, *for* $\bar{w}$ *read* $\bar{x}$.
— ,, — 6 from bottom, *for* $x\bar{w} + x$ *read* $x\bar{w} + \bar{x}w$.
— ,, — 8 from bottom, the letter w imperfect, like v
— 129, — 2, *for* $x\bar{y}z$ *read* $\bar{x}\bar{y}z$.
— 221, — 18, *for* vy *read* $\bar{v}y$.
— 231, — 10 from bottom, *for* vz *read* $v'z$.
— 261, — 2 from bottom, *for* p, q, r, *read* p', q', r'.
— 262, — 22, for p' and q' *read* p and q.
— 270, — 6 from bottom, for Xy *read* XY.
— 274, — 11, for $t_2\, t_2$ *read* $t_1\, t_2$.
— 282, — 10, *dele* (1) gives.
— 291, — 1, *for* $y + \bar{y}\bar{z}$ *read* $y + \bar{y}z$.
— 297, — 10 from bottom, prefix =
— 308, — 4, *for* limit *read* limits.
— 309, — 7, *for* $sq\,(1 - t)$ *read* $sy\,(1 - t)$.
— 313, — 13 from bottom, *for* $s_1 = 0$ *read* $\bar{s}_1 = 0$.
— 314, — 9 from bottom, *omit* the comma.
— 315, — 4 from bottom, *for* $\bar{s}_1\, s_2 \ldots s_n$ *read* $\bar{s}_1\, \bar{s}_2 \ldots \bar{s}_n$.
— 322, bottom line, *read* the second term as $\bar{s}\bar{t}x\bar{y}$.
— 330, line 6, for ν' *read* n'.
— 331, — 5 from bottom, *for* $p'm$ *read* $h'm$.
— 334, — 16, supply the letter *s*.
— 351, — 10 from bottom, *for* v_1 *read* x_1.
— 364, — 21, *for* 91° 4187 *read* 91°·4187.
— 373, — 7, *for* $p = r$ *read* $p - r$.
— ,, — 3, 5, and 6 from bottom, *for* 1 *read* 0^m.
— 385, — 6, *for* x_m *read* x_n.
— 386, — 13, w imperfect, like v.
— 388, — 16, *for* t_{m-1} *read* $\bar{t}_{m-1}$.
— 389, — 7 from bottom, *for* $t_1 \ldots t_m$ *read* $\bar{t}_1 \ldots \bar{t}_{m-1}$.
— 391, — 5, *omit* namely.